Islands at Risk?

Islands at Risk?

Environments, Economies and Contemporary Change

John Connell

Professor of Geography, University of Sydney, Australia

Edward Elgar

Cheltenham, UK • Northampton, MA, USA

Published by
Edward Elgar Publishing Limited
The Lypiatts
15 Lansdown Road
Cheltenham
Glos GL50 2JA
UK

Edward Elgar Publishing, Inc.
William Pratt House
9 Dewey Court
Northampton
Massachusetts 01060
USA

A catalogue record for this book
is available from the British Library

Library of Congress Control Number: 2012954985

This book is available electronically in the ElgarOnline.com Social and Political Science Subject Collection, E-ISBN 978 1 78100 351 0

MIX
Paper from
responsible sources
FSC® C018575
FSC
www.fsc.org

ISBN 978 1 78100 350 3

Typeset by Columns Design XML Ltd, Reading
Printed by MPG PRINTGROUP, UK

Contents

Preface vii
List of abbreviations ix

1 **Introduction** 1
SIS and SISI 2
Low-lying and high islands 14
A short synopsis 15

2 **Islands and political economies** 17
Economic development 18
Trade in patchwork economies 27
Aid 39
Politics and problems of the possible 43
Vulnerability and resilience 52

3 **The historic core: agriculture and fisheries** 55
Agriculture 56
Forestry 69
Fisheries 72
Out of resources? 85

4 **Towards modern economies?** 90
Manufacturing 91
Mining 94
Tourism 97
Services 110
Employment and incomes 116
Discontents of modernity 122

5 **Urban futures?** 124
Populations 124
Rural–urban migration 127
Resettlement 133
Urbanization 137

Urban environments 140
Beyond the city 148

6 **Leaving the islands: international migration** 150
Contemporary migration 152
Migration rationales 156
Remittances 159
Brain and skill drain 169
Return migration 174
National interventions 177
The outward urge 181

7 **Environmental change** 185
Climate change 187
Sea-level rise 190
Coral reefs 195
Cyclones 201
Floods 205
Drought 207
Water 209
Disease 213
Biodiversity 215
Catastrophes: volcanoes, earthquakes and tsunamis 222
Environmental futures 226

8 **Islands at risk?** 230
Black swans, uncertainty and insecurity 233
Struggling economies, challenging environments? 238
In search of hybridity? 245
From the ends of the world 254
Warm winds of change 258

References 263
Index 339

Preface

> To be forgotten is to die a little. It is to lose some of the links that anchor us to the rest of humanity
>
> Aung San Suu Kyi, 2012

As the twenty-first century has evolved, economic and political issues that hitherto dominated debates over development have increasingly taken on environmental themes. Nowhere is this more evident than in small island states. The aim of this book is to examine how the interactions between economics and environment play out within the changing geographies of small islands and small countries. It focuses on those island realms conventionally perceived as 'developing', rather than 'developed', which has therefore also meant a focus on tropical islands. It is precisely these tropical islands (alongside continental deltas) that are usually considered at most risk from environmental change, notably global warming and sea-level rise. This book seeks to respond by examining these risks, alongside other forms of vulnerability, and the extent to which even in and between small islands risks and outcomes vary enormously. It also attempts to shift from more conventional approaches to the political economy of island development towards a more political ecology, geography and ethnography.

The book was being completed at what seemed an auspicious time for considering environmental changes in small islands: greenhouse gas emissions had increased by a record amount in the previous year, even in the wake of the Global Financial Crisis. The financial crisis and subsequent Eurozone crisis brought both regulation and a new bout of Western introspection. In what seemed a curiously Melanesian acronym, the political economies of the PIGS (Portugal, Ireland, Spain and, above all, Greece), struggled to survive. Both financial crises contributed to global recession and concerns over aid and remittance flows, migration, investment and tourism. In 2011 global aid spending fell for the first time in 14 years. Meanwhile the Food and Agriculture Organization of the United Nations (FAO) announced that a quarter of the world's agricultural land was degraded, and the world population reached 7 billion. After the Japanese tsunami the future of nuclear power had become more uncertain, the melting of Arctic sea ice at an unprecedented rate was being blamed on human activity, extreme temperatures in the United States were being seen as the 'new normal' and United Nations talks on a global treaty on climate change had stalled in the empty promises of Durban. The Rio+20 Earth Summit was written off before it had begun. Island states were frustrated at the inaction that measured their marginalization, and fearful of the consequences of environmental change, perceived as a transnational threat to national development and security.

While this book primarily focuses on tropical and independent island states, it emerged from an earlier and much briefer study undertaken as a consultancy to the British government. This had a particular interest in the UK's overseas territories, most of which are themselves tropical, but which include the distinctly cool Falkland Islands and Tristan da Cunha, so they have also been included here in passing, as have other dependent territories. There is some Pacific bias, partly because of greater personal familiarity, but the Pacific has been rather better served by research in relevant areas than the Caribbean and Indian Ocean. The breadth of coverage has meant that the book is necessarily synthetic and second-hand in places, but consistent trends frequently recur.

At one time or another I have been fortunate enough to be able to visit all but two of the small island states (and many of the individual islands) that are discussed here. I am grateful to many people on many islands, from the Cayman Islands to the Maldives, from Bermuda to Rapanui and from Woleai to Aniwa, for their wisdom and information, their cups of tea, bowls of kava and sometimes stronger forms of sustenance. I am grateful to Neil Adger for inviting me to undertake the earlier project that enabled me to think more widely about the relationship between environmental and economic change. I am also grateful to Conway Pene and Olivia Dun for the maps, Cluny and La'avasa Macpherson for loaning a title, James Terry and Randy Thaman for their comments on an earlier version of Chapter 6, Wes Morgan's corrections on trade, Mark Hampton's thoughts on tourism, Robert Aldrich's scalpel to, and Godfrey Baldacchino's detailed reading of, Chapter 8, and Amelia Roberts' and Yayoi Fujita's patient reading of so much of the book. The usual disclaimers of course apply. They are neither responsible for the intermittent acronym soup, nor for my wish to cite a myriad sources and predecessors and include mini-vignettes. Ultimately, in that vein, it would be remiss of me not to thank Google for being there when all else had failed.

Abbreviations

ADB	Asian Development Bank
ANU	Australian National University
AOSIS	Alliance of Small Island States
BIOT	British Indian Ocean Territory
BVI	British Virgin Islands
CARICOM	Caribbean Community
CSME	CARICOM Single Market and Economy
EEZ	Exclusive Economic Zone
ENSO	El Niño Southern Oscillation
EPZ	Export Processing Zones
EU	European Union
FAO	Food and Agriculture Organization of the United Nations
FSM	Federated States of Micronesia
GATS	General Agreement on Trade in Services
GFC	Global Financial Crisis
HDI	Human Development Index
IAS	Invasive alien species
IDA	International Development Association
IMR	Infant mortality rate
IMF	International Monetary Fund
IPCC	Intergovernmental Panel on Climate Change
IVF	*In-vitro* Fertilization
LDCs	Least Developed Countries
MDGs	Millennium Development Goals
MIRAB	Migration, Remittances, Aid and Bureaucracy
MPA	Marine Protected Area
NAFTA	North American Free Trade Agreement
NCDs	Non-Communicable Diseases
NGO	Non-governmental Organization
NICs	Newly Industrializing Countries
OECS	Organisation of Eastern Caribbean States
PICTA	Pacific Islands Countries Trade Agreement
PNG	Papua New Guinea

PRC	People's Republic of China
PSWPS	Pacific Seasonal Worker Pilot Scheme
RAMSI	Regional Assistance Mission to Solomon Islands
RSE	Recognized Seasonal Employer
SAPs	Structural Adjustment Programmes
SAWP	Seasonal Agricultural Workers Programme
SDGs	Sustainable Development Goals
SIDS	Small Island Developing States
SIS	Small Island States
SISI	Small Islands and Small Island States
SITES	Small Island Tourist Economies
SLR	Sea-level Rise
SOPAC	South Pacific Applied Geoscience Commission
SPARTECA	South Pacific Regional Trade and Economic Cooperation Agreement
SPC	Secretariat of the Pacific Community
SPREP	Secretariat of the Pacific Regional Environmental Programme
UKOTS	United Kingdom Overseas Territories
UNCLOS	United Nations Convention on the Law of the Sea
UNCTAD	United Nations Commission on Trade and Development
UNDP	United Nations Development Programme
UNEP	United Nations Environment Programme
UNESCO	United Nations Education, Scientific and Cultural Organization
USP	University of the South Pacific
USVI	United States Virgin Islands
WHO	World Health Organization
WTO	World Trade Organization

1. Introduction

> The pattern of growth in a very small country must of necessity be different from that in a large continental country
>
> Demas, 1965: 22

This book examines relationships between socio-economic and environmental change in small island states, dependent territories and other small islands. It considers change and stability, in island environments and socio-economic systems: the evolving geographies of island regions. Islands, and especially coral atolls, have often been regarded as the 'canaries in the coalmines' of climate change – a particularly inapt metaphor for places without canaries, coal or mines – where environmental change will be experienced first and most catastrophically, because of low elevations and extensive and economically significant coastlines. Indeed tropical islands are near the frontline and many have economies that make coping difficult. This book seeks to examine the extent to which islands are particularly vulnerable, whether that vulnerability is environmental or economic (or social and political), and why and how islands and island states vary considerably. While vulnerability has been defined in many ways, like risk, it is simply used here in a common-sense manner to refer to the extent of exposure to changes that affect livelihoods. Resilience becomes the ability to both withstand and cope with vulnerability.

A major problem for comparing and contrasting socio-economic and environmental changes in so many diverse places is the absence of adequate data. This has necessitated an unusually qualitative approach, where such over-used adjectives as 'some' and 'most' and adverbs such as 'usually' and 'increasingly' pass for quantitative analysis. In some countries little reliable data of any kind exist. By contrast, in particular places excellent detailed studies have been undertaken, but with no necessary relevance for islands less than 50 km away. Indeed various data issues cannot be resolved, especially and crucially those relating to the 'science' of environmental change, where a more detailed assessment of many issues is necessary.

SIS AND SISI

No agreed definition of a small island state exists, let alone one that includes the various entities that are not quite states. Even defining islands is notoriously difficult. Numerous attempts have been made to conceptualize small size but ultimately smallness is a 'comparative and not an absolute idea' (Wood 1967: 29) and there is a continuum. Definitions have focused primarily on population, but often with some reference to land area and national income. Small may even be seen as a pejorative term, in relation to larger islands or mainlands, where smallness equates with unimportance. At the 2012 Pacific Forum, it was suggested that Pacific entities were 'large ocean' rather than 'small island' states.

When small states were fewer, Kuznets (1960) used an upper limit of 10 million people. In the formulations of Demas (1965) and Jalan (1982), small size was under 5 million but smallness was revised downwards as smaller independent microstates emerged. The Commonwealth Secretariat, after once using 500 000, settled on a threshold of 1.5 million people (but included much larger member states, such as Jamaica, Lesotho and Papua New Guinea, some of which were not islands). The World Bank has used a variant of this while numerous other definitions exist (Hein 1985, 2004, Crowards 2002, Liou and Ding 2002).

This book primarily focuses on some of those states that have been classified as Small Island Developing States (SIDS). The United Nations identifies 52 such states (38 of which are independent) most of which are included here. However, several have been excluded – Bahrain, Belize, Cuba, Dominican Republic, Guinea-Bissau, Guyana, Haiti, Jamaica, Mauritius, Papua New Guinea, Surinam, Timor Leste (East Timor) and Trinidad – either because they are not physically small, nor with small populations (all having over a million), or are not islands (cf. Maass 2009, Sutton 2011). That is not to say that they have very different characteristics from the places discussed here. Singapore has been excluded as a rich world state, as have relatively 'developed' small island states in temperate regions (Cyprus and Malta). Consequently the island states that are subsequently referred to in this book are the 24 residual independent island states, in plain type in the list below and referred to here as small island states (SIS) (see Table 1.1). The list of countries included here has some similarities with membership of the Alliance of Small Island States (AOSIS), a coalition of small island and low-lying coastal countries that regard themselves as sharing similar development challenges and environmental vulnerability, which functions as a lobby group and negotiating voice for SIDS within the United Nations system. AOSIS has a membership of 42 states and observers, including such large island states as Cuba and Mauritius, but excluding dependent territories such as Anguilla or Aruba.

A range of dependent territories, with political ties to France, the UK, Australia, New Zealand, the United States and the Netherlands, where the geographical and socio-economic context is similar, are also included here. However, one British overseas territory (Gibraltar), some temperate dependent territories (e.g. Greenland, the Faroes and St Pierre and Miquelon) and Puerto Rico and Réunion (with over 4 million and 850 000 people, respectively) have effectively been excluded. Little information is available on several such dependent territories (Aldrich and Connell 1998). Some reference is also made to islands in larger tropical states such as Papua New Guinea (PNG) and Mauritius, and to other islands (such as the Galapagos and Rapanui/Easter Island) that offer similarities and parallels in comparable contexts.

The book thus concerns islands and states where loose notions of 'islandness' or insularity are evident (such as the entire population identifying with the island and country, or the island being too small to influence its macro-climate significantly). That certainly does not mean that there are no divisions, tensions or climatic variations on islands. (Indeed coastal 'saltwater' people often differentiate themselves from those inland, while rainfall varies enormously on high islands.) This also accounts for the exclusion of larger islands, matching a distinct and convenient demographic break between Mauritius with about 1.3 million people, which is excluded, and Fiji with about 850 000 (though such a break has no necessary economic or environmental significance, and the two states share similarities). No definitions and categorizations are ever entirely satisfactory.

The book therefore seeks to consider as far as possible the following places:

Atlantic and Caribbean: *Anguilla*, Antigua and Barbuda, *Aruba*, *Ascension*, Bahamas, Barbados, *Bermuda*, *British Virgin Islands (BVI)*, *Bonaire*, Cape Verde, *Cayman Islands*, *Curaçao*, Dominica, *Falkland Islands*, Grenada, *Guadeloupe*, *Martinique*, *Montserrat*, *Saba*, Saint Kitts and Nevis, *St Helena*, St Lucia, St Vincent and the Grenadines, *Sint Eustatius*, *Sint Maarten*, São Tomé and Principe, *South Georgia and the South Sandwich Islands*, *Tristan da Cunha*, *Turks and Caicos Islands*, *US Virgin Islands (USVI)*.

Pacific: *American Samoa*, *Cook Islands*, Federated States of Micronesia (FSM), Fiji, *French Polynesia*, *Guam*, Kiribati, Marshall Islands, Nauru, *New Caledonia*, *Niue*, *Norfolk Island*, *Northern Marianas*, Palau,

Pitcairn, Samoa (formerly Western Samoa), Solomon Islands, *Tokelau*, Tonga, Tuvalu, Vanuatu, *Wallis and Futuna*.

Indian Ocean: *British Indian Ocean Territory (BIOT), Cocos-Keeling, Christmas Island*, Comoros, Maldives, *Mayotte*, Seychelles.

Some 59 small island states and territories are thus discussed, however briefly, alongside various other islands. Ultimately inclusion is arbitrary. Collectively the 59 entities are referred to here as small islands and small island states (SISI) although their diversity only occasionally enables generalizations at that level. Basic data for the distinct states and territories are provided in Tables 1.1 and 1.2. Providing comparable data for the various islands, under different jurisdictions, occasionally referred to in the book, is simply not feasible.

SISI are diverse, in size (of land area and population), geography, culture and history, in health status, income levels, life expectancies and so on. While they have small populations and are tropical, similarities often scarcely go beyond that. Indeed 'it is not always easy to disentangle the effects of smallness from those of remoteness or peripheralness … It may not in the end be especially profitable to do so' (Davies 1985: 248). A few, such as Nauru or Mayotte, are single island states or territories. Many, such as the Bahamas, Kiribati and the FSM, are multi-island states. Within these states there is often considerable diversity of lands, peoples, and economies. The FSM consists of four states, spanning over 1000 km of the western Pacific, each with separate languages, three of which states have a central 'high island' and a scatter of outer islands that are coral atolls, the definitive 'low islands'. Languages and societies and, in one state, ethnicities, also differ. Vanuatu has 60 populated islands and over a hundred languages. The nearby Solomon Islands has a mere 60 languages, almost as many populated high islands, and a cluster of outlying atolls that are home to ethnic minorities. By contrast the tiny British overseas territory of Pitcairn has fewer than 60 people, yet is more than 1000 km from the next occupied island. Tristan da Cunha is even further (2400 km) from any other inhabited place. Economies and political systems change and some regions are tectonically unstable. At least two new volcanic islands briefly appeared in Tonga in this century, while Urelapa island (Vanuatu) is uplifting at a rate of 3 mm/year (Cabioch et al. 2003). Such transformations, in a context where sea-level rise (SLR) dominates debate on environmental futures, emphasize uncertainty and diversity, both of which extend deep into the economic arena. Environments and livelihoods vary enormously.

Table 1.1 Small Island States

Small Island States	Population 2010	Annual Population Growth Rate (%) 2000–2010	Population Density/km²	GDP per capita (USD$) 2010	Life Expectancy (years) 2010	HDI (ranking) 2010	ODA as a % of GNI 2010
Atlantic and Caribbean							
Antigua and Barbuda	89 500[e]	1.3	201	12 600	73[e]	0.76 (60)	2
Bahamas	353 658	0.9	34	14 500	75[e]	0.77 (43)	0
Barbados	273 925	0.4	598	14 300[e]	75	0.79 (42)	0
Cape Verde	491 575	1.4	127	3 200	74	0.57(118)	21
Dominica	72 000[e]	0.2	99	7 000	76	0.72 (81)	9
Grenada	104 000[e]	0.5	305	7 400	76	0.75 (67)	6
Saint Kitts and Nevis	51 000[e]	0.8	192	10 500	73 (2008)	0.74 (72)	2
St Lucia	176 000[e]	0.4	282	6 700	74	0.72 (82)	5
St Vincent and the Grenadines	109 000[e]	–0.3	280	6 100	72	0.72 (85)	3
São Tomé and Principe	168 500[e]	2.0	168	1 300	64	0.51(144)	25
Pacific							
Federated States of Micronesia	102 360 (2011)	0.4	146	2 700	69	0.64(103)	41
Fiji	851 745 (2011)	0.5	47	3 500	69	0.69 (86)	3
Kiribati	102 697 (2011)	1.8	127	1 500	63[e]	0.62(122)	11
Marshall Islands	55 000 (2011)	0.7	304	3 100	71[e]	–	48
Nauru	9 947 (2011)	2.1	485	6 200	56 (2006)	–	–
Palau	20 643 (2011)	0.6	46	10 800	71[e]	0.78 (49)	20

Table 1.1 Continued

Small Island States	Population 2010	Annual Population Growth Rate (%) 2000–2010	Population Density/km²	GDP per capita (USD$) 2010	Life Expectancy (years) 2010	HDI (ranking) 2010	ODA as a % of GNI 2010
Samoa	183 617 (2011)	0.3	63	3 300	72	0.69 (99)	27
Solomon Islands	553 254 (2011)	2.7	18	1 200	67	0.51(123)	61
Tonga	103 682 (2011)	0.3	160	3 500	72	0.70 (85)	19
Tuvalu	11 206 (2011)	0.5	430	3 200	64 (2002)	–	34
Vanuatu	251 784 (2011)	2.6	21	3 000	71 (2009)	0.62(125)	15
Indian Ocean							
Comoros	753 939 (2011)	1.4	360	700	61	0.43(140)	13
Maldives	245 000[e]	1.9	1 040	8 400	75	0.66(107)	8
Seychelles	86 525	0.9	190	11 500	73[e]	0.77 (52)	7

Notes: [e]Estimate only.

Sources: Population and GDP per capita are from the United Nations (data.un.org), the CIA World Factbook (https://www.cia.gov/library/publications/the-world-factbook/geos/kt.html) and the South Pacific Commission's Population and Demographic Indicators (www.spc.int). Where states had their own up-to-date accurate data available this was preferred. Life expectancy data came partly from UNICEF (http://www.unicef.org/infobycountry/index.html). HDI rankings are from UNDP (http://data.un.org/DocumentData.aspx?q=barbados+hdi&id=269). ODA data were taken from Aid Flows (www.aidflows.org). While some data are formally regarded as estimates other data also fall into that category.

Table 1.2 Small Island Territories

Small Island Territories	Population 1985	Population 2010	Annual Population Growth Rate	Population Density/km²	GDP per capita (USD$) 2010	Life Expectancy (years) 2010
Atlantic and Caribbean						
Anguilla	6 680 (1984)	16 373[e]	2.1	164	11 200[e]	81
Aruba	61 400	108 000	1.4	594	22 850	76
Ascension	712	884	–	10	–	–
Bermuda	58 000 [e]	65 000[e]	0.6	1 210	69 900	79 (2009)
British Virgin Islands	10 985 (1980)	29 537[e]	2.4	149	39 000	77
Bonaire	10 791 (1990)	13 389[e]	–	46	–	76 (2008)
Cayman Islands	21 000	56 000[e]	2.2	212	43 800	81
Curaçao	147 388 (1981)	197620[e]	–	320	20 000 (2009)	76 (2008)
Falkland Islands	1 916[a] (1986)	3 200[e]	0.4	0	55 400	–
Guadeloupe	355 000	46 1000[e]	0.9	270	–	78 (e 2005)
Martinique	341 000	406 000[e]	0.6	360	–	80
Montserrat	11 000	5 500[e]	0.5	55	8 500	73[e]
Saba	1 000 (e1991)	1 737[e]	–	134	–	–
St Helena	6 000	4 000[e]	–0.3	35	2 500	–
Sint Eustatius	1 500 (e1991)	2 886[e]	–	137	–	–
Sint Maarten	13 156 (1981)	37 429[e]	–	1 100	–	–
South Georgia and the South Sandwich Islands	–	30 (2006)	–	0	–	–
Tristan da Cunha	296 (1987)	285	–	3	–	–

Table 1.2 Continued

Small Island Territories	Population 1985	Population 2010	Annual Population Growth Rate	Population Density/km^2	GDP per capita (USD$) 2010	Life Expectancy (years) 2010
Turks and Caicos Islands	11 465 (1990)	42 000[e]	3.2	45	–	75[e]
US Virgin Islands	105 000	108 000[e]	0.0	344	40 000	79[e]
Pacific						
American Samoa	39 000	66 692	1.2	335	8 000 (e2007)	74[e]
Cook Islands	18 000	15 576	–0.3	66	12 200	73
French Polynesia	174 000	271 831	1.2	77	25 000[e]	75
Guam	119 000	159 358	2.7	355	15 000 (e2005)	76 (2008)
New Caledonia	155 000	252 331	1.3	14	–	76 (2007)
Niue	3 000	1 400	–2.3	6	–	72
Norfolk Island	2 367 (1986)	2 302 (2011)	–1.2	67	–	–
Northern Marianas	30 000	63 517	–0.1	139	–	77 (2008)
Pitcairn	58	66 (2011)	–	15	–	–
Tokelau	1 577 (1991)	1 162	–0.2	97	–	70[e]
Wallis and Futuna	14 000	13 193	–0.6	93	–	74 (2008)
Indian Ocean						
British Indian Ocean Territory	666 1968)	4 000[e,b]	–	66	–	–
Cocos-Keeling	579 (1983)	600[e]	0.0	42	–	–
Christmas Island	–	1 500[e]	0.2	10	–	–
Mayotte	71 000	199 000[e]	3.0	532	6 500 (e2005)	76 (2010)

Notes: [e]Estimate only. [a]Excluding military personnel. [b]Military personnel only.

Sources: As Table 1.1

Data coverage similarly is extremely variable, and often poor, compared with that for large states. Basic data deficits – and uncertainties over the quality and reliability of available data – rarely enable assessments of even contemporary influences on ecosystems or economies, let alone temporal change. First, even basic population estimates and projections, especially in the Caribbean SISI, are far from accurate. Data on significant issues such as inequality and poverty are rare; as in the relatively large Seychelles and Mauritius, 'anecdotal evidence points to a strong feeling of increasing inequality, particularly among the poorest' (Kothari and Wilkinson 2013). Such anecdotes, 'strong feelings' and uncertainties are true elsewhere and for a range of variables. Other data may simply be inaccurate. In the Comoros 'the commonly quoted statistics are often contradictory, out of date or guess work' (Hauzer et al. 2013). Official statistics on national land areas, as for Wallis and Futuna, may be quite wrong. Economic trends are particularly poorly recorded. Second, SISI vary enormously – most obviously physically between high islands and atolls, but also in size, location, resource availability, population and political status. Conclusions valid for some SISI (or parts of SISI) are invalid elsewhere. Third, the underlying variables (such as climate) that influence environmental change are poorly recorded, and data sources are particularly fragmented and over short durations. Consequently regional climate modelling is an area of considerable uncertainty even where data are more readily available and climates less subject to local variation (Foley 2010). Moreover short-term climatic variations emanating from El Niño can oppose global trends in average temperature and demonstrate the need to integrate the effects of climate change across multiple time scales (Hilbish et al. 2010). Relative to larger states information on critical hazard events and their impacts are lacking (Méheux et al. 2007, Simpson et al. 2011). Research on soil degradation and soil fertility is conspicuous by its absence. In recent years research in numerous places and contexts has declined, through changing research objectives, funding shortages and difficulties of access. No social, biological or physical science has ever been conducted in some places, even on islands such as Aniwa (Vanuatu), a high island of about 400 people, an hour's flight from the capital city, where just two articles of no great significance were the sum total of intellectual and practical inquiry (Wilson 2010). Major hazard events such as Hurricane Lenny in 1999, distinctive because it travelled in an unusual pattern through the Caribbean, have barely been evaluated in any detail. Very little practical research has been undertaken in entire territories such as Anguilla and both Virgin Islands groups. What has been written of data on water quality and quantity in the Pacific region – 'there has been a post-colonial decline in the level and

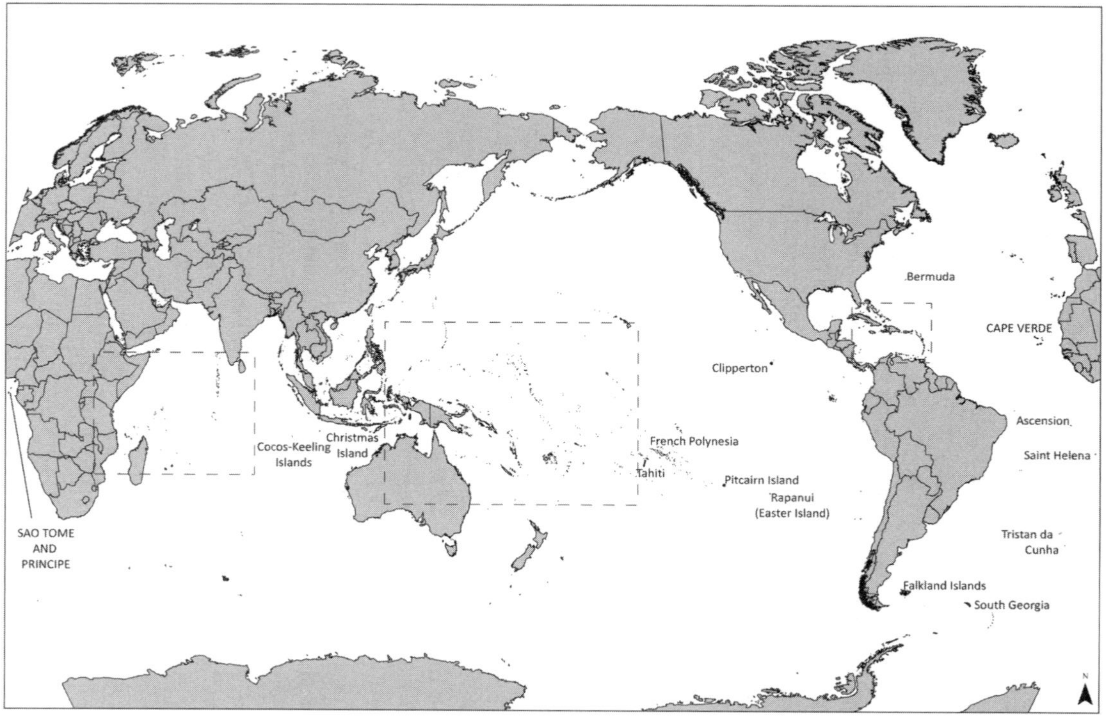

Figure 1.1 World Map of Islands

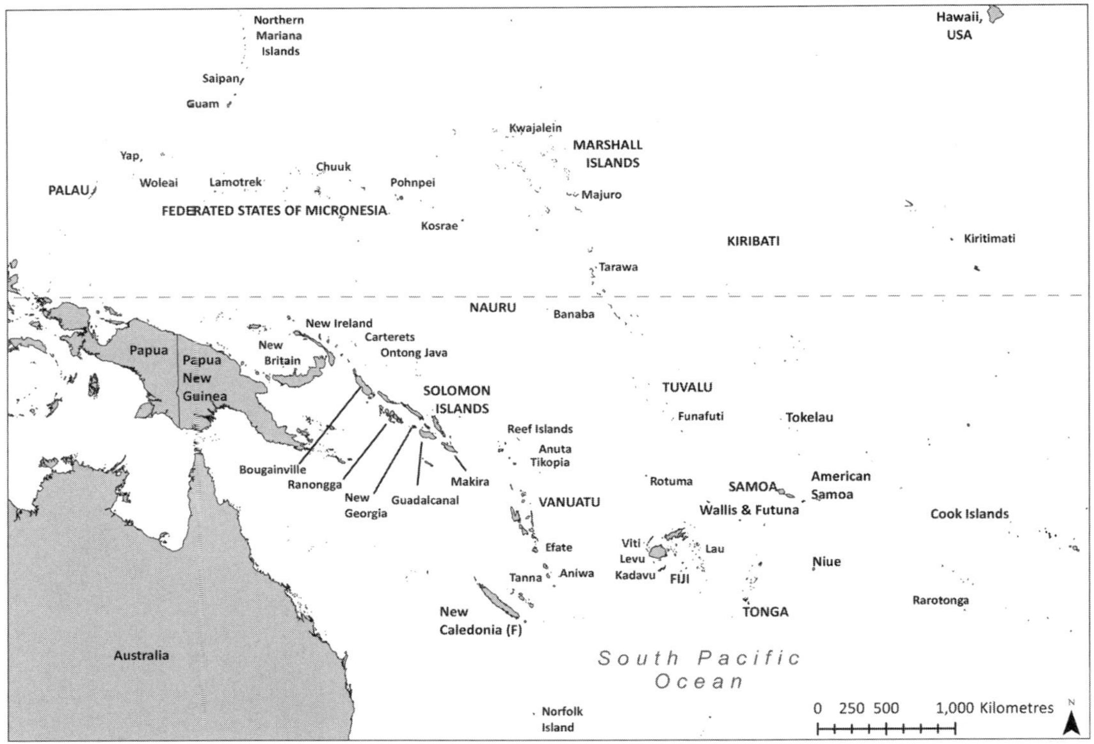

Figure 1.1a Pacific Islands

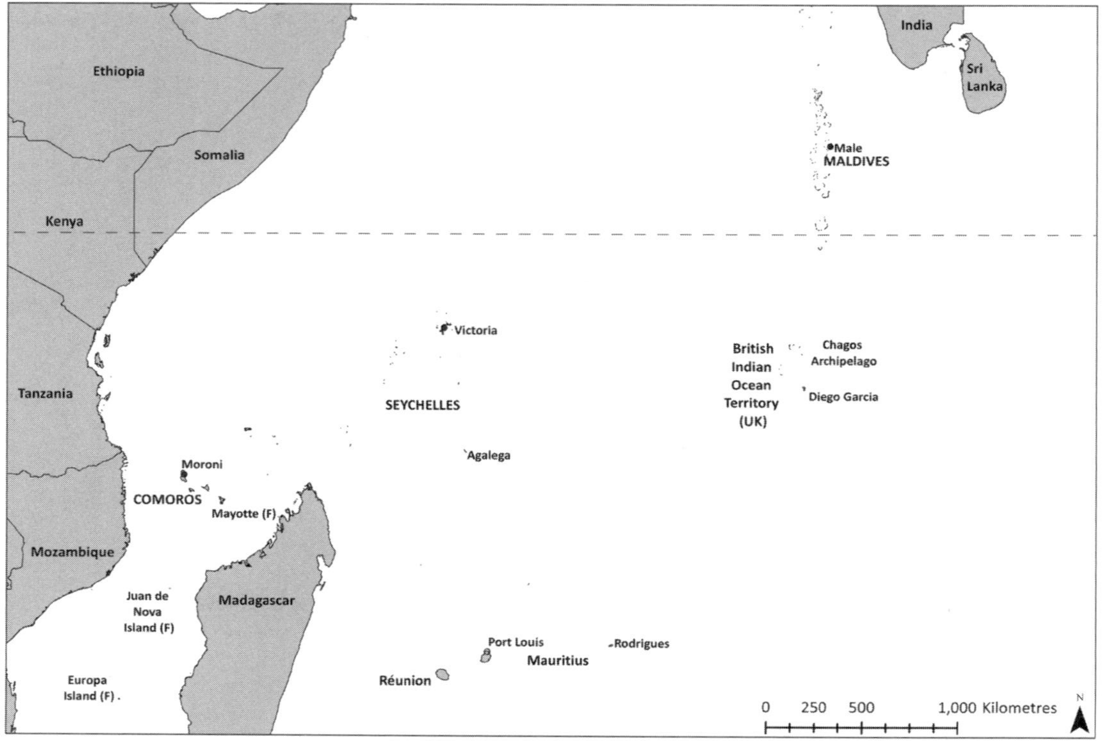

Figure 1.1b Indian Ocean Islands

Figure 1.1c Caribbean Islands

continuity of monitoring activity' (UNEP 1999: 11) – is just as true in most other places and for many topics. Unfortunately speculations and guesstimates have sometimes had to be made in what follows.

Some of what follows is based on 'spot' research in particular places, whose generality is sometimes limited and often impossible to determine. In terms of environmental change, much research, and even more media emphasis, has focused on 'problem areas', especially where SLR is involved, and this has biased analysis. Likewise much general information comes from more accessible centralized urbanized islands. Finally several states are still actively negotiating the extent of their coastal waters under the United Nations Convention on the Law of the Sea (UNCLOS). The task of tracing evolving relationships between environmental change and human activities and outcomes is particularly difficult. Relatively few detailed studies have been undertaken in SISI on changing ecosystems 'despite such countries harbouring the majority of the world's biodiversity and the high dependence of people on those services for their survival and livelihoods' (Kenter et al. 2011: 506). As the lengthy bibliography might suggest, this book is an attempt to provide a more comprehensive perspective on these relationships.

LOW-LYING AND HIGH ISLANDS

Conventional divisions distinguish low islands and high islands. 'Low islands' are synonymous with atolls, coral cays and reefs. Such islands are sinuous, rarely more (and usually rather less) than 3 m above sea level at their highest point, and consequently highly vulnerable to natural hazards. They are also usually seen as being in the frontline of climate change and SLR. A small number of independent states (Kiribati, Marshall Islands, Maldives and Tuvalu) and dependent territories (Tokelau, Cocos-Keeling Islands) are exclusively composed of such islands. Others, such as the Cayman Islands, the Turks and Caicos and the Bahamas are primarily low-lying limestone islands and at particular risk from storms. Because of the coral and limestone structures of low islands, most of which are also small in area, biodiversity is limited and agriculture constrained, because of poor soils and dependence on rainfall, while natural resources are particularly scarce.

High islands, by contrast, are generally regarded as those islands that are not coral atolls, but conventional classifications distinguish 'continental islands', formerly linked to a nearby continent, from 'oceanic islands', that are of independent volcanic origin. Continental islands are usually much larger, like those of New Guinea and Solomon Islands, with a much greater ecological diversity than oceanic islands, having once been connected to continental mainlands. Oceanic islands have emerged independently, often by volcanic eruption. They may not necessarily be 'high' above sea level (e.g. Bermuda), are often quite isolated and may also occur in association with fringing coral reefs (e.g. Aitutaki). They are distinct from atolls in having a developed soil structure, a more diverse flora and agricultural potential, natural resources (including minerals) and a reduced risk from storms and SLR. High islands also include such raised coral atolls as Niue and Nauru, whose soils are porous and whose biological potential is limited. Ultimately attempts to set physiographic criteria and distinctions are not particularly helpful for distinguishing categories of high islands, though atolls constitute a broadly similar and comparable category.

A larger number of small island states (e.g. Cape Verde, Comoros, Grenada, São Tomé) and dependent territories (e.g. St Helena and Ascension) are effectively exclusively composed of (populated) high islands. In most cases they are at less risk of negative impacts from environmental change, though many have fragile ecosystems and limited biodiversity. Several states, such as Solomon Islands and FSM, include both high and low islands, and thus a great range of ecosystems (including cloud forests in the montane Solomon Islands). In some islands the fauna and flora are so poorly studied and recorded that new discoveries, even of small mammals, are still being made. On the large island of Santo (Vanuatu), as many as a thousand new species were reportedly found in 2006 alone (Bouchet et al. 2011). In the Marquesas (French Polynesia) some 62 new vascular species, mainly ferns, were recently found in the high elevation cloud forests, increasing the number of known species of the island group by 20 per cent (*Radio Australia*, 2 August 2011). The extent of biodiversity, still being revealed, is much greater on high islands.

A SHORT SYNOPSIS

The book first sets the general background by examining the overall economic and political structure of SISI, the constraints to achieving development that they face and their vulnerability and resilience (Chapter 2). It then examines their changing agricultural and fisheries systems, once almost universal, but steadily becoming anachronistic (Chapter 3). The following chapter traces the extent of economic evolution towards a more 'modern' economy that embraces manufacturing, tourism and services, some of which have taken unusual forms (Chapter 4). Transition in economic development has almost everywhere accompanied urbanization, but an urbanization without industrialization, rather being associated with limited job opportunities, a 'youth bulge' and consequent high levels of unemployment and marginal living conditions for many (Chapter 5). Scarce local opportunities, for employment, incomes and access to services such as education, have stimulated an increasingly complex international migration to metropolitan states, and thus the emergence of a substantial diaspora, with island economies heavily dependent on remittances for social and economic development (Chapter 6). Socio-economic changes have been paralleled by less obvious but highly significant environmental changes, partly in the degradation of land and marine environments, notably reef systems, especially where there has been a coastal squeeze. A more insidious environmental change is linked to global warming and SLR, some evidence for and impacts of which are now perceptible (Chapter 7). The final chapter draws together the various threads that link SISI to wider worlds, and which tie social, cultural, economic and environmental stability and change together, and so reiterate the diversity of both islands and island futures. This then is not a book about economics, politics or the environment, but in a very loose sense purports to be a political ecology that provides an interdisciplinary perspective on several complex island realms.

2. Islands and political economies

> If São Tomé and Principe were a private company it would probably have long ago been declared insolvent. The country owes its continued existence to an international system which guarantees the survival of small internationally recognized, albeit unviable, state entities.
>
> Frynas et al., 2003: 60

Small island states and islands are and have been particularly disadvantaged in their efforts to achieve economic development, hence the emergence of the SIDS category. Most SISI are small and vulnerable, economically and environmentally. At the core of most debates about development are vague but long-established notions of viability, which usually conclude that small size precludes 'genuine' development. Indeed the discursive construction of smallness as synonymous with vulnerability is so pervasive that alternative perceptions pointing to smallness as no necessary barrier to development, and stressing the power of agency at local and national level, have been largely discounted (Lee and Smith 2010). However, rather more than remnants of agency exist in SISI.

Many states are isolated and fragmented, with numerous populated islands and, in Melanesia especially (from the New Guinea islands to New Caledonia), with many distinct language and cultural groups. Prospects for economic growth are limited, especially in Polynesia, Micronesia and the eastern Caribbean (Antilles), where small states are remote from international markets. Ten island states are officially classified as Least Developed Countries (LDCs). Some, like Nauru and Solomon Islands, have been described as failed states, in being unable to provide physical security and/or public goods and infrastructure to citizens, though that epithet is challenged (Chauvet et al. 2010). Others, like the Comoros, Fiji and the Maldives, have been characterized as having a propensity for coups. Politics is uneasy. Diverse cultures and economies, and a usually peaceful transition to independence, have posed problems for stimulating national identity and nationhood, especially in Melanesia and Micronesia. Secession struggles were once frequent, notably in Solomon Islands and Vanuatu, and islands like Nevis, Barbuda, Rotuma, Anuta, St Croix and Anguilla have had complex and contested political and administrative relationships with larger islands in the same region (e.g. Roopnarine 2011), just as Bougainville, the Galapagos and Rapanui have with distant mainlands. Only in Mayotte has secession from any SIS succeeded, with that island choosing to remain with France rather than achieve independence with the other Comoros. Both Tuvalu and Mayotte demonstrate that secession provides material rewards.

Political evolution is ongoing. Most French overseas territories and departments have changed status several times in recent decades. Saint Martin and St Barts separated from Guadeloupe to become French Overseas Collectivities in 2007. The Netherland Antilles fragmented late in 2010; two islands (Curaçao and Sint Maarten) achieved greater autonomy while the three others (Bonaire, Sint Eustatius and Saba) established stronger ties with the Netherlands. Bougainville (PNG) and New Caledonia are expected to vote on their political future in this decade. In most SISI government is highly centralized, a legacy of colonial times, despite token efforts at decentralization and regional government, and effective government reach has often weakened. This chapter seeks to examine the basic structures of economic development and politics in SISI and how these have evolved in recent years.

ECONOMIC DEVELOPMENT

In most SISI conventional economic development has been characteristically limited through a combination of now well-known factors including small populations, limited resources, remoteness, fragmentation (especially in the Pacific), susceptibility to extreme events (from cyclones to earthquakes), vulnerability to external economic shocks, excessive dependence on aid and imports, and fragile environments. Growth and development are constrained by high communication and energy costs, irregular international transport volumes (and uneconomic routes and loads), disproportionately expensive public administration and infrastructure, scarce local skills, problems of matching skills and jobs, limited domestic capital, small domestic markets and few opportunities for economies of scale. Manufacturing is similarly disadvantaged and without backward linkages to the agricultural sector. All such problems are compounded in remote archipelagic states. Consequently development has long been widely recognized to be constrained by small size (Demas 1965, Benedict 1967, Jalan 1982, Dommen and Hein 1985, Connell 1988a, McKee and Tisdell 1990, Fairbairn and Worrell 1996, Easterly and Kraay 2000, Armstrong and Read 2003, Winters 2005a, b, McGillivray et al. 2010).

Development constraints have resulted in narrow export specialization and market orientation, which 'exposes the economies of ministates to real shocks of an intensity unparalleled in larger countries' (Galbis 1984: 37). Export prices and foreign exchange earnings can fluctuate considerably and alarmingly. Transport costs are such a high proportion of import costs and export incomes that some crops are too expensive to market. Unlike larger neighbouring states Montserrat was unable to sell high quality bananas because it grew too few to pay for good harbour facilities, let alone afford disaster insurance, or to attract shippers bound by regular inter-island services (Lowenthal and Comitas 1962). The 'transport stranglehold' has been seen as the single most important constraint to island development (Brookfield 1987). Several states, including Kiribati and the Seychelles, have had to reclaim land for housing, since land resources are so scarce. Economic growth benefited from proximity to a global region, such as Europe, where trade and tourist potential were superior, and was weakest in regions with many other small states: the Pacific and the Caribbean. The most remote states, especially those in the Pacific, were particularly disadvantaged (Gibson 2007). Limited economic growth and development opportunities have resulted in substantial emigration, reducing the pool of qualified professionals. Relatively generous overseas aid and remittances have not compensated for such cumulative disadvantages.

Political sovereignty brought increased and disproportionate costs for state functions, including environmental protection, infrastructure development, national defence and international representation, otherwise undertaken by colonial authorities (Campling and Rosalie 2006). Yet several independent states are dependent on such external institutions as universities and banks for key services. Currencies are often linked to overseas currencies (as the Tongan pa'anga is linked to the Australian dollar, or the Eastern Caribbean to the US dollar), or overseas currencies are the national currency (as the Australian dollar is in Kiribati and Tuvalu) or in free circulation (as the US dollar is in Bermuda and Antigua). There is, therefore, limited scope for exchange rate and monetary policies, further emphasized by open economies, though this provides some insulation against external volatility. Even the largest SIS are heavily dependent on external influences and linkages. Fiji's economy, for example, is based on tourism (dependent on Australians and New Zealanders), sugar (dependent on currently suspended EU support), remittances (dependent on Fijian expatriates resident overseas or employed by US security companies and as UN peacekeepers) and aid (from various countries, including China). Fiji is also reliant on financial support from the International Monetary Fund (IMF), World Bank and the Asian Development Bank (ADB) (where the US and Australia wield considerable influence). While Fiji's range of external contacts is more wide-ranging than some SISI it emphasizes the growing diversity of external linkages. The much smaller state of Samoa is a member of some 34 international agencies (Macpherson and Macpherson 2009), even excluding significant religious and sporting organizations; a crude but impressive measure of the extent of globalization. One way or another, the fate and future of SISI are at least partly in the hands of others.

Classifications

Many studies have sought to classify countries according to various economic, political, environmental and social variables, with a growing degree of sophistication despite inadequate data. Most such categorization reached the conclusion that small size was disadvantageous, had similar ranking outcomes (Crowards 20002, Liou and Ding 2002) and failed to stimulate policy formation. Thus early studies from the Pacific region, adopting conventional perspectives on development broadly convergent with economic growth, simply concluded that the largest states had better development potential, if usually unrealized, than smaller more remote states and especially Kiribati and Tuvalu. Patience with classification quickly wore thin, and at the end of one of the first regional evaluations of development issues in the Pacific, Arnold Zable concluded that 'the categories of analysis were beginning to take over and the people … were disappearing' (1973: 430). Classification nonetheless continued with growing subtlety but largely predictable results and divorced from policy directions.

The emergence of newer smaller states and the gradual development of better data sets made classification and measurement more complex. Despite increasing subtlety, variables remained primarily economic. A group of LDCs was first distinguished by the UN in 1971 centred on three criteria (income, human development – nutrition, literacy, health and education – and economic vulnerability). Most were in sub-Saharan Africa but, as the list grew to 49, nine SIS were included (Cape Verde, Comoros, Maldives, Kiribati, Samoa, São Tomé, Solomon Islands, Tuvalu and Vanuatu). In 40 subsequent years only three countries (including Cape Verde and the Maldives) graduated to 'developing' country status, though in 2011 the UN suggested that Samoa, Tuvalu and Vanuatu might soon follow. In 1972 UNCTAD initiated studies of 'land-locked and island developing countries', that eventually became

SIDS, as it became apparent that such states faced distinctive challenges (Hein 2004). The emergence of a distinct group of SIDS at the United Nations Conference on Environment and Development in 1992 was followed two years later by the Barbados Programme of Action for the Sustainable Development of Small Island States – the first attempt to define policies specific to SIDS.

The dominant economic orientation remained evident in 2001 when the World Trade Organization (WTO) set up a work programme on Small Economies, in which the terminology gradually shifted from 'small state' to 'small vulnerable economy'. Formal definitions of disadvantaged countries were resisted since the evidence of disadvantage resulting from smallness or islandness was equivocal and could be challenged by larger developing countries seeking special consideration in the global economy (Read 2004, Sutton 2011). Thus a 'small vulnerable economy' was never defined and the WTO resisted special treatment for any group of countries (see below). This did not however prevent the emergence of either the SIDS, with 52 eventual members, although no definition was ever agreed upon, or the related AOSIS, established in 1990.

In 2000 the World Bank also began to take a special interest in small states, eventually identifying 'a diverse group of some 40 sovereign countries with limited populations for which diseconomies of scale and resource constraints constitute inherent challenges' (World Bank 2012: ii). That group excluded Tuvalu and Nauru who were not World Bank member states. Despite many SIDS being classified as 'middle income countries' efforts by the countries themselves, and organizations such as the Commonwealth Secretariat, prevented their 'graduation' from being recipients of International Development Association (IDA) resources, since it was argued that their economies shared many of the problems of low income countries and experienced particular additional problems such as vulnerability to national disasters, isolation and lack of natural resources. Small states thus perceived themselves as quite different from larger states and disadvantaged by their size.

Some classifications included dependent territories, to consider whether political independence conveyed economic advantages. Detailed analysis of 105 small states and territories, including both more developed entities, such as Bahrain, Iceland and Singapore, and political dependencies, unsurprisingly revealed considerable diversity. Dependencies, though usually smaller, benefited from colonial ties to the extent that their welfare and human development circumstances were often better than in SIS, because of extensive budgetary support, but economic growth was usually weaker (Armstrong and Read 2000, 2004, Read 2004, Dunn 2011). Their superior socio-economic status (and small size) partly explained why the push for independence effectively reached an infinite pause by the 1980s (Aldrich and Connell 1998). Islandness was a further distinct disadvantage, mainly because of the particular vulnerability of island states to external shocks, but also because of transport costs.

By the 1990s classifications placed greater emphasis on vulnerability, primarily economic (an openness to trade) and later increasingly environmental, with one such index placing six SIS (Antigua, Tonga, Seychelles, Vanuatu, St Kitts and St Lucia) as the most vulnerable of 114 countries (Briguglio 1995). Indices developed by the Commonwealth Secretariat centred on 'income volatility', which was linked to economic exposure (dependence on international markets, remoteness and insularity reflected through transport costs) and susceptibility to environmental hazards. Ranking countries by the resultant combined index placed São Tomé as most vulnerable and Barbados least vulnerable; ranked within over 100 developing countries, only 8 of the 26 most vulnerable states were not SIS. São Tomé was the most vulnerable to commodity price fluctuations and Vanuatu to natural disasters (Easter 1999, Atkins et al. 2000). Small states were disproportionately vulnerable to natural disasters with 13 of the 25 countries that had experienced the greatest number of natural disasters during the 1970s and 1980s being SIDS (Pelling and Uitto 2001). The UN Committee for Development Policy devised an Economic Vulnerability Index in 2000, as most analyses continued to centre on more easily measurable economic variables, partly because remedies to development problems appeared primarily economic. That index combined three 'shock indices' (instability of exports of goods and services, instability of agricultural production and a homelessness – due to natural disaster – index) and four 'exposure indices' (small population size, remoteness, export concentration, share of agriculture, forestry and fisheries in the GDP). With smallness and remoteness as variables SIS again fared poorly, with Kiribati and Tuvalu the two most vulnerable states (Guillaumont 2009). Such indices pointed to the impact of globalization and the marginalization of small states within the global economy, which led to an increasing focus on trade issues, alongside a more distanced interest in the mitigation of environmental hazards.

Although 'economic' variables continued to dominate most indices, growing evidence of the impacts of environmental degradation, hazards and change resulted in more sustained attempts to incorporate environmental factors. An Environmental Vulnerability Index was created in the early 2000s focusing on three concepts – the risks associated with hazards, the ability to resist hazards and the vulnerability caused by degradation of ecosystems – and with 50 indicators. It ranked American Samoa as the most vulnerable state, but nearby Samoa, with very similar characteristics, was ranked 81st. The challenge of creating and measuring concepts of environmental vulnerability at a national scale proved too great and the Index soon disappeared (Barnett et al. 2008). Another attempt to devise a measure of physical vulnerability for developing countries used four variables: a coastal index (the length of coastline related to total land area), a peripherality index (remoteness and insularity), urbanization and vulnerability to natural disasters (during 1970–96). While the results were somewhat tautological and predictably concluded that smaller island states were most vulnerable to biophysical, socio-economic, political and environmental hazards (Turvey 2007), it further emphasized the range of vulnerabilities that were particularly damaging for SISI.

More formally the extent of physical vulnerability was taken further in the United Nations University's 'World Risk Report' where a World Risk Index was compiled from four components: exposure to natural hazards such as earthquakes, storms, floods, droughts and SLR; susceptibility as a function of public infrastructure, housing conditions, nutrition and economic structure; coping capacities as a function of governance, disaster preparedness and early warning, medical, social and economic security; and adaptive capacity for future natural events (United Nations University 2011). Here more than in most contexts evaluations were subjective, since accurate data were scarce, social factors (such as traditional structures of self-help) could scarcely be measured and some countries were excluded. The outcome indicated that disaster risk was greatest in Asian and Latin American countries but three SIS – Vanuatu, Tonga and Solomon Islands – were among the 15 countries with the greatest disaster risk, with Vanuatu ranked the country most at risk (because of its exposure to earthquakes, and – somewhat improbably – drought and SLR).

As data improved and increased and more institutions took an interest in small states, acronyms proliferated. In the 2000s the World Bank recognized low-income countries under stress (LICUS), which included the Comoros, São Tomé and Solomon Islands. UNCTAD recognized structurally weak, vulnerable and small economies (SWVSEs) that involved many more countries, 29 of which were small island states, and several of which were also LDCs. More complex definitions were intended to focus on particular problems and thus solutions, but usually simply identified the same countries and variants of the same problems. No analysis could effectively define SIS (or small states, or SIDS) as a distinct group, disadvantaged in particular ways that might warrant some form of special policy formation.

The most utilized broadly based index has been the UNDP's Human Development Index (HDI), first established in 1990 and subsequently computed annually. The HDI ignores environmental variables but combines life expectancy, literacy and education, and standard of living (measured by per capita income). It both ranks and classifies countries into four groups: in 2011 one SIS, Barbados at Number 47, was the last country in the 'Very High Human Development' group, 11 were in the 'High Human Development' group, 7 (including Kiribati) in the 'Medium Human Development' group and three (São Tomé, Solomon Islands and Comoros, in 150th place amongst 187) were in the 'Low Human Development' group (UNDP 2011). The last ten countries were all sub-Saharan states – thus SIS fared reasonably well when more weight was attached to social rather than environmental factors. The relatively positive performance of SIS at least partly resulted from previous preferential access to developed country markets (Lewis 2005). While the HDI had some obvious implications for development policies, it implied no necessary strategies or goals. By contrast, the establishment of the Millennium Development Goals (MDGs) by the UN in 2000 set targets for all countries to achieve by 2015. The MDGs focused on three major areas: human capital and wellbeing, infrastructure and social, economic and political rights. While some were hard to measure, they gave new emphasis to environmental sustainability, gender and the relatively poor.

None of these classifications and directions directly concerned political matters, despite the role of governments and institutions in influencing change. States are also vulnerable to social upheavals that have resulted in many being classified as failed states. The economic and social costs of failure are high, especially in terms of lost production, and particularly high in SIS due to the greater openness of economies and the flight of financial and human capital (Chauvet et al. 2010). The journal *Foreign Policy* used 12 indicators covering a range of components of state failure (including extensive corruption and criminal behaviour, inability to collect taxes, sharp economic decline, severe demographic pressures, brain drain and environmental decay), to produce a list of 60 states that are variously 'critical, in danger or borderline' (Anon 2012). Only Solomon Islands and the Comoros were recorded as 'borderline' (though not all SIS are assessed) in a list which otherwise again centred on sub-Saharan Africa.

More philosophical and emotional issues were absent from indices. The Buddhist economics and anti-materialism of Schumacher's *Small is Beautiful* (1973) lay largely dormant until revived in the Earth Institute's 'World Happiness Report' of 2012, which focused entirely on social values, arguing that political freedom, strong social networks, trust and an absence of corruption were more important than incomes in contributing to wellbeing. The report further argued that global society had moved into a new era, the Anthropocene, where humanity, through its technological prowess and growing size, was the major driver of changes of the Earth's physical systems, including the climate, the carbon cycle, the water cycle, the nitrogen cycle and biodiversity: 'If we continue mindlessly along the current economic trajectory, we risk undermining the Earth's life support systems – food supplies, clean water, and stable climate – necessary for human health and even survival in some places' (Helliwell et al. 2012: 3). It suggested that the MDGs be followed by Sustainable Development Goals (SDGs) centred on environmental sustainability, social inclusion and good governance, arguably the indicators that were being developed and collected in Bhutan to measure Gross National Happiness. Its analysis of over 150 countries, based on subjective, self-perceptions of happiness, placed Denmark at the top, with other Scandinavian countries close behind and Togo firmly at the bottom, alongside several other sub-Saharan countries including the Comoros, the only SIS to be included. The Earth Institute's report was based on earlier studies that had included some SIS. Indeed in the New Economics Foundations's compilation of a Happy Planet Index in 2006 Vanuatu was placed top of some 178 countries, based on indices that ignored income but measured life satisfaction, life expectancy and environmental footprint. Dominica, St Vincent and St Lucia were not far behind (New Economics Foundation 2006). While the survey briefly made global headlines, and Vanuatu reaped three years of valuable tourism publicity as the 'happiest country in the world', when the second report appeared in 2009, it and several other SIS had been excluded because of the lack of adequate data. Rather later Vanuatu developed its own set of indicators that sought to measure happiness and reflect Melanesian values, notably resource access, cultural practice and community vitality (Malvatumauri National Council of Chiefs 2012).

Over time, data availability further improved. By 2011 the World Bank could list 331 indicators for 209 countries, compiling these data separately for 40 small states, including 22 SIS, although many were unavailable or of doubtful quality (World Bank 2012). With so many variables, the weight attached to particular variables was less significant while constructing indices became more challenging in terms of analytical or practical utility. Consequently, the same basic variables (notably national income) tended to recur because they could be measured and were readily comprehensible. Indices and classifications have remained largely conformist, despite a focus on MDGs, poverty and the environment that attached concern to sub-Saharan Africa. While SIS rarely featured in the worst categories, several sub-Saharan 'development traps' (Collier 2007), especially being 'landlocked' and experiencing 'bad governance in a poor country' had resonance for SIS.

Indices and the ranking of countries into development silos have been criticized because cut-off values seem arbitrary, they provide incentives for strategic behaviour in reporting official statistics (to obtain aid), statistics are compiled differently (and are less useful in countries, including several SIS, where subsistence economies are important), classification is of limited value for assessing change, is misleading without broader contextualization and is overly deterministic in ignoring cultural diversity (Wolff et al. 2011, Kothari and Wilkinson 2013). National averages disguise regional, social and structural differences (though this is less significant in SIS). Even the subtlest indices are of little use for policy formation. Simplification and subjectivity concealed tautologies and technical flaws. Better data enabled no greater clarity of categories and, even in areas where small states might have been expected to share common goals, as in trade or climate change negotiations, common objectives were absent. Thus while small states were initially united under AOSIS, as the climate agenda grew, differences between AOSIS members became more obvious than common elements, with individual members opposing some AOSIS positions (Betzold et al. 2011). Much earlier in 1978 the Niue government prepared a paper 'Developing island states: a sub-grouping necessary' that called for recognition of the distinct problems of the very smallest countries.

> Not only was the paper rejected by the bigger countries but it led to a very heated argument, confirming the contentious nature of 'smallness' which derives from its intrinsic advantage in a much wider international context. The bigger Pacific countries that rejected Niue's paper were fearful of its implications for their attempt to gain international recognition for the special problems of SIDS. They felt the acceptance of the paper would be harmful to their national interest as it would divert attention away to a special group within the 'developing island states' category, to their exclusion. (Neemia 1986: 119)

For many reasons, not least vested interests (or their acquisition), the creation of particular groups and interests posed problems. Devising criteria and classifications that were straightforward, acceptable, accurate and useful was impossible.

While better data and more sophisticated evaluations added complexity, the same old 'tired colonial and neo-colonial anthropological tropes of categorization and hierarchy, governance and technology' (Richmond 2011: 130) seemed to dominate and failed to capture achievements, freedom, capabilities and opportunities. Such features, let alone self-worth, dignity, equity and creativity, could not be 'completely defined by the rise and fall of national incomes, life expectancy and schooling' despite being more important than the commodities and income people happen to produce and possess (Høyland et al. 2012: 3). Not only did typologies merely emphasize what had become well-known by the 1960s, but also inadequate data resulted in countries like Vanuatu sometimes being placed in strange and fluctuating positions that had limited bearing on local circumstances and no possible practical value. The dead weight of variables ignored the human face and consequences of development and change.

Despite their limitations and biases, development indices and classifications repeatedly emphasized the particular challenges that face small states (from remoteness and susceptibility to natural disasters to limited economic diversification), but also the variations between them. They showed that SIS were rarely so disadvantaged as many sub-Saharan countries (for example, in terms of poverty and malnutrition). The more complex studies emphasized the multidimensional nature of development, and the connectivity and relatedness of variables, especially where resilience is hampered by small size, open economies and narrow specialization, so that natural disasters perpetuate development problems and increase vulnerability (Heger et al. 2009). A real assessment of development trends and status in SISI demands a more subjective focus on local and regional processes and the specificity of their outcomes, at a scale below the national. Ultimately the Bahamas are not Tuvalu, and Bridgetown (Barbados) is far from Adamstown (Pitcairn).

TRADE IN PATCHWORK ECONOMIES

Definitions and categorizations cannot represent the complexity and diversity of social and economic change and development in SISI. Between SIS such as the Comoros and Fiji there are worlds of difference and within them regions of difference, where urban centres are connected to global financial markets and remote islands close to subsistence. The SISI are patchwork economies, where some sectors, countries and regions thrive while others stagnate, and uneven development can be as substantial as in larger states. Yet few SISI exports succeed in global markets while overseas businesses outcompete local rivals. Diversity and patchwork economies raise complex questions for public policy, and demand a range of approaches. 'One size' does not fit all SISI.

Most island economies are weak and have tended to be further weakened through recent trade liberalization, multinationalization of production and capital deregulation. From the mid-1990s trade liberalization resulted in more preferential trade agreements worldwide but a steady erosion in the benefits of post-colonial trade preferences offered to SIS as developed countries liberalized their markets to competing exporters from other states, so disadvantaging small states. Since the early 2000s, the Seychelles, rather typically, experienced a decline in overseas aid, increased competition in tourism and fisheries and the erosion of benefits derived from preferential access to EU markets (Kothari and Wilkinson 2013). Political problems have accompanied and paralleled economic change. Low levels of investment, political instability and failures of governance, manifest in inert and dysfunctional public service systems, have blocked development. The UK government suspended the Turks and Caicos constitution in the 1980s following corruption and extreme political patronage, and again resumed more direct administration between 2009 and 2012 after further extensive corruption (Blom-Cooper 1986, Clegg and Gold 2011). Very similar problems in Sint Maarten had brought similar direct 'colonial' authority from 1993–96. Pitcairn too was effectively recolonized in the early 2000s (Farran 2007). Quite different military external interventions have occurred in several states, where external powers have perceived high levels of instability, from the American invasion of Grenada in 1983, and mercenaries toppling the Comoros government in 1978 to the long-term intervention of the Regional Assistance Mission to Solomon Islands (RAMSI) that began in 2003.

Economic growth has been disappointing in the era of independence, beginning around 50 years ago. It resulted in a series of externally imposed attempts at restructuring, including structural adjustment programmes (SAPs), urged by such institutions as the World Bank, the IMF and, in the Pacific, the ADB, and directed through a free market approach that sought to cut government employment and expenditure, through deregulation and privatization. SAPs were followed by Poverty Reduction Strategies, with more local input, but which emphasized macroeconomic structural changes with little consideration of productive activities including agriculture and fishing (Walmsley et al. 2006). In turn, they were followed by the IMF and World Bank's Heavily Indebted Poor Countries initiative in 2000s, mainly focused on sub-Saharan Africa (thus including only São Tomé and the Comoros), providing debt relief and low-interest loans. While SIS have rarely been considered 'heavily indebted' (or low income) the ten most indebted 'middle-income' economies (with public debt ratios of over 90 per cent of GDP) included four in the Caribbean (St Kitts, Grenada, Antigua and Dominica) and the Seychelles, with debt levels proportionately beyond those of countries such as Egypt and Argentina. Low income SIS such as Solomon Islands have similar high levels of indebtedness (Dodhia 2008). Such external global interventions, however limited in their ability to effectively restructure the economies of SIS in favour of a more sustainable and equitable development, were indicative of a context where economic growth had failed to occur and national debts were mounting.

Trade has long characterized island economies, many of which began as plantation systems for commodities such as sugar and bananas. In the sixteenth century São Tomé was the world's largest producer of sugar, with a subsequent history and political economy like that of many Caribbean states (Frynas et al. 2003). Four centuries later, it was the world's largest cocoa producer. Until recently, favourable trade agreements were invaluable in protecting and promoting such exports, despite some criticism that they maintained antiquated technologies and discriminated against efficient developing country exporters (Armstrong and Read 1998). The commodities in which SIS specialized, including sugar, were prone to unpredictable fluctuations in the world market and to competition from larger states, making them unreliable as sustainable sources of national income in conditions of limited diversity, without some protection, notably the 'Sugar Protocol', negotiated in 1975 alongside the EU Lomé Convention.

Export diversity in island regions is less than in other world regions, and has narrowed in recent years with small states playing a declining role in world trade. The relative importance of SIDS in global trade actually diminished by half (from 0.4 per cent of world exports of goods in 1980 to 0.2 per cent in 2000), although their share of global trade in services remained stable (0.7 per cent of world exports of services, notably tourism). No SIS had more than 0.01 per cent of world trade and most had considerably less (Davenport 2006). For the past 50 years, the rate of growth of exports from small states has been inferior to that of larger states. 'Dependence on a few agricultural products that experience either price or supply shocks on a regular basis is the principal reason for merchandise exports instability' though 'the dismal performance of manufacturing exports [has] contributed most to the declining relative importance of small states in world merchandise exports' (Grynberg and Raffique 2004: vii–viii, ix). Openness and a heavy reliance on international trade mean that even small trade shocks have a large impact, because of export concentration (Baldacchino and Bertram 2009, Guillaumont 2010, Read 2010). Set against the range of disadvantages, the comparative advantages of smallness and isolation are few, and openness provides no compensation. Yet small states as a group have been unable to demonstrate specific disadvantages in evolving global trading systems, and hence able to claim and enjoy special treatment.

More rapid growth and development, derived from free market reform, have been widely touted as justification for reducing trade barriers, and at a global level prosperity has followed liberalized trade. The objectives for admission to the WTO hinged on notions that 'where trade preferences are given for particular products … artificial comparative advantage is generated and economic activity concentrated in these sheltered activities. Resources are thus diverted from where there may be a natural comparative advantage, thereby reducing the much needed diversification in production needed to reduce market risk' (Mahadevan and Asafu-Adjaye 2013, Palmer 2009). However, while trade openness was once seen as an advantage of small size, enabling more flexible responses to change, and growth in small states (Easterly and Kraay 2000, Armstrong and Read 2002, Athukorala 2004), little empirical evidence exists of real advantages to SIS from such an open economy. SIS are marginal to the global economy, areas of comparative advantage at a small scale are few, the resultant diversification was minimal and such theory proved far from practice. SIS were particularly vulnerable to fluctuating international trade regimes. Despite an unprecedented rise in the volume of global trade 'some countries failed to derive significant benefits from the ongoing process of trade liberalisation and globalisation … particularly small states' (Grynberg and Razzaque 2004: vii). At the start of this century SIS were in a relatively more unequal position than two decades earlier.

Characteristically, SISI trade with distant metropolitan states rather than with neighbours (with similar exports), and along transport routes established in colonial times. Trade deficits, made possible because of aid and remittances, are normal. In most cases they have steadily increased for several decades especially alongside trade liberalization. Since the 1990s, various concessionary regional and global agreements, such as the Lomé Convention of the EU and the Caribbean Basin Initiative, have gradually been phased out. Though SIS are often perceived as having 'middle' incomes, a beneficial natural environment and considerable political stability – and thus without the challenges of many least developed states – adjusting from trade preference dependence to a liberalized global trading environment has been daunting (Grynberg 2006). Here above all SIS face serious economic challenges. Lack of complementarity, and greater competition rather than cooperation between states in a more deregulated context, has even resulted in localized conflicts such as the 'kava-biscuit war' between Vanuatu and

Fiji (Connell 2007b, Connell and Soutar 2007, Kothari and Wilkinson 2013). Imports typically far exceed exports, and are characterized by manufactured goods including food and fuel. Exports are centred on a small range of agricultural goods, in hitherto protected markets. While reliance on a small range of traditional exports limits the scope for exploiting comparative advantage, diversification into more diverse exports is not an option, due to narrow resource bases, diseconomies in investment and poor transport (Laurent 2011). Despite having less to trade, SIS are dependent upon it, having small domestic markets and consequently being more susceptible than larger, more diverse states to changes in international trade arrangements.

Asymmetry is central even to the negotiation of trade arrangements. SIS have few negotiators, and even fewer who are highly skilled and well informed. Tonga and the Maldives, for example, had little capacity to participate in the WTO accession process, since they usually had no permanent representation in Geneva; thus the Maldives 'has not been able to participate effectively in the rule-making process and to promote and defend its interests in the WTO in accordance with its national policy priorities' (Athukorala 2004: 1415, Wallis 2010). As a former Tongan Minister of Finance, Siosiua 'Utoikamanu, has said in a more general context: at international meetings 'you feel intimidated because you don't have the capacity to engage' (pers. comm., 2012). Skilled negotiators are few and may have to be drawn from other activities. SIS have often been advised by expatriates, who have not necessarily been dedicated to SIS objectives and have made commitments that should have been discouraged (Grynberg 2010, Wallis 2010). At best SIS 'struggle to negotiate effectively with big players' (Gay 2005: 602). Such significant tactical disadvantages are replicated elsewhere, for example in negotiations with large corporations. Thus the logging sector in Solomon Islands is virtually unregulated 'because of a lack of state capacity to keep track of corporate structures designed to hinder transparency and monitor tax evasion strategies, and in part because of the active collusion of politicians with agents of logging companies' (McDougall 2011: 132, Hameiri 2012). São Tomé's shortage of skilled human resources resulted in the government relying on external actors to develop its oil exploration policy, to the extent that an American company effectively 'usurped some of the country's sovereign rights [being] solely responsible for the development of the entire oil and gas sector' (Frynas et al. 2003: 62). Inexperience and inconsistency meant that eventual policy was not as favourable to national interests as in any other African context, with reference to foreign oil companies, bilateral boundary negotiations, and royalties and other economic benefits, while oil exploration involved few local personnel (and thus employment gains) since necessary skills were missing. However by the end of the decade no oil had been produced.

> Since the discovery of oil, São Tomé has been rocked by scandals involving allegations of government corruption. In 2003, São Tomé's government was briefly deposed in an abortive coup attempt that may have been triggered in part by expectations that with oil, the rewards of political office would dramatically increase … If these problems are not corrected before São Tomé finalizes the 2010 auction of offshore areas under its exclusive national control, São Tomé's government risks sinking into patterns of mismanagement and corruption while squandering a unique opportunity to develop its economy. (Human Rights Watch 2012: 4–5)

São Tomé appeared to be moving towards a catastrophic resource curse scenario, the outcome of scarce human resources, bargaining and monitoring capacity and of local venality: an extreme version of asymmetry, even complicit asymmetry.

Asymmetry has made it difficult for SIS governments to formulate effective courses of action to negotiate global pressures, when spearheaded by such formidable global organizations as the WTO and by large nations who are also investors and aid donors. Increasingly free trade, centred on notions of comparative advantage, threatens countries producing agricultural commodities, particularly where diversification from traditional export staples is difficult (Murray 2000). Consequently 'the risk of increasing marginalization from the global economy in the context of trade liberalization is one of the greatest challenges that SIDS face' (United Nations 2005, quoted in Connell and Soutar 2007). Small states have some room to 'manoeuvre at the margins' despite human resource constraints (Jones et al. 2010) but the greater openness of global trade structures has eroded previous political influences, while the loss of preferences proceeded faster than anticipated (Dodhia 2008), reducing the capacity to respond in a context of resentment, disappointment and diminished opportunities. Trade problems have intensified with the shift from an older reliance on commodities towards a necessarily more diversified but less protected economy, especially for states producing sugar and bananas, such as St Kitts, St Lucia and Fiji, where effective substitutes for once dominant crops have yet to be found. Loss of preferential markets extends to manufactured exports, notably textiles, particularly disadvantaging Fiji and the Northern Marianas, although most SISI have had few manufactured exports.

The overall impact of the WTO's Uruguay Round of trade liberalization, between 1986 and 1994, was negative for small states, because of declining protection for key commodities (Grynberg and Razzaque

2004). The same conditions have effectively remained in place in the Doha round from 2001. Under the Lomé Convention many SIS enjoyed preferential trade margins that are now incompatible with the WTO regime, including commodity arrangements with the EU, which covered exports that would be uncompetitive on world markets but were of major economic importance in SIS, such as sugar in Barbados and Fiji. In St Kitts the EU sugar protocol effectively protected fully 50 per cent of the country's GDP; closure of the industry in 2005 – despite its dominating both the physical and psychological landscape of the island – resulted in abandoned fields, higher unemployment (1400 lost jobs), social despondency and accelerated emigration (Richardson-Ngwenya 2010). Remarkably many Caribbean Community (CARICOM) countries have become sugar importers so reducing their food security (Hunte 2011). Banana production in the Caribbean also declined spectacularly during the 1990s, as quotas and preferences disappeared, forcing prices below production costs. Eastern Caribbean countries were disadvantaged by the loss of EU preferences in 1997 as the United States fought new 'banana wars' with the EU (on behalf of American multinational corporations) to prevent special access of Caribbean bananas to the EU. In St Lucia production between 1993 and 2002 fell from 123 000 tonnes to 48 000 tonnes. The 'banana war' was the first major trade dispute that the WTO had to handle and the eastern Caribbean islands were 'unequivocally the losers of this war' (Payne 2006: 26). Even Fair Trade farmers were 'embedded in a world market over which they have little control' though this 'represents the last, best hope for family farmers' (Moberg 2005: 12). The cost of the loss of banana preferences has been very high for Caribbean states such as Dominica, St Lucia and St Vincent.

Crops such as sugar and bananas were often produced inefficiently, because of limited modernization, lack of access to credit, weak management and use of marginal land, with high marketing costs, hence the need for protection. Preferences and protection provided the basis for the agricultural sector (and the acquisition of foreign exchange) in many SIS and, equally importantly, generated the majority of wage employment that was otherwise limited, especially for landless workers. Revenue from agricultural exports funded critical public sector investment projects that supported social and economic development. Without real possibilities for diversification, countries like Barbados and Fiji sought to maintain preferential arrangements and their accommodation within the multilateral trading system (Rawlins 2003: 63). Pacific SIS argued that preferences should be maintained because removing them was highly disadvantageous, while their small share of world trade would result in only trivial trade distortions and little impact on much larger trading partners (von Tigerstrom 2005). Despite minimal diplomatic power, in successfully dangling the threat of 'turning banana growers into marijuana growers and traffickers', Caribbean states were able to get the United States to retreat from its original aim of immediately eliminating the Lomé Convention (Potter et al. 2004: 374). SIS had little ability to influence global market mechanisms, and that rearguard action was ultimately unsuccessful in safeguarding valuable preferential trading ties.

Loss of preferential agreements left SIS to compete on the world market, a task that proved extraordinarily difficult, even for larger countries such as Fiji (Asafu-Adjaye 2007). For SIS this effectively meant the need to develop specialized niche agricultural products, such as kava, noni juice, spices and coconut oil (Chapter 3). Hitherto these had played a trivial role in national economies, dominated by crops with larger world markets, such as sugar and bananas. In Fiji it meant an intermittent focus on crops such as ginger and kava, the latter, briefly referred to as 'green gold', having considerable promise (Murray 2000) until it was banned in its main European market, officially on health grounds. In an open global economy SIS rarely had any comparative advantage in agricultural commodities. Even meeting global safety standards posed new difficulties. When Barbados implemented its WTO compliant regime in 2000 there was a negative effect on a range of local products, such as poultry, carrots, cabbages, pork, onions and sweet potatoes where Barbados had hitherto appeared to have a comparative advantage (Rawlins 2003). The critical issue for niche product production, however obscure the product, is that other countries are better placed than SIS to develop the same niches closer to markets (whether kava in Mexico or taro in Australia), while SIS face continued disadvantages, notably transport and economies of scale.

The erosion of preferences for such key commodities as sugar and bananas was much faster than any emergence of new niche products. UNCTAD concluded:

> The policy challenge facing the LDCs [and thus several SIS] and their development partners is how to promote development and policy reduction in a very open national economy situated in an asymmetrically liberalized international economy … There is rather a need for innovative thinking about how to promote development and poverty reduction in a newly liberalized economy. (2004: 214)

But 'innovative thinking' failed to materialize. The IMF noted both that the ten countries most negatively affected by trade liberalization (with the exception of the Seychelles) were all sugar and banana producers, and that 'candidates for the development of more competitive alternative industries are not obvious'. Nevertheless the IMF concluded that it would be inappropriate to make special arrangements for them because this would distort and delay global progress towards free trade, while granting preferences to one group might be at the expense of other groups of countries who were not significantly better off (Alexandraki 2005). It was therefore recognized that the smallest states would be the most likely to suffer from trade liberalization, but little would be done about it: the outcome of a generalized rules-based trading system with diminishing scope for flexibility. Small states, with inherent disadvantages, could not easily diversify from the production of primary commodities to compete in deregulated markets.

Just as niche agricultural exports are disadvantaged in a more free trade regime, the costs attached to manufacturing are such that it would be virtually impossible for SIS to diversify into manufactured goods even in sectors, such as electronics assembly and clothes, where small states might be expected to be competitive (Winters and Martins 2005; Chapter 4). Likewise diversification into financial services, recommended by multilateral institutions such as the IMF, was hindered by global regulatory regimes and standards (Grynberg and Silva 2006). The Inter-American Development Bank noted for Barbados, and implicitly for other SIS, that 'the creation of efficient social safety nets and other such services in order to soften the impact of adjustment will be critical' (Jessen and Vignoles 2004: 47). Moreover import duties fell and protection policies developed within SIS also had to be abandoned, hence Samoa's accession to the WTO in 2011 meant that it was no longer able to impose a ban on turkey tails and other problematic imports, despite the nutritional and health advantages (Thow et al. 2010, Snowdon et al. 2011). Substantial trade concessions were costly, the 'special and different' treatment provisions designed to assist developing countries were inadequate and new investment was not forthcoming (Wallis 2010), potentially further entrenching Tonga – like other SIS – as a remittance- and aid-dependent economy (Chapter 6). Openness had few real benefits.

For the smallest countries the costs of complying with WTO policy prescriptions (or regional free trade arrangements) exceeded the economic benefits, evident for Vanuatu, which withdrew from discussions because of the many commitments demanded from them. It finally became a WTO signatory in 2011; a move that has been seen as not so much 'an endorsement of the benefits of WTO membership but instead surrender to neo-liberal ideology' (Wallis 2010: 274), and thus resisted by many within Vanuatu. Although several Pacific SIS, sceptical of the benefits from participation, were reluctant to participate in talks with

WTO, the full weight of the EU, ADB, Australia and New Zealand – all major aid donors – and the Pacific Forum Secretariat (the regional grouping of independent Pacific SIS, PNG and Australia and New Zealand), not to mention the authority of the WTO, supported the free trade agenda. Pacific governments had little option but to conform. Parallel pressures existed in the Caribbean (Lewis 2005).

In both the Caribbean and Pacific regions the necessity of moving towards the more liberal global trading regime resulted in attempts to develop a two-stage process. Regionalization was to precede globalization. The independent Pacific Forum states committed themselves in 1999 to negotiating a regional free trade area, which took shape in the Pacific Islands Countries Trade Agreement (PICTA) that came into force in 2003. The CARICOM states (other than the Bahamas) are still in the process of establishing a CARICOM Single Market and Economy (CSME) region, to include the free movement of all goods and services and a common tariff and trade policy (Jessen and Vignoles 2004). In the Pacific two major trade agreements were developed: PICTA, followed by the Pacific Agreement on Closer Economic Relations (PACER), as interim steps towards integration into the global economy. PICTA's aim was to enhance trade between 14 Pacific island states and it set out a programme for regional liberalization, while PACER has a broader orientation, to promote greater cooperation between states in areas such as knowledge sharing, financial and technical assistance, trade promotion and structural adjustment. Both PICTA and PACER responded to demands from the more powerful members of the Pacific Forum, Australia and New Zealand: their negotiation was 'to reassure aid donors and the institutions of global governance that they are adapting to the new global trading order' (Firth, quoted in Connell and Soutar 2007). Pacific island states had little choice. Under a similar more localized Pacific trade agreement between Fiji, PNG, Solomon Islands and Vanuatu (the Melanesian Spearhead Group), trade agreements broke down as it was perceived that Fiji and PNG were flooding the markets of the two other countries, indicating the potential difficulties in adhering to agreements between countries of quite different sizes even within the Pacific. Far from furthering cooperation among Pacific Island States, PICTA encouraged competition between them, replicating similar problems within CARICOM (Jessen and Vignoles 2004). In turn these were a microcosm of the wider structures that pit SIS against larger metropolitan states.

Island states have largely failed to develop regional ties and organizations, partly because of the lack of economic complementarity, and of trade routes (and other communications) focused elsewhere. That has disadvantaged SIS where regional collaboration would strengthen the effectiveness of trade negotiations. Even over the short distances between eastern Caribbean states regional cooperation is limited. Tentative efforts at political federation in the 1980s were quickly abandoned. Equally tentative currency unions have likewise failed. Regional cooperation that 'demands some sacrificing of national interests' (Sevele 1983: 27) is absent. Organizations, such as the South Pacific Tourism Organisation, have had limited impact, compared with national offices, since regional competition is more important than cooperation. Barbados has been particularly antagonistic to tourism development in Grenada and St Vincent, where that would discourage visiting Barbados (Connell 1988a). Regional universities in the Caribbean and Pacific, once seen as rare successes, are being undermined by attempts to go it alone as quite small states (such as the Seychelles and Samoa) have created national universities. Common markets, from CARICOM to the South Pacific Regional Trade and Economic Cooperation Agreement (SPARTECA), resulted in minimal regional trade. Regional labour markets have similarly failed to emerge other than for some elite occupations. In the Pacific the Forum Fisheries Agency was initially effective in securing regional fisheries management agreements, but that has not been easily sustained (Sem 1986, Hanich 2012). Repeated attempts in the Pacific to coordinate and centralize procurement of oil and pharmaceuticals, and so reduce costs, have been undercut by national interests. Organizations such as the Indian Ocean Commission, the Secretariat for the Pacific Community, the Secretariat of the Pacific Regional Environmental Programme (SPREP) and CARICOM have provided valuable technical support and expressions of regional solidarity, but have played no significant economic role. The Pacific Forum has, however, played a key political role in regional military and police operations, notably in RAMSI. Notions of a 'Pacific Way' were significant in regional rhetoric in the post-independence years but did nothing to unite the region. Cricket, with a single West Indies international team, has created ties but not without friction, while the comparative rise of basketball (and television) in the Caribbean has oriented countries northwards. Regional cultural events and sporting competitions have promoted unity and a greater sense of regional identity where economic relations have been divisive rather than unifying.

Telephone charges to metropolitan states are cheaper than to Caribbean and Pacific neighbours. Regional shipping services have met with little success in every ocean (despite the early success of the Pacific Forum Line), because of the lack of complementarity between countries and the dominance of metropolitan companies. Airlines have likewise failed, because of low and irregular volumes of passengers and freight, high maintenance costs and metropolitan competition. Pride in having a national airline once outweighed economics, but many regional airlines, like Polynesian Airlines (once the flag carrier of Samoa), Royal Tongan Airlines (founded in 1985 and liquidated in 2004) and Air Aruba, have either disappeared or been reduced to local operations. Similarly Trinidad's British West Indian Airways, once the largest Caribbean airline, went out of business in 2006, after operating for 60 years, although Leeward Islands International Transport (LIAT) has survived after partial privatization. Massive losses have meant that most national airlines have given way to regional and international carriers, accentuating distant international ties rather than regional ones.

Despite the existence of regional bodies such as CARICOM and the Pacific Forum, and various others directed at particular themes such as environmental and fisheries management, regional integration has been weak to non-existent and has failed to create more formidable blocs that might protect regional interests and obtain concessionary goals. In large part that is because SISI have their own interests but, primarily, because these interests frequently overlap and compete. As the former President of Kiribati, Ieremiah Tabai, pointed out in 2012: 'even your big neighbours will take advantage of you' (pers. comm.). Not only have politically dependent territories in every region tended to seek stronger ties with metropolitan states, rather than a sovereignty that results in independence, but independent states have effectively followed similar trends rather than engage in processes where sovereignty is diminished through greater integration with regional neighbours. Rather than regional unity there has been a 'progressive balkanization of the islands' (Grynberg 2012). SIS have tended to go it alone: a tradition of divisiveness and distinctiveness, separation and pride in national identity. Fission and regional rivalry are more obvious than fusion.

So much pressure has been placed on SIS to enter a more liberal trading regime that this has been described, for the Pacific at least, as an imposition that was nothing less than 'the recolonization' of the Pacific by 'free' trade (Kelsey 2004). The global ideological status of 'free trade' is so powerful that there has been 'discursive closure' where no alternative seemingly exists (Wes Morgan, pers. comm., 2012), evident in one conclusion that 'Given the narrow resource base and small domestic market, openness to foreign trade and investment remains the "natural" policy choice for the Maldives' (Athukorala 2004: 1418). By contrast, and more generally, 'many small state representatives perceive themselves to be operating under a high level of threat from large states [including] fears of possible trade and aid reprisals, as well as of intimidation in the negotiating room' (Jones et al. 2010: xiv, Grynberg

2012). Antigua is the only small state that has successfully initiated and concluded a WTO dispute, through its support for internet gambling versus the United States (Bernal 2011, Cooper 2011, Laurent 2011): a unique exception to asymmetry. While SIS could make a strong moral and technical case for special and differential treatment, and providing distinctive market access for small states would have trivial consequences for larger states since SIS have a tiny share of world trade, they had an inherent negotiating weakness in the face of powerful opposition. Small states, particularly those that are geographically disadvantaged, cannot compete on a level playing field. Free trade has brought little of benefit to SISI, and has effectively emphasized and further entrenched asymmetry and uneven development.

AID

Most SISI have benefited very substantially from aid (officially 'overseas development assistance'), and also from migrant remittances, enabling them to run big current account deficits, maintain substantial bureaucracies and undertake relatively large public investment programmes of a kind that could not otherwise be financed. Mainly because of their small size, SISI are the most heavily aid-assisted parts of the world on a per capita basis. Were data available, most SIDS would be seen to receive far less support than almost all dependent territories, but they have long benefited from the 'small country bias' (based on perceptions of aid 'visibility' and the effective establishment of political relationships) that marks aid delivery from most donor nations (de Vries 1975, Connell 1986). Overall SIDS receive about three times as much aid per capita as sub-Saharan African states. However the enormous differences between the extent of aid in SIS (Table 1.1), with the northern Micronesian states (FSM, Palau and the Marshall Islands) and Tuvalu being particularly dependent, further emphasize their diversity. Over a 30 year period, aid to Tuvalu has been as much as 82 per cent of GDP, and 40 per cent of that of Kiribati, whereas for the larger and more developed Fiji it has been around 3 per cent (Fielding 2010). SIS with very low national incomes (notably São Tomé and the Comoros) have fared poorly (World Bank 2012). Even so São Tomé's overseas aid at its peak in the 1990s was twice that of its GDP and 97 per cent of public investment was externally funded; São Tomé was consequently perceived as 'an unviable state', since its domestic economy failed to generate enough revenue to sustain its import-reliant consumption patterns (Frynas et al. 2003). By that criterion, most SISI would be unviable.

With rare exceptions, such as relatively affluent Bermuda, politically dependent territories receive extremely high levels of aid, and budget support, from colonial powers, vastly in excess of aid to independent states. Despite United Nations pressure, many dependent territories have thus resisted movement towards independence, primarily to take advantage of 'aid', ensure an externally guaranteed internal stability and maintain migration opportunities, as in Niue and Tokelau (Connell 2008, 2009d) and some of the United Kingdom Overseas Territories (UKOTS), most obviously the Falkland Islands. The three independent states of northern Micronesia – the Marshall Islands, Palau and FSM – negotiated defence support, continued aid and unrestricted migration to the United States as they gained independence within Compacts of Free Association. Similarly Mayotte chose to remain with France rather than choose independence with the other Comoros Islands, just as three of the Netherlands Antilles have chosen additional political 'dependence' and migration rights to the Netherlands. At independence, Tuvalu separated itself from Kiribati in order to retain a separate identity and gain a separate source of income. However, just as some metropolitan states have experienced 'aid fatigue', so colonial powers have been reluctant to invest substantial sums on development in distant territories, unless there are special strategic or economic circumstances, as in Christmas Island (Australia) and the Falkland Islands. More frequently a version of aid fatigue has resulted in colonial powers even discouraging continued settlement in remote territories, such as Niue, and seeking to devolve administrative costs to local authorities, as in Norfolk Island and Tokelau (Connell 2009d). Islands that were once proud symbols of global empires have become millstones rather than jewels.

Aid delivery was once the province of developed western nations. More recently Japan, China and others have increased their aid to SIS. Multilateral organizations, including the United Nations family, the World Bank and the ADB, have become particularly prominent in SIS, with 'traditional' donors, such as Australia and the UK, committing a substantial proportion of their aid funds to such organizations. SIS now attract aid from numerous sources. In some there have been intense rivalries and fluctuating loyalties, for example, between Taiwan and the People's Republic of China (PRC) for recognition and therefore aid delivery (Frynas et al. 2003, Payne 2006, Atkinson 2010, Alexeyeff 2011), while the PRC is currently developing a strong aid (and trade, investment and migration) presence in the Caribbean and Pacific for economic and political reasons (Bernal 2010, Yang 2011). Strategic location and global rivalries have enabled some SISI to leverage aid and other forms of assistance from larger countries. Several Pacific states, including Solomon Islands and Kiribati, have repeatedly vacillated between recognizing Taiwan and the PRC as a means of securing aid, while Dominica has been accused of waging 'chequebook diplomacy'. Libya and Israel once similarly sought SIS support, while Russia has recently lobbied SIS to gain recognition for South Ossetia and Abkhazia. Nauru recognized both and gained Russian aid. In 2012 Tuvalu's recognition of Abkhazia resulted in Georgia breaking off its own single year of diplomatic relations. A North Korean mission reached Fiji in 2012: orientations and opportunities evolve.

Substantial aid and migration have resulted in many SISI being characterized as MIRAB (MIgration, Remittances, Aid and Bureaucracy) states and islands, where dominant sources of income are aid and remittances, and production is limited (see Chapter 6). The acronym both points to the significance of aid and aid dependence, but also to the manner in which aid – governed by urban bias – has spawned an urban bureaucracy arguably at some variance from local needs. However, for small poor countries, with few prospects of significantly expanding exports, 'aid dependence [is] a sensible income-earning strategy' (Moore 2011: 1771). Some SIS, notably Kiribati and Tuvalu, have been able to transfer aid funds into state-owned trust funds – sovereign wealth funds – designed to protect fluctuations in revenue and ensure a steady interest flow. Yet aid is so often conditional: tied to the products and personnel of donors, to support for them in international forums and to preferential access to such resources as forests and fish (e.g. Petersen 2003). It may result in unsuitable edifices – symbols of the limited national ownership of aid resources – and rarely supports necessary recurrent expenditure. Aid can distort national development strategies. Increasing demands from donors for consultation, accountability and engagement, alongside complicated adherence requirements, have created an 'inverse sovereignty effect' where recipient states are losing control over their development strategies, policies and programmes, as these become politicized (Murray and Overton 2011, Banks et al. 2012). That is complicated by the growing significance of aid from 'non-traditional' donors (such as China and Japan) that may be linked to natural resource use or global diplomacy, rather than local priorities or such goals as poverty alleviation.

Despite commitments to harmonization, notably the 2005 Paris Declaration on Aid Effectiveness, the proliferation of aid donors and projects has sapped the attention, energy and resources of recipient governments, and reduced the capacity for aid management, policy development and practical implementation, which in turn has resulted in donors focusing on 'good governance'. Pressures for 'good governance' have constituted a form of deterritorialization, partly dislodging linkages between territory and political authority by involving transnational forms of regulation and surveillance through military and police linkages and regional institutions: regulatory regionalism through 'capacity building' (Hameiri 2009, Harris and Goldsmith 2012). This represents a transition from the role of aid in economic and social development to the prevention of state fragility and failure, and enhancement of regional security.

Remarkably little analysis traces the relationship between aid delivery and development in SIS, although, as in larger states, it is probably most effective where good local policy contexts exist and where aid responds to local needs. That has been, as in Samoa, a function of high-level leadership in aid coordination, the ability to reject approaches that did not fit in with national priorities and the capacity and stability of key staff (Delay 2005). More widely the absorptive capacity of SISI has been questioned. When new sources of funding for climate change mitigation and adaptation, notably from the Fast Start Finance scheme that emerged from the Copenhagen Climate Summit in 2009, few SIS had the capacity to respond, develop strategies and absorb funds, on top of already complex aid programmes whose management demands were considerable (Nunn 2010). Aid cannot necessarily strengthen the policy environment, despite the focus on good governance, however, it can stimulate economic growth in small states but with diminishing returns (Feeny and McGillivray 2010). Data are extremely limited. Tonga is the only country in the Pacific where there is measurable evidence that aid improves productivity (Fielding 2010). Despite the direct impact of aid on economic growth being rarely measurable, the World Bank coined the phrase 'Pacific paradox' with reference to the combination of substantial aid and minimal economic growth in that region.

A quarter of a century ago one Pacific scholar concluded that:

> heavy dependence on foreign aid seems to be unavoidable if living conditions are to be maintained at tolerable levels. In the longer term the only meaningful course (leaving out emigration) against the threat of low level poverty calls for further population control and measures to restrain material aspirations to levels that can be sustained without undue dependence on international charity. (Fairbairn 1985: 57)

Restraint on material aspirations has been absent and SISI have both diversified the sources of aid and depended on migration and the human resources of the diaspora. The contours of future aid delivery are hard to envisage, in an era of changing global security and resource concerns. However, the small size of SISI means that relatively high levels of per capita aid are likely to be maintained and also that present donor nations, and others, are likely to assist in response to hazards. Aid fatigue is less evident for small states, and their location is one partial guarantor of continued support.

POLITICS AND PROBLEMS OF THE POSSIBLE

SISI are very much political economies but political and bureaucratic systems are weak. Several SIS, such as São Tomé, the Comoros and Solomon Islands, reached independence with minimal autonomy, and without colonial consideration of what institutions and human resources would be necessary for an independent state. At independence, Tuvalu had just two university graduates and in one ministry only the minister and his secretary had more than primary education (McIntyre 2012). Post-colonial SIS have often lacked the organization, political will and sometimes even legitimacy of their colonial predecessors and have held little sway over ordinary citizens, especially where local traditional authority retained vigour and authority, as it did in much of the Pacific. With rare exceptions, such as Vanuatu, there was no struggle for independence that might unify disparate populations, and independence was often granted willingly, or thrust upon reluctant 'nations', as in Kiribati and Tuvalu, by colonial powers anxious to extricate themselves from remote and unprofitable fragments of empire. States became independent without real nationhood, most problematic in archipelagic states.

With limited financial power and management capacity SISI are both weak and 'soft' states seeking short-term solutions, rather than committed to long-term development, and hence vulnerable to external political pressures. A few have been designated 'failed states', though such categorization has been rejected for Solomon Islands as being of little analytical value and inappropriate for a state that was 'never properly put together' in the colonial or post-colonial era (Wesley-Smith 2006). Kiribati gained a somewhat reluctant independence in the same year that its single real source of wealth, the Banaba phosphate mine, closed down. Its infrastructure was oriented to export not integrated national development. Some SIS might be better regarded as 'atrophied states' (Jon Fraenkel, pers. comm., 2011) where government functions have collapsed. Many SIS have ineffective governance through both the absence of social inclusion – gender equality, equity of public resource use, human resource development and social protection (Chauvet et al. 2010) – and the intermittent presence of practices such as corruption, media manipulation, disrespect for human rights and arbitrary application of the rule of the law. In Samoa, and implicitly for other Pacific SIS, such governance problems have been summarized as 'misunderstanding, frustration, alienation, migration, discrimination, malpractice, patronage and violence' that are at least partly a function of tensions between Samoan and imported practices and values (Huffer and So'o 2005: 311).

Radical political movements have been rare. At different times Cape Verde, São Tomé and the Seychelles have had rhetorically socialist governments but they practiced little. An exception was the People's Revolutionary Government of Grenada between 1979 and 1983, whose achievements have been interpreted in various ways (Thorndike 1985, Maingot 1986). A more general conservatism emphasizes the relative powerlessness of small states and an inability and unwillingness to successfully challenge conventional development contexts, evident in WTO debates, so that a more technocratic version of governance has prevailed (Roberts et al. 2007).

Most SIS and even small SISI have multi-party systems. Particularly in Melanesia, political parties are ephemeral and without distinctive ideologies and policies, but dependent on their leaders for charisma and promises delivered. Politicians are often elected on the basis of their personal standing, or their membership of a dominant clan or cultural group. In a few SIS, including Antigua and the Marshall Islands, a number of leaders absolutely dominated political systems for several decades. In the Maldives the President was in power for 30 years, until the first multi-party election in 2008, the democratic outcome of which was short-lived. In such circumstances, as in Antigua, government 'operated under a form of arbitrary rule with some degree of personal intimidation. Transparency and accountability were subordinated to the impulses of a personal mode of leadership' (Cooper 2011: 70). Yet in small places political strategies may be very transparent, centred on tangible material promises. Opportunism is more important than policy development. Distinctive philosophies rarely distinguish political parties and governments can be shifting unstable coalitions, with parties usually constituted only for elections and without local membership. In tiny Niue, as Cabinet Minister Pokotoa Sipeli has observed: 'we usually make policies after the elections' (pers. comm., 2012). As in São Tomé, differences between political parties are only superficial; 'personalities and personal connections matter most' (Frynas et al. 2003: 74). Consequently, in St Lucia, 'If you support the Party in Power, or the PIP as we call it, then you get everything. And if you do not support the party you get victimized. I'm not speaking about any one party … this is how politics works in general' (quoted in Moberg 2008: 14). In Dominica the businesses of opposition politicians might simply not be patronized (Olwig 2007). Factions may emerge but, as in the Marshall Islands, such factions 'have failed to combine or compromise in the interest of all Marshallese [hence] factionalism is quite possibly the greatest development challenge confronting the Marshall Islands and the Pacific' (ADB 2009b: 9). Politicians are under pressure from the groups that supported them to reciprocate, rather than support other constituents let alone the national interest. In parts of the Caribbean, such as St Lucia, trade unions have considerable political power, whereas in Solomon Islands: 'Local chiefs and clan elders support and demand respect from aspiring leaders who must commit themselves to promoting the interests of their locality. If they falter they will be cast aside and replaced by emerging rivals' (Steeves 2011: 344). Attention to such needs distracts ministers from their portfolios. Opposition parties are powerless and being in opposition holds no interest for most politicians. Loyalties are tenuous; in the 2000s one member of the French Polynesian assembly changed sides 17 times in 6 years. In the eastern Caribbean similar electoral volatility, through strong anti-incumbency attitudes, merely increased the numbers of politicians without experience (Payne 2006, Barrow-Giles 2011). Half of all Solomon Islands MPs lose their seats at every election. More generally, as in the Seychelles, government jobs, licences and investment are used as political patronage for ruling party supporters (Yoon 2011). In the Caribbean

> [s]ince the end of the ideologically based politics of the 1970s to the mid 1980s, politics in the Caribbean has tended to be fought along lines largely revolving around the material comfort of voters, state handouts to citizens, and contracts for large and small public works. The politics of money has replaced the politics of ideas, freedom, democracy, sovereignty and national self-determination. In this context men of ideas, skills, talent, honesty and true ability count for very little. (Joseph, quoted by Barrow-Giles 2011: 287)

And the Caribbean is little different from elsewhere. The turnover of politicians (and bureaucrats) is often rapid, hence the reluctance and inability to tackle long-term and difficult issues.

Democracy prevails in most SIS, successive governments achieve power through the ballot box, election results have rarely been questioned, and political dynasties are unusual. Government leaders are often short-lived. By the end of 2011, Solomon Islands had had six prime ministers in five years. In 1997 Nauru had four different presidents within four months and three in a single week in 2011. It also had 36 changes of government in the 42 years after independence (Ratuva 2011). Some SISI are over-governed. Niue has 20 politicians for a country of less than 1500 people, and Kiribati has 46 MPs for barely 100 000 people, but neither were willing to abolish themselves. Palau, with about 20 000 people, has two levels of government (and a Council of Chiefs). Nonetheless policy formation and implementation are dependent on the public service. Governments are not reform minded and enjoy too much discretionary power (Barrow-Giles 2011). Political instability and ineffectiveness is normal, undermining local trust in the outcomes of formal political processes.

Most politicians are men. The dual burden on women (especially their role as income producers), longstanding beliefs about women's role, the authoritarian nature of politics in several places and the expected financial commitment from aspiring and actual politicians have discouraged and prevented women from participation (Yoon 2011). Consequently, advancing women's rights within the socio-legal system is difficult, and implementing supportive policies even more difficult. In Samoa where few women are *matai* (chiefs), and only *matai* are eligible to stand for parliament, women's representation – already low – fell in the 2000s, but resulted in a constitutional amendment requiring at least 10 per cent of parliamentarians to be women. Yet in Caribbean SIS women have achieved a degree of political representation no different from that in larger states.

Small size creates inherent governance challenges. While size may enable better contact between governments and governed, it may also encourage nepotism and clientelism (Armstrong and Read 2004). Politicians are more likely to know constituents, which may facilitate corruption and the personalization of decision-making, so fuelling various forms of patronage and cronyism. Making delicate and demanding policy changes are difficult when they affect immediate neighbours and family. Anonymity and neutrality are almost impossible. Because many individuals may have direct access to ministers and senior bureaucrats, and use informal channels of communication, individual influence and the lack of recording of decisions make administration more likely to experience discontinuities and respond to inappropriate and unpredictable external influence (Jacobs 1975). Nepotism is common, and in smaller SIS almost inevitable, manifest in a predisposition to favour kin, relatives, political allies or people from the same tribal or language group in appointments, contracts and scholarships (Larmour 2009, Baldacchino 2012). Where the private sector is small, politicians and bureaucrats are particularly likely to seek to protect their own interests and positions. Exceptionally the creation of a centralized and politicized bureaucracy has resulted in the 'mass exodus of the most talented and qualified of the civil servants' (Hope 1986: 32). Accountability is limited because of a weak or constrained civil society.

Without policies and ideologies, elections centre around pork barrelling. Inadequate access to the barrel prompts dissent while excessive use of the barrel, and access to the perks and privileges of office stimulates corruption. Close linkages between elites facilitate corruption and a lack of accountability. Various forms of administrative and political corruption exist, ranging from misappropriation, embezzlement and abuse of power, to manipulations of budget processes to benefit particular interests (Larmour 2009), with often tacit and fatalistic acquiescence of something viewed as inevitable and, if not condoned, then expected (Alexeyeff 2011). Relying on external rent, whether from resource development or overseas aid, lowers the incentives for ruling elites to serve a domestic constituency and increases the likelihood of venality, evident in the licensing of overseas fisheries agreements (Hanich and Tsamenyi 2009). Weak governance has contributed to some degree of 'Dutch disease' (Fielding 2010, Moore 2011), but rather less than that in mineral exporting SIS, where it is also defined as a 'natural resource curse', where countries with abundant extractable natural resources are 'cursed' by a focus on them, becoming prone to such problematic phenomena as high income inequalities and inflated living costs, distorted and weak economies, authoritarian rule, poor governance and civil conflict. Corruption in São Tomé increased after the discovery of oil: an unusual example of elites having effectively captured the state (Frynas et al. 2003, Vicente 2010). While corruption can occasionally be rampant in dependencies, such as Pitcairn and the Turks and Caicos Islands, colonial mechanisms enable some control (Clegg 2009). Some have perceived 'the increasing criminalisation of the Caribbean state' (Barrow-Giles 2011: 300), though the UKOTS at least are 'safe, secure and governed by relatively democratic norms' (Hintjens and Hodge 2012: 208). Overall, SIS are no more corrupt than other countries, with Barbados being as incorruptible as the UK, and only the Comoros and the Maldives among the worst 25 per cent (Transparency International 2011). Corruption has decreased economic output, reduced capital formation and distorted economic development.

Natural resources are particularly vulnerable to mismanagement (in effect a 'political resource curse'), exemplified by logging in Solomon Islands (Hameiri 2012), while limited economic development opportunities have resulted in governments introducing activities such as offshore banking that are hard to monitor and engender distinct risks. Other activities that, almost by their nature, invite social problems, such as passport sales, as in Vanuatu, Dominica and São Tomé, or casinos in the

Marianas and Marshall Islands, have posed serious regulatory and management problems (Kupferman 2011). Corruption risks from poorly monitored offshore banking activities, rampant in the 1990s, have substantially declined following global concern with terrorism and money laundering. In the mid-2000s, however, Sint Maarten was regarded as not merely a centre of the cocaine trade but of laundering money to both Hamas and the Taliban. Some 400 shell banks were closed down in Nauru (Connell 2006a). Small states are not easily able to defend themselves against corrupt activity emanating from overseas, whether drug-running or illegal fishing (despite patrol boats being a common form of aid). As in Antigua 'there is no doubt that small size has attracted gangster and other undesirable elements who see small islands as admirable bases for their operations. They seem able to buy the occasional politician whose "street wisdom" is somewhat lacking' (Thorndike 1987: 110). Like many other new Pacific Prime Ministers before him, Darcy Lilo of Solomon Islands declared in 2012 that '[a]n overarching issue of governance is ensuring that all internal institutions must be fully up and running to fight against corruption, so that we can have a government that is operating transparently and effectively in delivering services, and also facilitating investment in the country' (*Radio Australia*, 12 January 2012). Internal anti-corruption efforts are hampered by weak government capacity, limited state penetration, unskilled and divided police forces, lack of civil society involvement and institutional inertia.

Few SIS have armies and even fewer have more than a thousand armed forces. Without navies several Pacific SISI have found it difficult to monitor illegal fishing. The Seychelles has been unable to ward off contemporary piracy in its international waters. Nor have SIS been able to prevent illegal drug running. Some Caribbean SISI, in strategic positions between the Americas, have become centres for drug running and laundering illegal gambling funds. Overall the value of drugs exported from the broader Caribbean is equal to the total value of all the region's legal exports (Moberg 2008). The Bahamas is 'the principle conduit for marijuana and cocaine into the United States' (Bartmann 2007: 301), Puerto Rico is a second centre while St Lucia is third in the Caribbean, after Jamaica and Trinidad, in the interdiction of drug mules heading for Britain (Henry-Lee et al. 2010). Martinique and Guadeloupe are valuable local markets for St Lucia and Dominica where marijuana fetches ten times the price in St Vincent (Moberg 2008).

> The mixture of corruption and drugs creates a vicious cycle of crime and violence that is engulfing many of the small Caribbean states. Corruption … allows the illegal drug trade to flourish. Drug money payoffs to government

> officials, police officers, and justice system officials cause these officials to 'look the other way' as the drug gangs go about their business in many Caribbean states. (Collier 2002: 7).

Expenditure of marijuana exporters in St Vincent was such that influential businesses were reluctant to support eradication programmes because their earnings were so dependent on this, and governments have been reluctant to oppose it (Payne 2006, Moberg 2008). More remote SISI are less well placed for a similar role, although the Northern Marianas is a smuggling hub for illicit drugs, sourced from both Asia and the United States. Drugs have also been shipped through such SISI as Fiji, Tahiti and Tonga. Endangered species, dolphins and birds, have been taken to Asia from Solomon Islands. In almost all such activities local connivance has emphasized the presence of weak administration and corruption, and external aid donors have provided patrol boats and stressed the need for good governance.

While corruption and weak governance have been common they have only exceptionally led to more serious crises. In Solomon Islands a combination of fragmented governance, corruption, self-interest, patronage and the criminalization of politics, economic stagnation, urbanization and high unemployment, poor law and order, and lack of international interest created an unusually fragile and divided state where external intervention eventually ensued after substantial civil unrest and violence (Fraenkel 2004, 2006, Connell 2006b, Allen 2012). In recent years such strife has otherwise been unparalleled in SIS.

Weak bureaucracies have limited management expertise and efficiency, tend to exhibit urban bias, especially in archipelagic states, are often highly politicized and easily swayed by commercial interests. The actual concept of public service is lacking. Public sector management – including that of the police forces – is weak because of a shortage of skills, cultural traditions that discourage individual initiative, and political incursions (Larmour 1998). Justice systems are flawed and slow, police are often ineffective and prisons overcrowded (UNDP 2012). Resource managers in SIS 'often have poor access to literature, are required to focus on day-to-day management, are often diverted to attend meetings called by international agencies, and often receive inadequate support from their governments' (Wilkinson and Salvat 2012). Regional pressures on limited numbers of skilled professionals echo international pressures; in the Pacific bureaucrats lament the proliferation of meetings, especially with regional agencies, that add to workloads, have unhelpfully mixed mandates, little continuity of staff attendance, duplicate activities and distract them from work (Negin et al. 2012). Problems may not be recognized, or ignored as beyond the realm of responsibility or resources; drawing attention to inadequate services and violence against women in urban settlements in Honiara (Solomon Islands) resulted in responses that 'ranged from outright refusal to acknowledge a problem to toothless gestures' (Amnesty International 2011: 15). A review of the Norfolk Island public service concluded:

> This Review found little evidence of a clear vision or sustainable long term plan for the future of Norfolk Island. The Norfolk Island public sector does not have the capacity to provide strategic resourcing, policy and direction-setting advice [or] to deliver the current range of services to the standards required. Poor work performance is a problem … The current middle level management culture breeds complacency and doesn't respect or understand contemporary management practices. The delivery of services is not coordinated. (Australian Continuous Improvement Group 2011: 5–7)

The limited penetration of the state implies that regulatory and compliance functions may also be weak or non-existent. Ombudsmen have no effective presence. Due to the lack of infrastructure, physical and human capital, central government often lacks the means to reach out to districts and communities outside the capital city (Barcham 2009). Decentralization is characteristically limited, with even development banks allocating most finance to urban areas (Connell 1988a), public services are generally contracting, while natural disasters place heavy burdens on states that are ill-equipped to cope with them.

Bureaucratic problems extend overseas. In international contexts small states 'generally lack the financial and human resources to properly staff and fund the required embassies and missions' (Laurent 2011: 219) so affecting their ability to negotiate and be aware of international trends. SIS are often represented by roving ambassadors. Representatives of small Pacific states in the United Nations experience difficulty in voicing their concerns, partly because they have so few representatives (averaging 2.5 per mission, compared with 21 for Western European and 8 for African missions) to enable constant vigilance, or give adequate focus to particular issues, even within blocs. Ambassadors themselves also saw this inability to influence opinion as a function of the Pacific being seen either as idyllic and peripheral or doomed to disappear, and in neither case likely to attract resources or concerned attention (McNamara 2009). Consequently small states usually exhibit limited participation in world affairs, and rely on larger states for protection and partnerships.

SISI experience shortages of specialists and technicians, especially where overseas training is required, partly because of migration and partly because of limited education systems. A lack of skilled human resources typifies many health care systems (Connell 2009c), with specialist functions such as ophthalmology routinely carried out by visiting overseas teams; part of a routine reliance on larger states to assist in some recurrent administrative tasks, from conducting censuses to ameliorating the effects of natural disasters. Most SISI have near universal primary education, and literacy rates are generally high, but secondary and especially tertiary education is attenuated, especially in rural areas, so that skilled graduates are fewer than required (Chapter 4). Education systems have become oriented to international curricula and scholarships, as expectations have risen, that might equip children to qualify and compete for the small proportion of jobs that require secondary education but, more significantly, gain access to overseas jobs where educational qualifications enable visas (Chapter 6). It has become conventional to argue that, as in Samoa, 'traditional institutions and customs pose a problem' for development, for example in the cost of text books in local languages and the operation of a national broadcasting system (UNCTAD 1985). However local languages and cultures enable improved literacy and act as a shield against inappropriate and too rapid change (Seers 1983), though that has rarely been favoured in policy.

Civil society is typically weak, with non-governmental organizations (NGOs) limited in scope and organization and, other than in parts of the Caribbean, trade unions are rare outside the mainly urban public service. Churches are often powerful (with the Methodist Church almost the only local organization willing to oppose the Fijian military government) often with a reach beyond that of the government. Policy formation and implementation are limited by lack of involvement of civil society. In the trade policy context, but certainly not only there, a constraint 'concerns the weak influence and engagement of trade unions, as well as academia and think-tanks' (Jones et al. 2010: xiii), all undeveloped in most SISI. Private sector involvement may be equally absent but may also be too involved where political, bureaucratic and business interests are entwined. The media are rarely independent – and choice is minimal – and have often been put under pressure to support governments. Restrictive ownership of the media constrains information and opinion, while local content can be very limited. Its role can be further hampered by government's attacks on the press. Stories have been blocked, advertising withheld and journalists deported in Fiji. Physical attacks on journalists and newspaper owners have occurred in Vanuatu without repercussions (Van Trease 2012). In the Seychelles radio and print media ignore opposition parties (Yoon 2011). The lack of access to independent newspapers, radio and television makes it difficult to promote accountability, citizen participation and investigative journalism. Self-censorship has enabled media survival but the media operate 'in an environment that is increasingly oppressive and restrictive and which has led to a "culture of fear" seen especially in a reluctance to speak openly on political issues' (Barrow-Giles 2011: 295). Yet while most SIS experience constraints on press freedom, and political cronyism and corruption are far from unknown, polities in the Caribbean at least 'have not experienced many of the antidemocratic practices and developments that have marred post-colonial states in the developing world' (Barrow-Giles 2011: 300). However basic checks and balances for good governance and accountability are weak in a context where social media are still emerging and the judiciary rarely neutral. In multiple ways small size, rather than valuing and enabling social contact, enhances flaws in governance, accountability and vulnerability.

VULNERABILITY AND RESILIENCE

However SIS and development are conceptualized and defined – and their diversity restricts easy categorization – most have long faced a range of challenges beyond those of larger states: 'vulnerabilities rather than opportunities … come through as the most striking manifestation of the consequences of smallness in global politics' (Payne 2004: 634). Development is inherently problematic. Basic economies of scale and transport costs disadvantage SISI in conventional development trajectories. Consequently small states have frequently been separately distinguished. Every classification reveals other disadvantages. Economic growth has been limited, with growing evidence of unemployment and poverty, though disguised by weasel words like 'hardship' (Abbott and Pollard 2004; Chapter 4). Being a political dependency has proved advantageous in economic terms, and enabled superior incomes, welfare provision and health status, partly because of their relative global insignificance. Yet where colonialism was, as in Vanuatu, 'government by benign neglect' (Tonkinson 1979: 106; Challinor 2012), it provided few foundations for effective participation in a post-colonial and increasingly global world. The accelerating rate and scale of change, broader forces of regional integration (and disintegration) together with widespread weakening of the apparatus of the state and the shift of economic initiatives to an unfettered private sector, weak locally but powerful at an international scale, have been destabilizing forces.

Policy directions are not easily developed in small states with small bureaucracies where political change is a regular phenomenon and political will elusive. The relative paucity and depth (both hierarchically and regionally) of skilled human resources constrain effective policy formation and implementation. With weak short-term governments, under-equipped and inexperienced bureaucracies, scarce resources and limited access to information, developing and implementing sound policies is a constant challenge, and one that has become more subject to international influence. External perceptions of 'failures of development' in SISI have gradually shifted from economic issues (such as lack of resources and constraints of distance) to politics and governance, so transferring the blame for development failure to the SIS and resulting in development acquiring its technocratic focus.

Alongside diverse (and sometimes uncritically welcomed) foreign investment, and the blandishments of entrepreneurs of a variety of persuasions and degrees of self-interest, island states are overwhelmed with advice from beyond, including well-meaning and sensible consultancy reports (though many diverge in recommendations or are insensitive to local circumstances). Advice and documents appropriate to the needs of large states may both overwhelm smaller states and disrupt planning, when, for example, SIS are part of regional organizations based in Asia, Africa or Latin America. Here too regional and global symmetry prevail. As has been politely phrased for trade negotiations, 'the research and information provided by external consultants are not always tailored to the needs of small states, and there are concerns that some information is biased towards the interests of donors' (Jones et al. 2010: xiii). Moreover, as one Barbadian has noted: 'Barbados … is currently engaged in too many international trade negotiations simultaneously, with the result that there is too little time to dedicate capacity and resources to defining national interests' (quoted in Lewis 2005). Complex international negotiations require complex technical knowledge, yet skilled human resources are scarce and recruitment, development and retention of high quality officials is a challenge. Poor career prospects and low pay, limited recognition, frequent reshuffles and lack of clear policy direction discourage the public sector. Where negotiating skills are as important as technical knowledge, in trade or aid deliberations, SIS are further disadvantaged by experience and language, while few have 'the analytical capacity to properly assess the trade-offs of different policy options' (Jones et al. 2010: xi–xii), at domestic and international level. Without the international or domestic ability to critically access external 'advice', the choice and implementation of development strategy are rarely easy. Data on which to make decisions are often inadequate (e.g. on poverty), those involved in decision-making may not always want data to be available, while those who are able to assess priorities disinterestedly may have no political influence. Nepotism is no great help.

Trade has become increasingly problematic, and aid unable to significantly influence development trends and enable a more effective sustainable development. Difficulties posed by liberalization have contributed to economic stagnation and political instability, notably in the banana- and sugar-producing states, while lack of economic power and authority has made it more difficult for small vulnerable states to exert an effective presence in trade negotiations. Economic activity thus takes place in a context of uncertainty. The management of development and policy formation is not merely challenged but threatened by global perturbations far beyond the ability of SISI to influence, let alone control. Politics and economics pose acute problems for SISI. How these play out in practice and how policy shifts, global trends and local distinctiveness have affected island development can now be examined.

3. The historic core: agriculture and fisheries

> Knowledge about fishing, as well as farming, hunting, medicine and navigation is disappearing because younger members of island cultures are often no longer interested in mastering it. They judge it no longer useful. Why learn to fish well when nine-to-five jobs in air-conditioned offices beckon in the district center and there is an endless supply of fish in cans? Why learn to build a canoe and sail it when fiberglass boats and outboard motors can be bought and operated with little detailed preparation or knowledge? This disdain is reinforced by well-meaning educators, for the exclusion of traditional skills and knowledge from westernized school curricula in many developing countries amounts to a constant, tacit assertion that such things are not worth learning.
>
> Johannes, 1981: 148

Virtually without exception SISI were historically characterized by agricultural systems, inevitably supplemented (or dwarfed) by fishing, a situation which continued until far into the twentieth century. Such dominance typified even relatively recently colonized islands, from Tristan da Cunha to Réunion. Yet in very few places, usually relatively remote islands, is this still true. However, there are great historic differences between the more autonomous Pacific islands, and those of the Indian and Atlantic Oceans and the Caribbean, where many islands were populated by European settlers and their slaves developing plantation agriculture oriented to distant markets and never having to adjust to a life of subsistence broadly within the ecological limits of island environments. Since that first wave of globalization, agricultural change has taken quite different directions, especially in the past half century, with external influences – trade policies, migration and so on – increasingly shaping contemporary agricultural systems. Market orientation, commodification, specialization, and shifts towards monoculture and individual land ownership, alongside movement out of agriculture, typify recent trends. Fisheries too have changed, but rather less dramatically.

This chapter examines what were once the dominant sources of livelihoods in SISI, and the marginalization of agriculture that has gradually ensued. Here, as much as in any other context, there is remarkable diversity between SISI, some of which never supported much more than a rudimentary agricultural system and others that continue to sustain both subsistence and an export economy. Broadly, Caribbean SISI have moved away from agriculture, in circumstances where plantations were once dominant – and disdain and contempt for agricultural labour linger on – the environment is less propitious and tourism seems to offer much more. Pacific SISI have retained and occasionally diversified longstanding agricultural systems, while Indian Ocean SISI have performed more like those in the Caribbean. Decline and disintensification have been the norm.

AGRICULTURE

Subsistence Agriculture

Subsistence agriculture is declining in most SISI as populations have become more urban, agricultural work offers fewer attractions (especially for youth), and food is increasingly purchased. In some places, such as Nevis, the Comoros and the Seychelles, it has long been true that 'there is simply not enough land to provide either subsistence or the cash crops to buy subsistence' (Benedict and Benedict 1982: 109). In the Caribbean especially, the area under agriculture and its diversity have steadily declined. Such movements towards monoculture increase the risk from hazard, with implications for residual food security. Deterioration of agricultural systems has been most evident in the Caribbean, especially where tourism is dominant, soils poor, degradation long established and rainfall low; beyond a few home gardens, agriculture is largely absent in Anguilla, Barbuda, Sint Maarten, the USVI, Turks and Caicos and the BVI (e.g. Potter and Sluyter 2010). Land shortages, tenure constraints, poor soils, inadequate rainfall, preference for wage employment, new tastes and poor marketing infrastructure have all discouraged production. Subsistence agriculture has been further threatened by shifting work preferences, migration, food imports (with imported foods acquiring more status) and pressure on available land. On Rendova (Solomon Islands), where population growth had reduced the area of agricultural land, one villager observed:

> The young generation are not planting ngali [almond nut] trees … The young generation are changing their food habits; they prefer rice over sweet potato

> and tinned tuna over ngali nut. [But] if the ngali nut is value added then young people will think it is a foreign food and buy it. (quoted in Grant 2012: 42)

Such transitions, and disintensification, have also usually involved a reduced agricultural area (both absolutely and per capita), the shift to less demanding crops (whether in terms of soils or labour) such as cassava (with reduced nutritional value), the disappearance of more demanding techniques (such as irrigation and mounding) and the loss of particular varieties of cultivars, as in Vanuatu and New Caledonia, that constitute a threat to conservation (Caillon and Degeorges 2007, Gaillard and Manner 2010). The relegation of agriculture to a reduced role in nutrition, work and local market orientation, is typical of most SISI, where off-farm income-generating activities exist.

If alternatives to subsistence agriculture (and fisheries) exist, whether off-farm employment or a dependence on remittances and state salaries, the transition away from agriculture has been more substantial. On atolls such as those of the Tuamotus (French Polynesia) and Woleai (Yap, FSM), where agricultural work is challenging, people prefer to 'eat from the store' and have no wish to 'return to the past', even though nutritional status has declined (Connell 1992, Chazine 2005). French overseas possessions, such as Martinique and Guadeloupe, have even been called 'consumer colonies' where production is no longer of consequence (Ormerod 1981), though it has certainly not disappeared. Where migration has been considerable, subsistence agriculture generally plays a reduced role, imported foods have become more important and agricultural land has gone out of use, as on Mokil/Mwoakilloa (Pohnpei, FSM) (Oles 2007a, b). All these atolls are 'government islands' (see Chapter 4) where external aid is substantial.

Agriculture is widely typified and derided as an activity of low status and prestige, especially for men. In the Cook Islands men who work on the land (or in other forms of manual labour) are described as *repo taro* (dirty taro), a derogatory term that also refers to 'nothing men', without land or capital, who are typically also denigrated as coming from outer islands (Alexeyeff 2008). By the 1970s in Tonga 'one often hears parents expressing the wish that their children "work at something better than agriculture" [which] invariably refers to white collar jobs which carry with them a lot of prestige' (Sevele 1973: 16). In Barbados agricultural employment has given way to tourism, and working in agriculture, even commercial agriculture (especially cane cutting), is disdainfully regarded by the young especially as 'degrading' (Gmelch and Gmelch 1997). Half a century ago in Samoa, declining interest in agriculture and the presence of uncultivated land was the outcome of the 'low social status accorded to skill and diligence in agriculture' (Ward 1959: 126). Plantation agriculture is particularly derided and avoided, especially in the Caribbean and São Tomé, where it remains redolent of slavery and forced labour. Universally, farmers are steadily aging.

Where 'modern education' has both taken youth from fields, gardens and lagoons and inculcated some opposition to 'traditional' knowledge, indigenous technical knowledge has declined and disappeared and agricultural diversity contracted. More demanding components, particularly taro pit cultivation and taro terracing, have almost universally declined, even when not threatened by hazards, as a labour-intensive activity of low status (e.g. Momsen 1986, Marshall 2004a, Caillon and Degeorges 2007, Oles 2007b, Rudiak-Gould 2009). Nonetheless, in most places, new crops have been introduced because of their higher yields, ease of cultivation or preferred taste. On Bellona (Solomon Islands) broad patterns of land use have remained much the same for 40 years, but crop combinations changed with the more easily cultivable sweet potato and cassava (which demand less labour input) becoming more dominant, as elsewhere in Melanesia, resulting in some reduction in crop (and nutritional) diversity. Greater access to cash and physical access to the national capital resulted in agriculture becoming more a supplement to a diverse 'portfolio' of livelihood strategies (Thaman and Thomas 1985, Reenberg et al. 2008, Birch-Thomsen et al. 2010). Quantitative measures of declining diversity are rare, but in New Caledonia, where yam cultivation is significant, only ten of a former 38 species are still cultivated (Gaillard and Manner 2010).

Yet significant agricultural decline is not universal. Where migration is uncommon and population pressures on land have intensified, agriculture has become more difficult, notably in Melanesian states, which continue to experience 'the "classic" challenge of sustaining rapidly growing populations in the context of limited agricultural land and natural resources' (Schwarz et al. 2011: 1129). Pressure on land has usually come primarily through the extension of cash crops. On high islands, such as Samoa and parts of Fiji, where agriculture is less challenging, and where subsistence agricultural systems have not greatly changed, the local economy has been relatively resilient to both global market forces and natural disasters (Paulson and Rogers 1997, Kuhlken 2007). In particularly remote islands, such as Anuta, Tikopia and New Georgia (all in Solomon Islands), agricultural diversity has been retained (partly out of necessity as transport links with the capital city have worsened), which has both strengthened food security and enabled the islands to be more resilient to cyclone (and other hazard) damage (Furusawa and Ohtsuka

2009, Mertz et al. 2010). Islanders have also revitalized subsistence agricultural systems when cash incomes have declined for some reason, as on Ontong Java atoll (Solomon Islands) (Bayliss-Smith et al. 2010), if migration is not a viable option. Where externally oriented economies have collapsed dramatically, as on Nauru, with the effective demise of phosphate mining at the end of the twentieth century, agricultural systems have been slowly revitalized even in hitherto unpropitious circumstances.

In some remote island contexts, such as Ifaluk (Yap, FSM) and Aniwa (Vanuatu), islanders have deliberately retained food production strategies on land and sea that have opposed maximization of production and commodification, but enabled greater equality of access (Sosis 2002, Wilson 2010). In a very different way subsistence agriculture on the even more isolated island of Tristan da Cunha, where population has not grown, has similarly scarcely changed. All Tristan families are farmers, owning their own stock and tending potato patches and settlement gardens, on land that is communally owned. Stock numbers are strictly controlled to conserve pasture and prevent better-off families accumulating wealth. But this sort of continuity is rare and usually only in particularly isolated contexts. Temperate Tristan has unusual similarities with tropical Ifaluk.

Agricultural areas in some SISI, especially in the Pacific, have usually expanded with population growth but also with the establishment of cash cropping alongside subsistence agriculture. Agriculture may consequently have extended on to marginal land that is more susceptible to erosion or with lower natural fertility, as in Fiji where sugar cane farms extended into marginal hilly areas. After the 1987 coups, as the price of sugar declined and farm labour left, contour planting was no longer practiced, farm husbandry worsened, soil erosion was exacerbated by pre-harvest burning of cane which exposed bare fields to rainfall; soil erosion and landslides became more common, alongside the silting up of drains, waterways and rivers which contributed to greater flooding in lowland areas (Lal 2010; Chapter 7). Likewise, the relatively new niche crop of ginger, farmed on steep slopes, further contributed to erosion (Barbour and Terry 1998). As the value of agriculture declined, key components of the agricultural system thus became more vulnerable to hazard, and their vulnerability contributed to the worsened effects of heavy rain in low-lying areas. In Barbados a broadly similar evolution might have happened in this century, but the government chose to subsidize and retain sugar cultivation (and employment) rather than countenance unmanageable erosion (Richardson-Ngwenya 2010). In the Comoros and São Tomé erosion has long been problematic since better land was reserved for plantations (Newitt 1992), with small farmers forced to use marginal land.

'Traditional' subsistence agriculture, often replicating the structure of rainforests, has been better able to cope with environmental hazards. Modifications and simplifications of such systems have increased exposure to risk, especially in high rainfall environments, and where agroforestry has declined. The loss of diversity in food gardens, and the loss of tree species, has reduced productivity, cultural utility and ecological stability (Clarke and Thaman 1997). Burning has contributed to species reduction. Where livestock numbers have increased substantially or where former agricultural areas have been used for livestock, as in parts of the Caribbean, erosion has followed. Land degradation has occurred in many places, especially in the more mountainous and drier islands of the Caribbean, and in Barbados, where 'soil erosion was rampant 250 years ago' (Potter et al. 2004: 123) and in upland Pacific SISI following new land clearances (UNEP 1999). Until quite recently SISI economies, in the Pacific at least, were characterized both by limited monetization and by the existence of an agricultural and fishing system which offered both a 'subsistence safety-net' and an 'exit option' from inadequate cash crop prices (Rodman 1987a, b). As subsistence economies have weakened, market orientation strengthened, and land degradation occurred, safety nets have withered, making both sustainability and withdrawal from an urban and distant economy more difficult to achieve.

Characteristically accounts of rural economies in most SISI now exclude any mention of hunting and gathering. Hunting is infrequently practiced because of the loss of many hunted species through improved hunting techniques (mainly acquisition of guns), growing population numbers in most areas where hunting played a significant dietary role (notably Melanesia), the decline of habitat, especially forests (mainly because of clearance, but also replacement with alien species), and also the lack of adherence to traditional conservation practices (see below), and thus the growing scarcity and extirpation of hunted species. Where forest areas remain, especially in Melanesia, hunting of various species, including wild pigs and possums, provides valuable nutrition. Coconut crabs are widely sought even when protected. Parallel changes have resulted in the gathering of edible species also declining in significance. Changing tastes have also influenced these trends, with some local species no longer regarded as appropriate nutrition. In a number of SISI, again especially in Melanesia, forests have been valuable sources of medicinal plants. Deforestation, the reduction in biodiversity, and declining knowledge of plants and their utility, as in Bellona and Kolombangara (Solomon Islands) (Thaman et al. 2010, Katovai et al. 2012), have reduced their significance. In many Pacific SISI forests retain a role in contributing to health care, not necessarily only for poor households. Reduction in the significance of species hunted and gathered (and fished from small river systems) increases nutritional vulnerability and dependence on the market economy.

Commercial Agriculture

Only in the most remote islands are agricultural systems without some commercial component with food crop surpluses traded in local markets (or involved in social exchanges), and crops such as cocoa, coffee, chillies and coconuts being specific cash crops. Large-scale estate agriculture continues to characterize some SISI despite commercial agriculture having had a long, often unpleasant, history in many SISI, with periods of boom, collapse, investment and disillusion. Slavery characterized the genesis of commercial agriculture in much of the Caribbean and Indian Ocean and in São Tomé. Commercial agriculture was most significant in islands that were larger and closer to western markets – notably those of the Atlantic and Caribbean, though, like the Canary Islands, prone to boom and bust cycles as cash crops (sugar, grapes, cactus – for dyes – and bananas) boomed and collapsed, or St Lucia, where sugar was followed by coal and bananas (Baldacchino and Bertram 2009, Walker and Bellingham 2011). Plantations reached the Pacific only belatedly and small, remote islands such as the Lau group (Fiji) were written off for plantation agriculture, as were the Cook Islands (Simpson 1990). The Gilbert and Ellice Islands (which became Kiribati and Tuvalu after independence) were merely unwanted 'Cinderellas of empire' (Macdonald 1982). Most coral atolls had little to offer other than labour, hence colonial intervention eschewed agriculture and remote islands, such as Lau, were supposedly 'spared the journey through Hell' (Knapman 1976: 83) that transformed larger SISI.

At independence not only did several SISI have agricultural systems inappropriate for independent states but many were bequeathed very little. In São Tomé particularly, the economy and infrastructure were oriented to export, not integrated national development. Estate agriculture raised questions over monoculture, management, environmental degradation and the marginalization of poorly paid workforces. It languished or collapsed in the post-colonial era and resulted in governments struggling to manage nationalized systems (as in Grenada), achieve land reform, seek to diversify cash crops and reach greater levels of food production. The colonial imposition, most marked in the Caribbean, where the best agricultural land was used for export agriculture and the poorest for food production, was never conducive to self-sufficiency in food production. Yet some SIS remain dependent on scarcely transformed agricultural systems. Until the 1980s at least 80 per cent of São Tomé's export earnings came from cocoa, an exceptional degree of specialization, historically linked to various forms of forced labour, especially where food production was trivial. Such extreme specialization ended with the decline and collapse of the plantation economy, accelerated by the onset of independence in 1975 and low prices in the 1970s and 1980s (Frynas et al. 2003). Ironically, despite the collapse of the plantation economy, drought and mismanagement, cocoa contributed over 90 per cent of national exports in 2010 as agricultural diversification failed.

As São Tomé demonstrated, achieving a successful transition towards a smaller scale agricultural activity, that was more efficient and diverse, employed more and was better integrated into local market systems, proved extremely difficult. While food production barely grew, estate crops began to disappear. In the Caribbean and Pacific the estate sector slumped and plantation cultivation for copra ended in small, remote islands such as the Cocos-Keeling Islands and the Line Islands (Kiribati). Sugar disappeared particularly quickly in countries like Antigua, where tourism became a new 'quasi-staple' (Henry 1985). In St Kitts agricultural production fell from 40 per cent of GDP in 1964 to less than 6 per cent by the end of the century, and in Barbados the comparable decline was from 38 per cent to 4 per cent (Potter et al. 2004). Outside the Caribbean, failures of past schemes have not prevented some experimentation with new plantation-type schemes, such as oil palm and rice in Solomon Islands (e.g. Juvik 1987, Fraenkel et al. 2010), but where local participation and land ownership are more evident.

Commercial agriculture remains the primary source of livelihoods in many SISI, especially in the Pacific, and in São Tomé and the Comoros, often from small-scale household production, since agricultural employment itself is rarely sought after. In the Caribbean particularly, both food and export crop production have declined as a result of multiple factors: droughts, floods, fires (especially in sugar), cyclones, crop diseases, declining soil fertility, industrial disputes and labour shortages, foreign exchange problems, lack of access to credit, high costs of other inputs, shortages of chemicals and machinery, loss of agricultural land, political instability, aging farmers, lack of physical planning and technical services (including irrigation), inadequate prices and marketing structures, 'parcelization' into tiny plots through inheritance, price fluctuations, 'modern' education, along with higher wages in tourism and disdain for agricultural employment (Axline 1986). With such a litany of problems, many replicated elsewhere, smallholder commercial agriculture struggled to compete with larger producers elsewhere with superior market access.

Production of many long-term staples, such as sugar, copra and bananas, has declined; several are no longer traded since prices in remote places are too low to justify marketing. Bananas (the first real smallholder market crop) have gone from Samoa and Tonga as export crops, outcompeted by South American countries, and copra likewise from most atolls in the Marshall Islands, FSM, Tonga and Tuvalu (Perminow 1993, Pollock 1996, Stevens 1999, Chambers and Chambers 2001), remaining only where there are no other local income earning opportunities (van der Velde et al. 2007, Rudiak-Gould 2009). A corollary has been a greater specialization in a smaller number of niche export crops, which has contributed to greater susceptibility to external price shocks and market shifts.

The greatest shifts in agricultural production and marketing have been the outcome of changes in the global trading system and thus world prices (Chapter 2). In the eastern Caribbean the Windward Islands (between Martinique and Grenada) have experienced much reduced banana production, especially in the wake of the twentieth century 'banana war' where United States pressure ended EU preferences for Caribbean bananas, with disastrous consequences for such Caribbean states as Dominica, St Vincent and St Lucia (Grossman 1998b, Mycoo 2005, Klak et al. 2011, Rose and Mathiason 2011). The economy of Dominica, which depended solely on the export of bananas through a single company to the UK, imploded (Bernal 2011). St Lucia saw its banana industry collapse between 1992 and 2005, when exports were a quarter of what they had been; since banana farming was the main source of national employment, increased unemployment, crime and a greater proportion of people living below the poverty line were outcomes (Jones et al. 2010). In St Kitts the once dominant sugar industry has disappeared. The EU sugar protocol effectively protected fully 50 per cent of the country's GDP; closure of the industry in 2005 resulted in 1400 lost jobs, an enormous number of formal jobs in a small economy (Richardson-Ngwenya 2010). Barbados, Fiji and other sugar producers have likewise experienced similar problems dependent on price fluctuations – a general but more widespread 'Caribbean sugar crisis' (MacDonald and Demetrius 1986, Tubuna et al. 2007) – and have experienced recent contractions in the sugar industry. Sugar made up 31 per cent of exports in 1980 but only 9 per cent in 2009. The end may be imminent. No SISI have found comparable substitutes for sugar and bananas, despite attempts to develop niches, and agricultural processing has similarly contracted. After trade liberalization in the mid-1990s many

SISI experienced a surge of cheap food imports, including items produced locally, making local farmers even more vulnerable and marginal.

The belief that free trade would benefit all participants relied on the theory of comparative advantage, suggesting that countries should specialize in the production of commodities for which they had a superior resource base, and stressed 'advantages' that were nowhere apparent (Chapter 2). For SISI this has meant specialized niche agricultural products, such as vanilla, kava, ginger, noni juice, coconut oil and various spices. Yet, by definition, niches are small, and have only ever played a slight (and often temporary) role in economic development. Vanilla has boomed intermittently, notably in the Comoros, but world prices have fluctuated dramatically. Between 2004–07 prices fell from $600 to $20 per kilogram, while competition from low-cost Indonesia and Madagascar intensified. Fiji has experienced an intermittent focus on crops such as ginger and kava, until kava was banned in its main European markets. It has survived and been viable in some islands, notably Pentecost (Vanuatu), Kotu (Tonga) and Kadavu (Fiji), because it commands a high price on local markets (Perminow 1993, Sofer 2007, 2009). Papayas from the Cook Islands to New Zealand have been outcompeted by the Philippines. Ylang-ylang in the Comoros is an outstanding and unusual exception, since the Comoros supplies about 80 per cent of world demand (much of which becomes the main component of Chanel No.5), a rare example of an island state not being disadvantaged by intervening marketing opportunities. In the late 1980s Tonga captured a significant squash market in Japan (seizing a seasonal gap between Californian and New Zealand production) but high quality requirements made market production difficult (because of transport infrastructure problems), resulting in a shift to monoculture, increased individualism, overproduction, reduced soil fertility, pest and environmental damage (either from 'complacency' or 'imperfect understanding') from excessive fertilization (James 1993a, van der Grijp 1999, 2004). Ironically, in those parts of Tonga where commercial agriculture prevailed, lack of labour raised its price above that elsewhere in the Pacific, and reduced the competiveness of agricultural goods (James 1993a). By the late 2000s, after price falls, little remained of the squash industry. For various reasons niche agricultural products have only exceptionally been successful, and have not replaced traditional commodity exports.

Inability to achieve economies of scale (albeit less significant for niche products), production costs (including labour) and the cost of credit affect competitiveness, while larger countries are often better placed to develop niches. While greater integration and free trade theoretically provide strong incentives for local producers to become more efficient, diversify and specialize, this has largely failed to occur. Any future shift to new niches, that might include organic production and fair-trade products (such as the banana industry of the Windward Islands), is unlikely to significantly counter the costs of free trade (Murray 2000, Moberg 2005). Even more critically, 'unable to compete in deregulated markets, household-based agriculturalists have steadily withdrawn from commercial farming, opting instead for emigration, seasonal reliance on wages and, in many instances, the production of illicit drugs' (Moberg 2005: 7). In St Lucia, St Vincent, Vanuatu and Fiji (and also Jamaica) farmers have turned to illegal crops, such as marijuana, because of their high value and the ability to compete in the market. Trade in marijuana, and even cocaine in parts of the Caribbean, has boomed. Following the collapse of the banana trade, St Vincent became the largest marijuana producer in the eastern Caribbean, second only to Jamaica in the Caribbean, with marijuana being the most valuable export (Chapter 2). By the 1980s, it was also the most valuable export from Palau (Marshall 2004b) and is widely if clandestinely traded in Vanuatu, Fiji, Samoa and elsewhere. In St Vincent it was said that its most important role could be in 'fostering a reduction in the stigma attached to working land among young people' (Rubenstein 1988: 130), but that never spilled over to less valuable and less 'cool' crops.

In hundreds of SISI villages households produce a combination of food crops for local consumption, cash crops (such as cocoa and coffee) for distant markets and a range of products – vegetables, fruit, fish and also goods such as baskets and mats – for local markets. A few coastal areas and valleys, like the Sigatoka Valley (Fiji), have become valuable sources of food crops for local markets, but much commercial production is export oriented. A particular and perhaps ironic external market has been the diaspora, and very small states such as Niue have marketed small quantities of local produce, such as taro, exclusively to that market (Murray 2000). Markets thus provide a potential source of income for villages not too distant from urban centres, and such 'subsistence incomes' are often the dominant source of income where wage employment is scarce. Even within 25 km of Suva, the capital of Fiji, most village households retain a semi-subsistence existence, relying on agriculture and fishing for their livelihoods and main incomes (Zann and Zann 2008; see Chapter 4). More distant villages are similarly becoming market oriented because of the need for income. In three remote Lau (Fiji) islands, fishing has given way to the production of relatively durable copra and handicrafts for distant markets, and greater consumption of purchased food (Turner et al. 2007). Distant villages may be increasingly disadvantaged; in Paonangisu village, no more than 50 km from the capital of Vanuatu, rising fuel prices have made it five times as expensive to market produce from there as in 1976, hence returns from crop sales have not increased (Connell 2010a). Almost universally a greater dependence on the market economy, even in remote islands such as Makira (Solomon Islands) has resulted in people having less resort to nearby ecosystem services (as local provision of building materials, water purification and soil fertility declined). By becoming more dependent on markets, pressure on land resources has intensified further, and local values (such as asking permission, sharing and care) have been eroded with a resultant loss of community and social cohesion (Kenter et al. 2011; Chapter 8). More generally subsistence and cooperation have evolved towards commercialism and individualism.

Migration has contributed to some decline in the use of marginal and distant land. By the 1950s parts of Fiji and the Caribbean had already experienced contraction of cultivated land areas (Ward 1961, Gmelch 1980, Rubenstein 1983). Migration, remittances (and off-farm employment, often in tourism) emphasized the trend towards disintensification. In Cape Verde a combination of remittances and expanded tourism has enabled the country to become more viable at the expense of an unproductive agricultural sector, where once it faced recurrent famine problems, and was a major recipient of food aid (UNCTAD 1985). Alternatively, where populations have increased and alternative employment or migration opportunities are scarce, agriculture has intensified. Fallow periods have declined, where populations have grown and cash cropping expanded. In Tonga that has occurred across the country but particularly in peri-urban areas (James 1993a). In north Malaita (Solomon Islands), between 1971 and 1985 the area of tree crops (notably cocoa) doubled, which resulted in land shortages, a reduction in bush fallow periods, decline in crop yields and land degradation (Frazer 1987). Similarly in Samoa and Pohnpei (FSM), in the 1990s, increased demands for market participation resulted in the extension of cash crops such as kava to more upland areas, resulting in the clearance of steeply sloping areas (partly because kava required direct sunlight) and intensified erosion (Shankman 1999, Harmsworth and Raynor 2004). While farmers well understand sustainability, as in Tonga.

> Many farmers remained largely unconcerned with decreasing soil fertility and saw no relevance in loss of biodiversity. The possible consequences of tractor tillage, pesticide application, and market crop production … were insignificant compared to the consequences of unmet family obligations, and failing in one's *fatongia* (duty) was far more serious than environmental disruption. (Stevens 1999: 8)

More intensive agricultural practices and shorter fallow periods in Tonga reduced secondary growth to grasses rather than trees, while tractors and ploughs destroyed seedlings and mature trees, in what were once 'agri-forest' systems (James 1993a). Agriculture has placed great pressure on land in some areas, in extending into 'problem areas', for example those prone to flooding or landslips (as in the case of sugar in Fiji). Agricultural development has resulted in more rapid deforestation, aided by the recent proliferation of chainsaws, and consequent sedimentation of streams and rivers, disturbing riverine habitats and reducing forest cover. Changing social, political, technological and economic relations have made some practices obsolete and new practices unsustainable. However, just as subsistence agriculture centred on diversity, so commercial agriculturalists have sought to eschew monoculture and retain various options for generating cash incomes, continuously experimenting with possible crops and varieties. Ironically, by maintaining a variety of sources of food and income 'this strategy of individual self-reliance inhibits the achievement of national self-reliance and constrains capitalist development' (Rodman 1993: 181). Yet, in so many places, agriculture is part of an 'old order' that is gradually giving way, for many reasons, to other means of livelihood wherever they are possible.

Land

In most SISI, especially in the Pacific, land is bound into 'tradition' and is collectively owned. This has slowed some forms of development, including both urban infrastructure and cash cropping, as ownership complexities constrain leasing or selling land, resulting in tensions (especially at mine sites and on urban fringes). Frustrations over access to land, disputes over ownership (the outcome of traditional systems being flexible), alongside growing populations, have threatened security of tenure and usufruct. Village 'communities' are increasingly dispersed, leadership is fragmented, putative 'real owners' multiply and, via marriage and migration, once distant 'others' may claim some form of ownership.

> The fact that there are multiple classes of claimants to a particular parcel of land, no strong tradition of delegated authority, and no statute of limitations with regard to customary claims, means that it is extremely difficult … to come up with schemes by which landowners can pool resources to convert customary land into modern alienable, property. (Fukuyama 2008: 21)

Such complexities can seem insuperable and have posed severe problems for urban development and resettlement schemes (Chapter 5). As cash cropping and tourism have increased and populations grown, land has acquired more obvious economic value. Typically, where commercial agriculture has expanded significantly, land tenure has become less flexible and more individualized, contested and unequal, as in Ambrym (Vanuatu) (Rodman 1987a) and Grenada (Thompson 1985). Even on islands like Taveuni (Fiji), where population densities are low, inequalities of ownership have resulted in a 'severe land shortage and a large unsatisfied demand for land' (Brookfield 1979: 45). Close to urban areas tensions over land ownership have sometimes been considerable, even violent, as on the fringes of Honiara, since land has acquired a high market value. In such localized contexts population pressure has created tensions where land is not freely available but zealously guarded by its traditional owners. Conflicts around Honiara in the early 2000s were partly a consequence of Malaitans leaving their own densely populated island and settling on the land of the local Guadalcanal population who resented the loss of their resources (and the greater competition for scarce urban jobs); a problem anticipated 20 years previously (Muria 1979). Relationships between land, migration, population pressure on resources and political tensions are increasingly complex.

Again, primarily in the Pacific, in societies where land is much more than an economic asset, were that not enough, and society is 'written on the ground', land ownership is complex, embedded in cultural and personal relations and not easily amenable to translation into western codes and conventions. Thus in Ranongga (Solomon Islands), where for 50 years outside agencies consistently called for the clarification of property rights as the necessary starting point for 'modern' economic development, and local residents were eager to have their rights recognized by outsiders and by other islanders, transforming complex, cross-cutting localized relationships into abstract rights that were commensurable, predictable and knowable to outsiders raised major political and ethical dilemmas for Ranongga leaders. Claiming exclusive rights to land for oneself or one's group would negate longstanding elements of reciprocity and the welcoming of outsiders, so effectively alienating others who are essential for the proper functioning of the local polity. Clarification of land tenure, however necessary for the working of a capitalist economy, thus threatened the tenuous achievement of unity that Ranonggans regarded as the prerequisite to peace, prosperity and, as they understood it, proper development (McDougall 2005). In other words it was more effective and advantageous for local development that land tenure remain cloudy and thus flexible, and even subject to competing claims, rather than be finalized and fossilized and a source of overt contention. Social relations were not abandoned for greater efficiency. In some cases where societies have for whatever reason adopted or had a more 'modern' individualized system thrust upon them, they have sought to revert to the old order, as in Rapanui (Gonschor 2011). Likewise in Caribbean societies, where family land tenure has been disparaged by some, it is socially efficient and from the viewpoint of the rural poor, in Gordon Lewis' memorable phrase, a system 'without which they would be naked against the blasts of life' (quoted in Moberg 2008: 28). Small proprietors, even when unable to survive from agriculture, are loath to sell their land at any price (Moberg 2008, Olwig 2007). Societal goals were more important than economic objectives, yet here as in most contexts external pressures have sought to transfer land ownership into the market economy, despite the social consequences.

FORESTRY

Only the larger islands of Fiji and Solomon Islands have significant timber industries, the former based on renewable species but the latter exploited by outside interests, at unsustainable rates, with damaging economic and environmental consequences. Reduction of forest cover has been one cause of reduced biodiversity (especially of larger bird and mammal species) including various iconic birds, such as hornbills in the western Solomon Islands (Chapter 7). Vanuatu has been able to minimize environmental problems by banning the export of logs, but also through the fortuitous geographical fragmentation of its forests (Regenvanu et al. 1997). Growing interest in the planting of tree species and especially sandalwood, largely wiped out by nineteenth century trade in the western Pacific, has been revived on a handful of Vanuatu islands, such as Aniwa, with valuable local results (Wilson 2010). Otherwise forests are being cleared rather than planted. External demand for forest products is undiminished, with the relative deforestation and more rigorous legislation of Asia having preceded that in the Pacific SISI, and the world on the verge of 'peak timber' (Shearman et al. 2012). Developing appropriate forestry management practices has proved difficult.

Logging activities through much of Melanesia, but particularly in Solomon Islands, where the economy has been highly dependent on timber for half the country's export income, have been undertaken with minimal environmental management. Whole logs are exported without downstream processing, creating few local products or significant employment. Policies to reduce the extent of logging in Solomon Islands have largely failed because of widespread corruption and landowners with few alternative sources of income to the windfalls from logging (Kabutaulaka 2006, Hameiri 2012). Forest exploitation continues especially through the larger western islands, both legally and illegally, although some of the worst excesses in Kolombangara and New Georgia have been curtailed (UNEP 1999, Hviding and Bayliss-Smith 2000, Pikacha 2008). So many of the large forests of Solomon Islands have been effectively worked out that the industry will have dwindled significantly by 2015, as production is about six times the estimated sustainable yield. Soil erosion and the silting of rivers and coastal waters have reduced water quality and damage to some coral reefs and coastal fisheries. Deforestation has predictably degraded river habitats, as in Fiji and Palau (Haynes 1999, MacKenzie 2008), but has been more devastating in Solomon Islands. Legislation to regulate forestry operations is unwieldy, and provides no compensation to landowners, while local agreements are often abused, exploitative or take little note of broad local interests. Governments have neither had the capacity to enforce the legislation, nor, in the case of Solomon Islands, 'the will to ensure that its citizens get a good bargain from the sale of their resources', hence illegal logging and corruption became 'rampant' (Danielsen et al. 2010: 110, McDougall 2011: 139). Outside Solomon Islands, forestry is too limited to be economically significant or environmentally problematic.

Most SISI have limited forest cover and, independent of commercial forestry, this is often declining because of growing agricultural pressures and the use of timber for construction and firewood (and copra or cocoa drying) (e.g. Shankman 1999, Garonna et al. 2009). Especially in poorer SIS, wood constitutes the main source of fuel, and is a significant marketable good. Widespread pressure on forests has depleted mangroves and coastal woodlands. Small, dry, cool and remote islands have been most prone to deforestation, irrespective of recent human activity. Perhaps typically, crude estimates suggest rates of deforestation of about 2 per cent per year in Samoa in the early 1990s: about the global norm for tropical forests (Shankman 1999). On the main island of Pohnpei (FSM) forest cover fell from 42 per cent of the island in 1976 to 15 per cent in 1995 (UNEP 1999, Harmsworth and Raynor 2004). In Samoa deforestation followed growing populations and demand for cash resulting in more rapid clearance for agriculture and the need to claim new land and avert future land shortages, which also marked a subtle but significant shift towards more individual land tenure (O'Meara 1990). More generally, and less dramatically, in most Pacific SISI the rate of deforestation exceeds the rate of reforestation (UNEP 1999: 8).

Compared to the extensive deforestation throughout the region in the early twentieth century, forest cover has increased in most Caribbean SISI in the last half century as agricultural lands have been gradually abandoned and left to naturally regenerate or be replanted: a revival from a limited base. In Barbados, most of the original forest cover was removed as early as 1827, a situation typical of those SISI with plantation economies, so that the wooded area of Barbados represents about 2 per cent of total land area. Between 1990 and 2000, the whole Caribbean averaged an annual increase in forest cover of approximately 0.7 per cent of the land area. The rate of increase grew between 2000 and 2005 (to 0.9 per cent annually). SISI with the greatest annual increases from 1990 to 2005 included St Vincent, where increased forest area followed the establishment of plantations for water, soil and coastal protection, and some timber and fuelwood production. Despite overall increases some islands have experienced continued losses. Between 2000 and 2005, the USVI lost an annual average 1.8 per cent of forest cover, while Dominica lost 0.6 per cent. Losses in the USVI were attributed to tourism and in Dominica to agricultural expansion, although deforestation was slowed by mountainous terrain and inaccessibility (McGinley 2007, Peteru et al. 2010). Similarly Saba retained much of its forest, because of very steep slopes, until a cyclone in the 1960s. In the Caribbean, therefore, deforestation has largely 'bottomed out' with increased areas of forest gradually becoming more common, contributing to greater resilience, but as the outcome of declining agricultural extension rather than adoption of conservation practices.

The environmental consequences of deforestation are wide-ranging. Mangroves alone, and to a much lesser extent sea grasses, being rich in both aquatic and terrestrial biodiversity, play a vital role in ecosystem stability, and in sequestering carbon. They provide protection against floods and storms, habitat for juvenile fish (nurseries for fish, conch, sea eggs and lobsters), other marine species and both sea and land birds, nutrient sinks and sources of firewood, food and building materials (Warren-Rhodes et al. 2011). Mangroves protect the land and the sea from each other: the land is protected from wave action during high tides and storms, while the marine environment is protected from soil runoff during heavy rainfall by the mangrove root structure. In the past 20 years the world has lost almost half its mangrove forests, making them one of the world's more endangered landscapes (Osti et al. 2009). They have largely survived where alternative sources of wood are available or where populations have only slowly increased (e.g. Naylor et al. 2002, Warren-Rhodes et al. 2011). Depletion and degradation rates are probably increasing, since there is still limited recognition of the need for preservation. Urbanization and tourism have been the principal threats, alongside agricultural expansion, road construction, waste disposal and removal for firewood, as people depend more on coastal regions and resources. Population increases and cash income needs have resulted in the sale of mangroves for timber and firewood, resulting in the degradation of mangrove ecosystems. In the BVI the development of coastal properties, including marinas, hotels and private residences, has annihilated several mangrove forests and the habitats they provided, resulting in a reduction in fisheries yields, a general loss of diversity in coastal areas and the loss of shelter functions (particularly significant during Hurricane Hugo in 1989 when over 200 boats sought shelter in one such mangrove environment). Destruction and degradation of forested shoreline habitats have increased the vulnerability of coastal areas to floods and erosion. The considerable social and economic value of mangroves to local people, in an immediate and long-term sense, has thus declined. Clearance of coastal mangroves in Fiji, and their replacement with sugar cane, has exacerbated the impacts of flooding at times of high rainfall (Lal 2010; Chapter 7), and broadly analogous situations occur on the fringes of other Pacific towns such as Apia (Samoa) and Nuku'alofa (Tonga). In Tonga, too, mangroves (and coconuts) have been depleted for firewood around Nuku'alofa, with the loss of 'natural fuels' making urban residents more dependent on higher cost imported fuels and on imported foods that require less cooking (James 1993a; Chapter 5). Like mangroves, sea grasses (especially in lagoons) provide habitats for juvenile marine species, and may be threatened by changes in water temperature, rather than by commercial exploitation or deliberate destruction.

FISHERIES

Fisheries have never realized the potential for SISI that maps of islands, maritime regions and exclusive economic zones (EEZs) imply. The seas around many SISI, notably in the Caribbean, are limited in biomass and fishery potential is slight. Furthermore, most commercial fishing is undertaken by fleets from distant nations, and the value of fish taken by these fleets from Pacific waters, for example, is considerably more than the licence fees paid to island states (Hannesson 2008, D'Arcy 2009). Revenues are also reduced through inadequate monitoring. Maximizing economic benefits from fisheries is crucial since the total world catch from marine and freshwater wild stocks has peaked, the majority of global fish stocks are intensively exploited and no significant global resources remain undeveloped. Although some of the world's major fisheries continue to produce substantial sustainable yield, others have been severely overfished, and many stocks appear to be heading toward depletion (Hilborn et al. 2003, Morton 2011). Most Caribbean SISI lie in a region where this is true. The depletion of Caribbean and Atlantic stocks, including the collapse of the cod and Atlantic bluefin tuna fisheries through overfishing and poor management (Safina and Klinger 2008), has put pressure on the Indian and Pacific Oceans, and provided crises for Newfoundland, St Pierre and Miquelon and the Faroes. It has placed increased pressure on other species, driving them closer to extinction, as 'rogue nations' ignored both management wisdom and legislation. Many marine ecosystems have been profoundly changed by fishing and other human activities. More responsible distant water fleets are being replaced by less responsible, low-cost fleets, to the detriment of island states (Morton 2011, Gagern and van den Berghe 2013). Mackerel are in danger of following the fate of cod, where a combination of overfishing and ocean acidification (from rising concentrations of carbon dioxide in the atmosphere) has created ideal conditions for the shift from a system dominated by fish to one dominated by jellyfish (Monbiot 2011). Sustainable management of deepwater fisheries is crucial given limited alternative regional economic options, especially for Pacific atoll states, but international regulation has long been weak. Inshore fisheries are under greatest pressure with coastal population increases.

Greater pressure has been placed on tuna fisheries in the Pacific region, as fisheries' potential elsewhere has declined. While the Pacific tuna fishery, with the exception of bigeye, was 'in good health' at the end of the 1990s (UNEP 1999: 14) this is no longer true, as illegal fishing has increased and proved difficult to regulate (D'Arcy 2009, Gillett 2009, Gillett and Cartwright 2010). Marine diversity has similarly come under pressure. The western Pacific has the greatest marine diversity in the world, with up to 3000 species on single reefs (SPREP 1993). Yet the extent of biodiversity in the Pacific, especially in marine environments, is unknown, with perhaps as few as 10 per cent of tropical marine species having been described. Even international agreements on the conservation of threatened species have been difficult to achieve. Turtles are still taken and consumed in some areas, notably in parts of Micronesia (e.g. Oles 2007b), and dolphins have been exported from Solomon Islands (Parsons et al. 2010). Dugongs have been extirpated around Mauritius and probably the Maldives, and are threatened elsewhere. Despite being legally protected in many countries, the main causes of declining numbers of marine mammals include hunting, habitat degradation and fishing-related fatalities. Whales are, however, no longer hunted near SISI (other than in the south Atlantic and the Faroes).

Further economic pressures, climate change and El Niño Southern Oscillation (ENSO) fluctuations will affect fisheries stocks and their locations, and be particularly critical for tuna on which the economies of many Pacific and Indian Ocean SIS depend (Miller 2007, Lima and Naya 2011). In the Maldives variations in tuna catches are greatest during La Niña and El Niño years when migratory fish patterns change and fish shoal at different depths (Mimura et al. 2007). Similarly in the Seychelles, during the peak of El Niño, the tuna fishery underwent a transformation, especially when the fisheries shifted several hundred kilometres eastwards, resulting in economic decline of the national fishing industry (compounded by the erosion of the competitive position of Seychelles in European markets, because of reduced EU tariffs for Asian countries with superior economies of scale and lower wages). At a different scale, coral reef fisheries, important for food security and employment, are also threatened by climate change (Graham et al. 2007). More generally, fisheries will be affected by ocean warming, its impact on the presence of phytoplankton and their subsequent impact on the food chain (Barange et al. 2010). Climate change will influence fish distribution, both tuna fisheries and local species composition, and restructure marine food chains, because of the sensitivity of marine species (and corals) to temperature change (Bell et al. 2011). Most of the benefits from fisheries for local people – such as nutrition, incomes and employment – come from coastal resources that are under greatest pressure.

Oceanic Fisheries

SISI lack the financial and technical capacity to engage as directly in oceanic fisheries as larger nations, though several have small fleets, since entry into the industry demands technically complex and expensive vessels, skilled labour and distant market access. SISI have no comparative advantage (Hannesson 2008). Much of the revenue from oceanic fisheries comes from the leasing of EEZs to fishing fleets from larger countries, notably in the Pacific where most fishing is undertaken by vessels from Asia and the United States. Some of the largest multinational companies are involved in tuna fishing and processing in the Pacific, and have long disrupted the efforts of SISI to develop national industries (Doulman and Kearney 1987). Licence fees represent a small proportion of the value of the fisheries and little processing is undertaken locally to add value. Pacific island nations as a whole receive about $70 million annually in fees, but that represents only 3.5 per cent of the value of the catch. In FSM licence fees still make up between 20–30 per cent of national income (Cassels 2006). In the Maldives, where fisheries employ more than a quarter of the workforce and account for two-thirds of exports, the rent from fisheries paid by foreign-owned vessels within its EEZ represents little more than a quarter of the value of the catch, and the resource is being depleted through economic overexploitation (Sinan and Whitmarsh 2010). For Nauru such leases have nevertheless been absolutely crucial as virtually the only source of national income after the collapse of mining. Value has been lost in countries like Tonga and Samoa where a medium-scale local tuna fishery existed for some years before largely being displaced by foreign-owned longliners in the 2000s (Barclay 2010). Fisheries collapsed in Cape Verde around 2011 following the incursions of factory ships from Europe and elsewhere, many fishing illegally, after the depletion of north Atlantic and Mediterranean fisheries. For most Pacific SISI, with limited land and fresh water, that are distant from trade routes, land-based fisheries activities (other than bunkering vessels) have effectively been impossible, and no local processing occurs. Minimal local employment is created and Pacific islanders are not employed on the ships (Pretes and Petersen 2004). Seychelles is a unique SIS where substantial value is added by local processing. Leasing is also significant on a smaller scale for Indian Ocean and Atlantic SISI.

The future of oceanic fisheries is vital for several island economies, especially in the northern Pacific, in the face of intensified global demand and competition. Eight Pacific island states (including PNG) have worked collaboratively, under the 1982 Nauru Agreement, to manage western Pacific tuna stocks (responsible for 25 per cent of the world's tuna catch), initially by restricting the number of overseas vessels operating in their waters and subsequently by implementing a management regime restricting the number of days fished by tuna purse seiners in their waters, in order to sustain bigeye and yellowfin and increase the return from licence fees (Shanks 2010). Formal negotiations have proved difficult. Since mid-2011, the eight Pacific Island nations have sought to re-negotiate the Nauru Agreement which failed to meet either their economic or conservation goals (especially the threat to bigeye), by giving unlimited access to American fishing vessels. Boats from other countries commonly paid $3000 or more for a fishing day, whereas the United States only paid $300. Late in 2011, three countries (Tuvalu, Nauru and Solomon Islands) closed their waters to foreign fishing vessels in an attempt to increase lease fees and maintain a sustainable tuna fishery. Effective and equitable revenue generation from deepwater fisheries has proved elusive.

Management is constrained by problems of data collection, monitoring and surveillance, weak communication between fisheries and other departments, corruption, variations in experiences and aspirations, the unwillingness of SIS to compromise their individual interests and the lack of a transparent framework for distributing the burdens of conservation, all of which have limited the collective management necessary for mobile fish and fleets (Hanich et al. 2010, Parris et al. 2010, Hanich 2012). Long-line fishing boomed in Samoa (and PNG) in the 1990s only to collapse a decade later. Bigeye tuna is now overfished, despite precautionary measures, and skipjack and yellowfin have reached sustainable limits (Harley et al. 2011). The conservation challenge that this implies is complicated by mobility and by the multispecies and multinational characteristics of tuna fisheries. Moreover the biological characteristics of each tuna species are quite different. Predicted climate change is likely to result in significant concentrations of the main tuna species further east in the Pacific (Bell et al. 2009), so benefiting fewer countries, especially as fuel prices increase.

Villagers have scarcely benefited from licence fees, but have gained employment where canneries have been established, as on Ovalau (Fiji), where a high percentage of village women and a handful of men are employed, and in Majuro (Marshall Islands), Noro (Solomon Islands) and American Samoa (Barclay 2010). The industry has been criticized for exploitation, of people and of local baitfish, and for low wages (Meltzoff and LiPuma 1983, Emberson-Bain 2004). High wages and diseconomies of scale make canneries uncompetitive with Asian states such as Thailand. The principal American Samoan cannery closed in 2009, mainly because American legislation raised the minimum wage to that of the US mainland and it was no longer economic. (The cannery employed 2000 people and was both the largest employer, after the government, and one of the largest private sector employers in any SISI, but 4000 people lost their jobs, since a substantial part of the American Samoan economy was linked to the cannery and its wages.) The Seychelles has retained the second largest tuna processing and canning plant in the world (40 per cent government owned), despite concerns over the loss of preferential access to EU markets, employing approximately 2400 people, around 16 per cent of the national workforce and accounting for 15 per cent of GDP (Campling and Rosalie 2006, Clifton et al. 2012).

Exceptionally, fishing is the core of the commercial economy. The crawfish (Tristan rock lobster) has been exploited commercially since the first canning factory was built in 1949. There is a catch limit quota for local and distant fishing boats, and sustainable management has been maintained despite commercialization. Isolation may have advantages. Since 1986, the Falkland Islands have become financially self-sufficient through leasing their fisheries rights, with the principal fishing fleets coming from as far as Poland and Taiwan, and fisheries becoming vastly more important than land-based economic activities (Kerr 2005). Until the 1960s dried fish was the main export from the Galapagos, but improved air transport enabled lobster sales and the fishing industry quickly became exceptionally valuable and (alongside tourism) resulted in a relatively remote group of islands going from largely uninhabited volcanic outcrops to the richest department in Ecuador, though over-fishing subsequently ensued (Kerr 2005, Edgar et al. 2010, Castrejón and Charles 2013). In absolute contrast, where small territories are not so isolated, and are close to fishing grounds where competition has been so intense that the fishery has effectively collapsed, as in the north Atlantic, the damage to an economy where diversity was almost non-existent has been enormous. That threat is not yet hanging over most SISI.

Inshore Fisheries

Inshore fisheries are of considerable economic and nutritional significance, especially on small islands where resources are otherwise limited. In most Pacific SISI, notably in atoll states and outer islands, inshore fisheries continue to have a vital role in ensuring subsistence livelihoods, and involve most people. On Woleai (Yap state, FSM) men fish daily, women fish almost as frequently – mainly on reef flats and in shallow water – and children play a regular part. Fish are consumed daily, but remoteness and the ubiquity of fishing means that they were not sold. By contrast in Nadoria, on the main island of Fiji, seafood provides the main protein component of household diets, and revenue from the sale of marine products covers school fees and household necessities, especially for more than half the village households where no-one had paid employment (Fay-Sauni et al. 2008). In similar small island contexts, as in the Lau lagoon (Solomon Islands), fishing and fish consumption are regular and ubiquitous, and inshore fish are a valuable but small-scale commodity. The more productive inshore fisheries are often associated with coral reefs, whose management is crucial.

Some inshore fisheries are highly specialized. Pacific island states supply a significant proportion of the world's trade in ornamental fish, and a quarter of the trade in live corals (Leh et al. 2008). Pearls have been grown successfully in French Polynesia and Manihiki (Cook Islands) but with mostly intermittent success, though it became the second most important industry in French Polynesia after tourism, until prices collapsed in the 2000s after over-production and poor management (Macpherson 2000, Andréfouet et al. 2012). Trochus is commodified in more remote islands where alternatives are few, because the shell is durable. Seaweed is similarly of significance where other opportunities are restricted, as in Kiribati and Solomon Islands, but is a low-value product sensitive to market price fluctuations (Thomas and Tonganibeia 2007, Kronen et al. 2010a). The market collapsed in Kiribati in the 2000s after EU protection disappeared and government subsidies ended (resulting in some seaweed choking coral reefs). Fishing (or diving or trawling) for sea cucumbers (bêche-de-mer) is valuable in several SISI, including some remote parts of the Pacific and the Galapagos Islands (Bremner and Perez, 2002), mainly for Asian markets. In the Galapagos external demand for bêche-de-mer brought uncontrolled migration to key harvesting areas, overexploitation in some areas (aided by superior harvesting techniques, and access to credit and markets) and new pressures on adjacent national parks, which in turn led to closing of the fisheries and subsequent regulation of seasons, resulting in violent protests by fishermen (Bremner and Perez 2002, Castrejón and Charles 2013). Similarly in the Marovo lagoon (New Georgia, Solomon Islands) the bêche-de-mer fisheries have had to be intermittently closed because of overexploitation, and 'boom and bust' cycles (Buckius et al. 2010). In several parts of Fiji bêche-de-mer, commonly harvested in the 1980s, could no longer be found 20 years later, having been extirpated by excessive harvesting, reducing local incomes (Leh et al. 2008). Tonga similarly banned bêche-de-mer exports for ten years, but found the ban unenforceable, though the trade largely collapsed through overharvesting. In Ontong Java (Solomon Islands) depletion of stock in nearby and shallower areas resulted in exploitation at greater depths and in more distant areas, partly because harvesting restrictions were no longer enforced, whereas on the neighbouring and similar atoll of Nukumanu (PNG) a different local management regime (not imposed by the central government) enabled conservation to remain effective, and sustainable bêche-de-mer exploitation to continue (Christensen 2011). In virtually identical atoll ecosystems, and in comparable high islands contexts in New Georgia, Fiji, Tonga and the Galapagos, management regimes (see below) and their relationship to market forces have been quite different.

Inshore fisheries, including bêche-de-mer, especially around urban areas, have generally come under pressure from growing populations, the disappearance of traditional management strategies, market pressures (for consumption and aquariums), pollution and more efficient but less sustainable practices – such as dynamiting, more widespread use of outboard motors to access fishing grounds, improved technology, spearfishing, ice box storage and better marketing (Brewer 2013). Even in some relatively isolated islands, such as Wagina (Solomon Islands), Rodrigues (Mauritius), Yap (FSM) and the Cook Islands, growing populations, the need for cash and improved techniques have begun to result in reduced catches and less income (Kronen et al. 2010a, Gemenne and Magnan 2010, Houk et al. 2012, Wichman 2012), posing both nutritional and economic problems where alternative sources of cash incomes are few. Some common species, such as giant clams in Moorea (French Polynesia), had disappeared by the 1980s (Walker and Robinson 2009). Only exceptionally, as in Ngazidja (Comoros), where technological change has been slight, and village fisheries associations are effective local institutions capable of achieving gear restrictions, has local solidarity enabled effective management of local resources (Hauzer et al. 2013).

While much of the Caribbean has experienced stock depletion the relatively poor Caribbean island state of Dominica, where tourism is limited, has retained a significant local fishing industry – with fish constituting more than twice the dietary intake of the Caribbean generally (Peteru et al. 2010). In Samoa, as some species have been depleted and catch sizes declined, the local inshore fishery has gradually gone further offshore and been taken over by foreign-owned longliners (Barclay 2010). In Micronesia, not untypically, the catch value of inshore fisheries has declined relative to inflation and the cost of commercial goods (Houk et al. 2012). In the Galapagos, despite efforts to establish management regimes, fishers competed with each other for larger catch shares every season – a 'race for the fish' – resulting in over-capitalization in larger, faster boats and high technology equipment, in the search for quick profits, increasingly becoming illegal, compromising management and the long-term recovery of fish stocks, and endangering their own health and safety – evidently an example of the 'tragedy of the commons' (Castrejón and Charles 2013). Versions of most of these trends exist elsewhere, hence the widespread need for effective management. Technological changes and market expansion have influenced patterns of exploitation, distribution and consumption, and greater ease of access and 'modern' education have tended to result in the attenuation of local knowledge of fishing techniques and fish behaviour, even in remote atolls, such as Etal, Kapingamarangi and Eauripik (FSM) (Nason 1975, Lieber 1994, Scourse and Wilkins 2009). In Lau lagoon villages (Solomon Islands) fishing has become an individual activity rather than a communal one as modern fishing gear is easily handled by individuals, and more accessible and affordable (Molea et al. 2008). Women play an increasing role in commercial fisheries once considered the domain of men (Novaczek et al. 2005, Walker and Robinson 2009).

Pressures on inshore fisheries are greatest around growing urban areas, especially where the poor take advantage of this economic and subsistence resource. In several areas this has resulted in conflicts over access to marine resources and fishing grounds, and reduction in species availability (Aswani and Sabetian 2009, Cinner 2009). Fishing has become less productive around some urban areas, but temporal data are rare, other than for western Madagascar and around the towns of Gizo (Solomon Islands) and Majuro (Marshall Islands). In some urban contexts, such as Noumea (New Caledonia), fishing has become more recreational than economic, but with no less pressure on environments (Jollit et al. 2010). Perhaps unusually, subsidies and loans from government and from the national development bank have exacerbated overcapacity and over fishing in Fiji, eventually pushing even experienced fishermen below the poverty line (DeMers and Kahui 2011). In the Galapagos, overfishing, species decline, ecosystem stress and distance to ports are correlated (Edgar et al. 2010). On parts of Upolu (Samoa) and in Rodrigues (Mauritius), declines in inshore fish catches have been attributed to sedimentation of reefs as a result of deforestation (including mangrove removal) and agriculture; a situation certainly more widespread. In several places local people recognize that their own practices are harmful to marine species and reefs (UNEP 1999, Gemenne and Magnan 2010). The impact of many different pressures on inshore fisheries has been poorly documented, but it has been predicted, at least for the Pacific islands, that over the period 2010–35:

> The amount of fishery products originating from coastal fisheries that is accessible to urban residents will decline sharply due to overexploitation and habitat destruction. A growing proportion of the population will not be able to catch sufficient fish to provide for household consumption, and purchased fish will become relatively expensive. Food security issues will grow tremendously in importance. (Gillett and Cartwright 2010: 7)

This scenario is probably already in place in several Caribbean SISI, as it appears to be in Rodrigues (Mauritius), where weak enforcement of fisheries regulations and a lack of income-generating alternatives 'have created a history of noncompliance and degrading fisheries' (Peterson and Stead 2011: 2). Fisheries on remote Caribbean islands, such as Navassa (Haiti), have similarly come under increased recent pressure, and that pressure has been exceptionally exploitative without any local management (Miller et al. 2007). By contrast in some other island contexts, including the Seychelles, there has been a less obvious decline in artisanal fish catches over the past decade, though interpretations vary (Daw et al. 2011, Clifton et al. 2012). While inshore decline could be compensated by offshore line fishing, fuel costs (that rose threefold between 2009 and 2011 in the Seychelles) are a disincentive (Clifton et al. 2012). Geographical variations in the extent of depletion of inshore fisheries are considerable.

The greatest decline of inshore fisheries has been experienced in Caribbean SISI, strongly associated with population density, increased fishing and human impacts on reefs. Larger fish, once abundant in the Caribbean, are increasingly absent from high-density areas, especially in the Antilles, but remain near uninhabited islands. Larger fish are more likely to experience declines since they are targeted by fishers, mature more slowly and often form aggregations, which increases their vulnerability. Tourist fishing and consumption of fish is less significant than local consumption in most contexts even when tourist numbers are considerable (Stallings 2009). In small Pacific SISI local fish consumption is also the main driver of decline of inshore fisheries, but especially where fish are a significant and (for some coastal people) a unique marketable commodity. Fish marketing has, however, been problematic in many places, and especially outer islands, because of the absence of freezers and other infrastructure (notably regular transport), and the lack of capital for commercial orientation. In both regions exploitation of fisheries is closely related to population density, population growth and the existence of alternative means of income generation.

Aquaculture

Aquaculture has not compensated for the decline of inshore fisheries. It is rare in SISI, because of its demands on freshwater resources, and because it is both labour- and capital-intensive, and high risk. It has rarely achieved commercial success. Almost the entirety of aquaculture production in the Pacific region comes from the two largest French territories of New Caledonia and French Polynesia (Gillett and Cartwright 2010). Technically successful aquaculture, as in New Caledonia, has demanded considerable state subsidy, is restricted to a limited number of species and contributes only 0.3 per cent of GDP (David et al. 2010). Aquaculture in the Caribbean has been conspicuous by its absence. Unless the economics of aquaculture change it is unlikely to become of great significance, even for small-scale subsistence production, despite its environmental impact usually being less than raising animals such as cattle, pigs and chickens.

Management

The widespread existence of traditional fisheries protection strategies (such as closed seasons, restricted areas, constraints on techniques, etc.)

is indicative of the perceived need for conservation in the past (at times of smaller populations and greater necessity for self-reliance). Social change and diverse market pressures have meant that various conservation practices have disappeared, even in relatively remote areas where populations are not growing quickly, such as Woleai (FSM), Etal and Kapingamarangi. Management practices have come under increased pressure in the recent past with sustained entry into the cash economy, increased capacity for exploitation through improved gear technology (e.g. nylon nets, spear guns, snorkelling masks, waterproof torches), changing social, political and demographic circumstances, weakened leadership and reduced respect for custom (Veiteyaki 1997, Johannes 1998, 2002, Ruddle 1998). In south-east Tanna (Vanuatu) coastal management declined because of lost beliefs in the efficacy of ritual prohibitions (assisted by religious teaching), night time harvesting (aided by modern torches) when surveillance is difficult, population growth and mobility (hence less community cohesiveness), modern education (and reduced environmental knowledge), greater demand for cash (and few alternatives), conflict between local leaders and remoteness from (weak) government intervention (Winthorpe 2004). In Fiji young people place less value on management and view the sea through a commercial rather than 'traditional' lens (Veiteyaki 1995, DeMers and Kahui 2011). Even in remote atolls superior access to modern technology (from steel hooks to outboard motors and freezers) has resulted in greater efficiency and greater individualism (Scourse and Wilkins 2009). Blasting has killed off significant proportions of reefs in some areas, such as Chuuk (FSM), and also led to an increase in ciguatera (see below). Children have become adept at breaking coral reefs with iron bars. By contrast, in some densely populated islands that are not far from markets, such as Dominica, traditional techniques appear to have largely been maintained, contributing both to biological sustainability and to the sustainability of the labour force, including significant numbers engaged in such secondary industries as boat-building (Peteru et al. 2010). In coastal PNG communities and the Tongan island of Lofanga, access to markets has been a more important influence on reef depletion and overfishing than population size and growth (Cinner and McClanahan 2006, Kronen and Bender 2007). Management practices, and their absence, vary considerably.

Traditional fisheries systems have generally shifted towards more open-access commons (Aswani 1999, 2002, Christensen 2011), with fishers becoming characterized as more rapacious, individualistic and competitive (Oles 2007b), resulting in widespread efforts to introduce marine protected areas (MPAs) to reverse resource depletion. MPAs vary in size and policy, and from absolute no-catch zones to regulations on technology and catch limits. In numerous places, people have been reluctant to welcome marine reserve creation, where their fishing livelihoods are perceived to be at stake. In more tourism-oriented communities, where alternative livelihoods are possible and dive-tourism is important, such as the Cayman Islands, acceptance has usually been greater (Trist 1999, Broad and Sanchirico 2008), but that was not true in the Bahamas and Galapagos. Similarly, at least in the case of Réunion, elite local people with limited connections to the island were more in favour of establishing an MPA than longer-established, lower socio-economic status residents who had both a more positive perspective on the status of the marine environment and a vested interest in utilizing its productivity (Thomassin et al. 2010). Only exceptionally, as at Ahus island (Manus, PNG) and near smaller towns in Solomon Islands are there perceptions that customary management is beneficial in terms of superior catches for all (Cinner et al. 2005, Brewer 2013), since such outcomes are normally long term. Most MPAs are small scale and tend to be linked to no more than small clusters of villages and limited areas of lagoons, although an increasing number of MPAs are extremely large, partly in response to global pressures to cover 20–30 per cent of marine habitats by 2012. Some have been imposed from above without local involvement, notably the controversial Chagos (BIOT) marine protected area, the largest in the world, and large enough to potentially protect migratory species (Koldewey et al. 2010, Sheppard 2010, De Santo et al. 2011). A general trend to scale up marine conservation and management has occurred, partly as 'pro-nature' approaches have displaced 'pro-people' approaches, reflecting debates about the role of science and local participation, but also as governments have taken a greater interest, raising questions about social justice and sustainability (Sievanen et al. 2013). In some contexts where government regulation has failed, private initiatives have created entrepreneurial MPAs, as has occurred at dive sites in Curaçao and Bonaire (de Groot and Bush 2010, Thur 2010). Future MPAs may need to be mobile to protect vulnerable migratory species such as turtles, sharks and seabirds.

Some relatively remote islands where populations have not grown significantly have both retained a greater suite of traditional management practices and have more effective MPAs (Aswani 2002, Aswani and Vaccaro 2008, Aswani and Sabetian 2009). Where population density is greater and population growth faster, establishment and compliance have been weak. In Rodrigues high levels of non-compliance with a variety of management regulations were evident at least by the early 1900s, since alternative incomes to fisheries were largely absent. More recently, the creation of a large marine park resulted in fishermen reluctantly contemplating either migration to distant Mauritius or salt production as alternative means of livelihood, but there was also substantial non-compliance because of scarce alternative income-generating opportunities for individuals whose skills were few (Gemenne and Magnan 2010, Peterson and Stead 2011). In areas with a significant dependence on fisheries (for nutrition and commerce) effective MPAs and other regulations have not been easy to establish and implement. Poaching in MPAs is common. On unoccupied Tetepare (Solomon Islands) significant factors influencing resource extraction from the MPA were fluctuations in local markets; when copra prices fell coconut crabs were hunted and fishing increased, while 'logistical challenges and isolation have to date limited the opportunities for Tetepare harvesters to access large-scale, permanent markets [but] high population growth … more portable refrigeration and aspirations for a higher standard of living are likely to intensify pressure on marine and terrestrial resources' (Moseby et al. 2012: 234). Similar pressures are widespread.

Some localized and sometimes short-lived success stories have succeeded in protecting endangered species and reducing use conflicts, but medium-term outcomes remain uncertain or limited (Johannes and Hickey 2002, Tupper and Rudd 2002, Bartlett et al. 2009, 2010, Clarke and Jupiter 2010, Dumas et al. 2010). Tourism benefits may be short term, the economic and nutritional benefits medium term while the indirect benefits from coastal protection and the 'bequest value' from preserving ecosystems lie far into the future (O'Garra 2012). Management is more difficult when indigenous ecological knowledge varies in even relatively isolated, culturally homogeneous communities, and local people cannot easily comprehend the scale, direction and causes of change in marine environments (Lauer and Aswani 2010). Most MPAs and coastal management projects have been developed without any detailed assessment of the value of such areas to local communities. In the Caribbean, MPAs have been more affected by terrestrial activities, have largely failed to prevent coral mortality and have not been designed to account for the effects of warming (Mora 2008, Selig et al. 2012). In Vanuatu also MPAs are too small to be able to adapt to climate change (Bartlett et al. 2010). MPAs may have helped fish populations but have done little to sustain corals (Mora 2009). However, as is true of Fiji, and of most SISI, 'the current status of reef-associated fisheries at the national level is still uncertain due mainly to the lack of dependable data on subsistence fisheries' (Teh et al. 2009). Despite the role of fishing in nutrition, and the crucial significance of nearshore environments, too little is known for effective management.

Reviews of the impacts of MPAs have been inconclusive over their role in influencing reef and fisheries' recovery rates, and on the relationship between effective management, compliance rates and nearby population density and thus on the resilience of marine ecosystems (Côté and Darling 2010, Pollnac et al. 2010, Graham et al. 2011). Unsurprisingly, therefore, 'Continued degradation of coastal and nearshore marine resources is expected to continue over the next decade unless deficiencies in management are addressed' (UNEP 1999: 15). However, poverty, 'inadequate opportunities' and uneven development, potentially exacerbated by migration (Cassels et al. 2005), constantly trump management and conservation whether of forests or coastal waters (e.g. Foale 2001, Cinner 2010) so that sustainable management is rare. The few broadly successful MPAs are in developed countries; those elsewhere epitomize the 'tragedy of the commons' with inadequate regulation of areas where self-interest has triumphed over cooperation (Wilkinson and Salvat 2012). Climate change will further emphasize existing stresses on MPAs (Keller et al. 2009) and put greater pressure on fisheries.

Economic and environmental issues are routinely intertwined. In the Lau islands of eastern Fiji, as people sought cash incomes they became less dependent on fishing and exerted less pressure on the marine environment. Indeed fresh fish consumption was inversely correlated with household income (Turner et al. 2007, see Kronen et al. 2010b). Isolation from markets prevented fishing down the food chain, as occurs closer to markets. Reduced dependence on fishing has resulted in the erosion of 'social memory and local ecological knowledge', reducing opportunities to fall back on marine resources while simultaneously reducing the likelihood of their conservation (Turner et al. 2007). Ecological knowledge declines as it becomes less essential and, paradoxically, the potential for marine management may be greatest precisely where that environment is no longer so crucial to livelihoods. Where poor people depend on marine resources for food and income, especially in peri-urban areas where social controls are weakest, management is least effective.

OUT OF RESOURCES?

However much has changed in recent decades, the majority of SISI populations remain in rural areas, and agriculture and fishing provide valuable livelihoods. Beyond that they contribute a significant proportion (sometimes the entirety) of exports, generate lease fees (in fisheries), enable local market provision and are the key to food security (especially as imports are costly) and nutritional wellbeing. Yet agriculture has largely disappeared in a handful of SISI, and dwindled in many, while the same kind of cumulative downward spiral that has affected agriculture has also affected fisheries, though rarely to the same extent. Pressure on forests has been just as great; few forests remain, biodiversity has declined sharply and wood has gradually become a scarce commodity. Pressure on coastal land – a 'coastal squeeze' – is matched by pressure on coastal fisheries, yet effective management is rare, with MPAs often rejected or ignored. Both agriculture and fisheries have been threatened by 'more sophisticated resource extraction technologies' such as chainsaws and diving torches, without similarly more sophisticated management mechanisms (Fazey et al. 2011). Conservation cannot be understood simply as good practice, but as involving issues of identity and politics that go far beyond that. Complex human ecologies have become fragile with greater individualism (and rising populations) resulting in more disputes over tenure and less social cohesion. Likewise management of oceanic fisheries has come under pressure as some oceans are depleted. Depletion and overexploitation of global fisheries has put greater pressure on SISI resources. Pressures are likely to increase further with the continued rise of a middle class, especially in Asia, and greater demands for fish consumption. Increased pressures, some illegal, threaten future sustainable incomes from agriculture, fisheries and forests.

Food security is threatened by changing production practices and land losses, where food production is growing slower than the rate of population increase. Changing nutrition is partly associated with declining food production and reduced diversity, mainly derived from changing tastes, low incomes, the inability to purchase adequate diets (at high market prices), a growing dependence on store foods, disdain for cooking and the poor quality of cheap imported processed foods and drinks. Exclusive consumption of local foods even hints at poverty. Imported foods range from corned beef, turkey tails and meat flaps to instant noodles, biscuits and soft drinks (e.g. Grossman 1998a, Evans et al. 2001, Dixon and Jamieson 2005, Cassels 2006, Oles 2007b, Foster 2008, Rudiak-Gould 2009, 2010, Gewertz and Errington 2010, Thow et al. 2010, Seiden et al. 2012, Wichman 2012), a function of cost, status, taste, convenience and accessibility. Tinned fish is a remarkably common import in many island contexts, with imported fish and meat preferred to local versions (e.g. Dixon and Jamieson 2005, Potter and Sluyter 2010). In Woleai, where more than 200 species of edible fish inhabit the lagoon, the best-selling item in most stores was tinned fish. Rice and flour have become food staples even in quite remote communities, sometimes eaten without accompaniments (Rudiak-Gould 2009, Schwarz et al. 2011).

Even in the most remote places stores are dominated by imported goods, mainly tinned foods. On Vatulele (Fiji) more than a third of store goods were foods and more than half of them came from outside Fiji (McInnes and Connell 1988), much as in Martinique, where 'the products on its dusty shelves form a startling microcosm of the world system, an astonishing testimony to the history of colonialism and the more recent organization of international commerce' (Price 1985: 119). Imports, especially food, have embedded islanders in wider worlds. The humble corner store is the ubiquitous symbol of globalization.

Food and beverage imports are consistently high, often being more than a third of the value of all imports, so contributing significantly to trade imbalances. The inability to retain policies banning inappropriate imports as WTO regulations are enforced (Chapter 2) may worsen nutrition, dental health and the balance of trade. Rising food prices have undermined children's nutrition, especially amongst the urban poor, where people are turning to cheaper but less nutritious food. In urban areas poverty partly explains such shifts, thus 'any government that heavily taxes these products or bans them will suffer the ire of the working poor at elections and so the issue is largely avoided' (Grynberg 2010: 34). In Kiribati, the former President, Ieremiah Tabai, observed: 'If the shops are without rice and sugar it is a big political issue' (pers. comm., 2012). Surprisingly food security may be inadequate in rural areas where transport infrastructure is poor and on isolated islands with high population densities (Schandl et al. 2011). The shift away from traditional staples poses particular problems for food security after hazard events. Root crops such as taro are more resistant to cyclones, hence when Fiji experienced a series of cyclones in the mid-1980s there was a greater subsequent dependence on imported foods than might hitherto have been expected, since less taro was being planted (Chung 1987). Hazards have reduced the availability of local foods and hastened the transition to imported foods; a process that might be accentuated with climate change (Barnett 2011). Food security is also threatened by the volatility of international commodity prices.

Just as many individuals have moved out of agriculture and fisheries, where they could, in favour of other economic activities (or, in some cases, into unemployment), so governments have not always looked favourably on rural people. In Mauritius, for example, there was reportedly an 'island-wide mentality to view traditional fishermen with contempt and blame their distress and misery on insufficient volition and salubrity' (Paul 1987: 144). Agricultural workers, like the *repo taro* of the Cook Islands, sometimes fare worse. In St Lucia the cultural and linguistic gap that separated bureaucrats from farmers saw the latter viewed as illiterate, ignorant and incapable of innovation (Moberg 2008: 89–90). Concerted efforts to develop and implement effective integrated rural and regional development policies have generally been lacking. It is not only in the Caribbean that 'the policy of most governments on fisheries has appeared to be in token form' (Walters 1984: 95). Rural bias and decentralization of services in favour of fisheries, agriculture and integrated rural development is implausible. Households themselves have had to diversify to achieve development.

As agriculture and fishing have become less likely to satisfy local needs for income generation, diversification has brought greater occupational multiplicity, increasingly a characteristic of SISI, in order to 'construct a portfolio of livelihood activities and social support capabilities in their struggle for survival or to improve their standard of living' (Ellis 1998: 4). On three tiny Lau islands, with a hundred households averaging four people each, there were as many as 4.7 occupations per household (Turner et al. 2007). Rural Tahitians have similarly developed 'creative solutions that allow them to maximize their economic options at the local level' whether in agriculture, handicraft production or local employment (Kahn 2011: 74; see Lockwood 1993). On Nevis multiple economic activities have similarly extended across agriculture, fishing and handicrafts (Olwig 2007). On the tiny Tongan island of Kotu, where people have long practiced a mobile economy making uses of resources on different islands and the lagoon – a long 'history of flexible adaptation' – diversity gave them an advantage over other islands in responding to the vicissitudes of the market economy, so making them less dependent on migration and remittances (Perminow 1993). Livelihood diversity also benefited biodiversity. Diversity and multiplicity provide a contemporary 'subsistence safety-net'.

Fewer SISI are now dominated by agriculture, forestry and fishing, either for employment, income or livelihoods. Most households and most SISI have long diversified away from natural resource based activities. Local socio-economic dynamics are complex, and household livelihood strategies vary enormously even within villages (according to demography, needs, access to land, labour, technology, credit and markets, etc.). Poorer households may seek to minimize risk, while meeting basic food needs, while wealthier households maximize profit. Other households may choose to migrate, in whole or in part. As 'traditional' agriculture and fishing have stagnated and dwindled, sustainable management lapsed, and 'modern' commercial agriculture and fishing competed with trade restrictions and external pressures, so the need for a more diversified economy has grown, and contributed further to agricultural decline. In most SISI at least a basic service sector is invariably present, tourism plays a part and manufacturing and other 'modern' activities have emerged. The extent to which SISI have diversified away from the 'old economy' and the impact of this can now be examined.

4. Towards modern economies?

> When we look at income distribution and the evidence of growing inequality in Fiji in terms of housing, wages, education, health care and access to opportunities we must admit that we are in danger of creating and accepting in Fiji – the Fiji of the rich, which belongs to a small number of elite, and the Fiji for the poor, which belongs to a vast majority who are struggling to keep their heads above water.
>
> Quoted in *Fiji Times*, 17 September 2006

Conventional theory has long argued that economic growth and development follow the structural transformation of economies away from the primary sector – agriculture, fisheries and forestry – towards higher value manufacturing and services. Evidence is drawn from the industrial revolution, the rise of the newly industrializing countries (NICs) of Asia, the more recent, if tenuous success of the BRICs (Brazil, Russia, India and China, and now South Africa), and even parts of the Maghreb (until the 2010 'Arab Spring' re-emphasized that economic growth required a distributive element for development). For several decades before and after independence island states were repeatedly advised to similarly shift national economic structures to embrace industrial development and other facets of the modern economy. Curiously, tourism, seemingly a prerequisite for modern development in places where sea, sun and sand were virtually ubiquitous, and foreign exchange might be generated through capitalism and competition, was rarely given particular priority either by national governments or external advisors. It smacked of frivolity, unwanted social change, and also of an inappropriate focus on the needs of well-off foreigners (and perhaps a lingering notion that this was not self-reliance), and it had certainly nothing to do with either the industrial revolution or the NICs. Yet here at least geography actually favoured SISI.

This chapter examines what is at best the mixed success of a 'modern economy' that has enabled a transition away from agriculture and fishing – which is neither to imply that island agriculture and fishing are entirely 'traditional' (and the previous chapter should have demonstrated otherwise) nor that they should be ignored. But industry and services have diversified island economies and provided new employment. Economic growth and employment in manufacturing and services are almost entirely urban, and where they have grown (or failed to grow) urban populations have transformed island landscapes and geographies. Even the 'modernization' of agriculture and fisheries have had urban components in financing, processing, packing and marketing. The second half of this chapter examines the impact of the modern economy on employment and incomes, wealth and poverty.

MANUFACTURING

Invocations to industrialize came early in the post-war Caribbean, spearheaded by Sir Arthur Lewis, the Caribbean's Nobel prize winning economist, but the minimal success of 'industrialization-by-invitation' might and should have been a warning for latecomers elsewhere. Yet, typical of late twentieth century invocations to achieve industrialization was that of a World Bank report on manufacturing prospects in Fiji which stated in its opening credo:

> Neither geography nor culture nor political forms necessarily preclude a country from sharing in the fruits of economic advance; what matters is the skill with which a constellation of policies mobilize domestic resources for development ends and enable the nation to grasp the trading and technological opportunities that have underwritten development gains most notably in East Asia. (World Bank 1987: 1)

Yet geography (including human and physical resource endowments), culture and politics clearly hindered industrial development, and the success of newly industrializing Asia was not globally transferable, other than in a dilute, minuscule and unrecognizable form. Indeed the Bank study went on to recognize many constraints to industrial development in Fiji, including the lack of capital, skills and technological development. Since one of the largest SISI was thus ill equipped the prospects for export-oriented industrial development elsewhere were slight.

Manufacturing is a small-scale urban phenomenon in most SISI where it typifies import substitution industrialization in its early phases, being concentrated on light manufacturing and food production – the agri-food industries – once derided as the 'Coke, Curtains, Coconut Cream and Corn Curls' phase of development (Theophilus 1987, quoted in Connell 1988a: 53). Limited resources, small domestic markets, scarce skills, high (and rising) costs of energy and insurance, inadequate access to technology (including spare parts) and investment capital and the lack of economies of scale have hampered even this development. For years

Tuvalu had nothing more than a bakery as an indicator of industrialization, since few local materials were available for processing. In Samoa and Fiji agricultural processing firms operate at less than full capacity because of small markets and the seasonal supply of agricultural produce. Downstream processing in the timber industry had no more success. Higher wages than in Asia, and open economies (enabling cheap untaxed imports and discouraging local investment) and consumer preferences for overseas products, including food and drink, have further restricted industrialization.

The establishment of export-oriented manufacturing production has been even more difficult, because of inadequate access to markets and high labour and transport costs, other than in a few processing industries. Conventional disadvantages for export production, notably transport and labour costs, are accentuated for industry especially in SIS (Winters and Martins 2005). Even handicraft production in most SISI is constrained, so that 'local' handicrafts are sometimes 'made in Asia'. Where industrial investment was most successful, as in Barbados, the factors that contributed to that were good labour relations, a skilled and pliable labour force (usually female), low (usually minimum) wages and weak unions, financial incentives, political stability, suitable real estate (for expatriate owners and managers) and ease of access to foreign markets (Sunshine 1985, Long 1987). Such conditions were similar to those in Tonga during its brief phase of industrial development and usually included subsidized infrastructure, the ability to expatriate profits and avoid exchange regulations, concessionary trade agreements and committed government support (Taylor 1987, McGregor 1989). In most SISI manufacturing wages remain low (and unions non-existent) compared with public service employment. Footloose industries benefiting from such suppliant regimes were usually outposts of multinational corporations that were able to migrate easily if conditions, such as 'tax holidays', changed: a phenomenon of 'corporate ephemeralism' (Taylor 1986). Several countries established export processing zones (EPZs), with special land and legal regimes, financial incentives and attractive taxation policies. Industries moved in search of cheap, compliant labour, but as EPZs became increasingly global the particular attractions of those in SIS were less obvious and footloose industries moved on. At the apogee of Caribbean industrialization in the late 1990s, St Lucia had 70 factories employing over 3500 workers – almost 10 per cent of the formal workforce. Almost all the workers were women, 'explained by their willingness to work for lower wages and to be more tolerant than men' but who could not survive and raise families on manufacturing wages without additional income-earning opportunities (Gmelch and Gmelch 1997, Potter et al. 2004:

340). In the 1980s an electronics industry was successful in Barbados, St Lucia and St Kitts, and for a time exports thrived (Long 1987, Goss and Knudsen 1994) but remnants of the industry survive only in St Kitts. A similarly fast growing industry in the 1980s in several Caribbean SISI was data processing, where at its peak the Barbados industry had several hundred workers, almost all women, processing airline tickets and medical records, but it too has failed to survive. Though conceptually data processing should be easily outsourced to SISI, it suffered from lack of state support for local firms, the narrow role allotted to companies, and paid low wages (Potter et al. 2004). Even in the heyday of the manufacturing industry in St Lucia, Tonga and probably elsewhere, factory workers in such low wage circumstances saw the principal goal of employment as that of obtaining valuable skills for subsequent emigration (Kelly 1986, Connell 1988a). In such circumstances, and because tax breaks and other concessions were offered to industries, the benefits were relatively few, and tightened global trade restrictions effectively ended them.

In several SISI, for example Tonga, de-industrialization continues, despite industry never becoming significant. An infant industrial estate, protected by tariff walls, once had 21 firms which exported footballs, woollen garments and watches; little remains beyond a brewery and a small fish-exporting business. The high import component of industrialization weakened local gains, and at its peak it employed no more that 1 per cent of the workforce and contributed 5 per cent of the value of exports (Connell 1988a). Certainly, as elsewhere, an inefficient and even inappropriate industrialization, dependent on imports of capital and intermediate goods such as wool, was sheltered behind tariff walls and barriers (Kaplinsky 1983). Assembly operations exist in a few states, including a car components factory in Samoa, employing 950 people in 2012, which has survived only through special amendments to Australian trading regulations on 'rules of origin'. Ironically, only in Fiji has a manufacturing industry been relatively successful, but largely within the protective barriers of global trade agreements, such as the Lomé convention, or regional agreements such as SPARTECA (with Australia and New Zealand), but mainly confined to the garment industry. At its peak towards the end of the 1990s, the clothing industry employed about 19 000 people (and garment exports were valued at 33 per cent of all exports in 2000, a drought year) but, since the 2000 coup and the gradual reduction in protection in Australia and New Zealand, several factories have closed, wages have stagnated and by the mid-2000s employment had fallen to around 8000 people (Storey 2006). Declining numbers of small clothing factories still exist, mainly employing women on low wages. Employment in manufacturing is falling rather than growing, as it is in other SISI.

In the Northern Marianas a similar textile manufacturing industry developed in unusual circumstances, using cheap (and often illegal and exploited) Asian workers and exporting finished goods to the United States (while being inside tariff walls). After the United States effectively imposed minimum wage legislation to Saipan in the mid-2000s, making wages comparable with those in the United States, the industry collapsed almost overnight, resulting in outmigration, population decline and the parallel collapse of the government. The demise of the Marianas textile industry emphasizes in extreme form the fragility of industrial sectors and the vulnerability of governments when protection disappears and industry, as in the Caribbean, engages in 'island hopping' (Kelly 1986) in flight from regulation and declining profitability.

The establishment of export-oriented manufacturing largely occurred through protective trade agreements and widespread local concessions, and collapsed when these conditions disappeared. The optimism that once prevailed over industrialization in Fiji and in the rather larger Mauritius and Trinidad (Grynberg 2006), that sometimes made them seem models for SIS, has been vanquished in more liberal trading conditions. Successes, such as the global marketing of the 'prestigious' Fiji Water (Connell 2006c) and tennis racquets from St Vincent, are rare exceptions with small niches and even the former's spectacular success has become threatened by concerns over its environmental impact. Likewise, import substitution industries have been constrained by a range of factors, not least small local markets and absent economies of scale. The World Bank's optimism and enthusiasm have been unfounded, and models are now infrequently offered.

MINING

Several SISI, even such tiny islands as Clipperton, Redonda, Christmas Island, Angaur (Palau), Fais (FSM), Makatea (French Polynesia) and Farquhar (Seychelles), have a colonial history of small-scale mining. Whole ecosystems have been lost in rare cases where mining has overwhelmed particular islands, notably in Nauru, but also in Banaba (Kiribati) and Angaur (Palau), to the extent that the resettlement of the entire population of Nauru seemed a possibility at one time (Tabucanon and Opeskin 2011) and the entire Banaba population was resettled in Rabi (Fiji). Where mining now exists – primarily in New Caledonia, Fiji and Solomon Islands – there are economic benefits but negative environmental consequences on land and sea. New Caledonia holds about 20 per cent of the world's nickel reserves. Mining has continued since 1874, now employing more than 10 per cent of the island workforce. Expansion of mining has been so significant in this century that its future is a key component in the determination of the political future of the territory, as it was in the twentieth century, and a source of tension between local people, mining companies and the administration (Connell 1987, Horowitz 2004, 2009, 2010). Outside Melanesia, mining is of no great significance for SISI.

Nauru, once said to be the richest country in the world, eventually became the ultimate victim of the 'Dutch disease' or 'resource curse'. No other country has gone so drastically and dramatically from riches to rags. By the time that almost a century of phosphate mining effectively closed at the turn of this century, no other economic activity existed (with agriculture long abandoned) and other than a thin, residential coastal strip the environment was devastated. Much of the income had been squandered on an unprofitable airline and other rash investments, no alternative sources of income were evident, Nauruans had eschewed education and employment, and sedentary lifestyles had contributed to world record levels of diabetes, alongside obesity and other non-communicable diseases (NCDs; Connell 2006a). During the present century, Nauru has become dependent on aid, with residual mining, and struggled to establish the basis of a new economy in a particularly unpropitious context.

Environmental degradation has otherwise been most severe in New Caledonia where mining has been and remains the principal disturbance to the environment. Biologically, where soil fertility is extremely limited, the extinction risk for many taxa is considerable (Pascal et al. 2008, David et al. 2010). In some places the extent of environmental degradation has resulted in violent conflict (Horowitz 2009). Several river systems and wetlands have been degraded. Erosion has been considerable from mined hillsides, with severe soil losses from the mountain tops, due to high precipitation, steep slopes and belated revegetation. Bed levels in some rivers have risen with consequent flooding of adjoining agricultural land. Polluted river delta areas have caused changes to marine flora and fauna, mangroves have been silted over and reefs and lagoons polluted. Environmental management problems, for both river systems and coastal fisheries, resulted in the closure in 1988 of the giant Panguna mine (Bougainville Island, PNG) after violent landowner protests and conflicts over compensation for environmental degradation and over resettlement failures (Connell 1991b). Few contemporary mines have been established on small SISI, so that mines on Lihir and Simberi islands (PNG) are special cases. Both dispose millions of tonnes of waste – including cyanide and heavy metals – into the sea. Ocean disposal was chosen because of high rainfall, rugged landscapes, high seismic activity, strong community support and proximity to deep water. While ocean disposal resolved immediate issues, long-term marine impacts are uncertain.

In recent decades mines have closed in Vanuatu and Fiji, but Gold Ridge reopened in Solomon Islands in 2011. It employs about 550 people, mostly local people, and was expected to account for a third of the country's GDP, over an anticipated life of ten years (as it did between 1998 and 2000 until closed by civil unrest). Within a year tensions had arisen over land tenure, compensation and retained profits. The Vatukoula gold mine in Fiji, probably the largest current operation in any SIS, employs over 900 people, but briefly shut down for a year in 2006 due to the high costs and the level of capital required to sustain the operation, and it has rarely contributed more than 1 per cent of the nation's GDP or provided 1 per cent of employment. A bauxite mine opened at Bua in 2011, extending and diversifying Fiji's mining sector. Various future prospects, such as copper and magnetite in Fiji, nickel, gold and cobalt mining in Solomon Islands, are feasible and prospecting continues elsewhere, mainly in the Pacific. Most new mines will not be coastal but, as elsewhere in SISI, waste disposal and the construction of mining infrastructure will affect coastal environments.

Deep-sea mining of sea-bed manganese nodules in the eastern Pacific has been anticipated for at least a quarter of a century but it has been delayed by technological, environmental, legal and political complications. PNG has granted permission for prospecting but since 2008 local communities, particularly in New Ireland and New Britain Islands, have opposed it. Nodules in the Cook Islands have the highest cobalt content in the region, and may become valuable, but mining technology is currently unavailable, and mining tailings could be harmful to poorly understood deep-water marine ecosystems. Nonetheless at least ten Pacific SISI have issued prospecting licences.

The absence of mineral resources in most SISI is paralleled by the virtually total lack of fossil fuels or other domestic energy resources. With the sole exception of São Tomé, where valuable discoveries were made in the late 1990s, oil has been entirely absent from the SIS. While São Tomé may become a petro-state, it has yet to develop an effective regime for oil production and management (Chapter 2). The presence of oil off the Falkland Islands has given that small UKOT a new economic importance and also re-awakened strategic political issues (Dodds and Benwell 2010).

When combined with various degrees of isolation, growing populations, urbanization, greater transport needs (dependent on fossil fuels rather than wind or muscle power) and the switch from wood fuels, most SISI face high and rising energy costs. Oil is one of the most costly imports in SISI, and the first two global oil shocks were severe economic blows. In 1980, for example, two-thirds of Samoa's foreign exchange earnings went on imported fuel (Harden 1985). By 2010 oil imports were equivalent to 29 per cent of GDP in the Cook Islands and 15 per cent in Tonga. Attempts to develop local resources, whether of ethanol or hydro-power, have had only limited success, and are impossible on most low-lying islands, hence substantial and costly energy dependence is likely to continue. Inadequate policy support, the absence of proper financial evaluation and technical expertise have all slowed development (Barrett 2008). However Guadeloupe obtains 8 per cent of its power needs from geothermal energy and also has some wind energy, St Lucia, St Kitts, Dominica, Martinique and Vanuatu are currently developing geothermal energy, with the Comoros and other SISI in volcanic regions likely to follow. A growing number of projects are being designed for outer islands. Fiji, a rare SISI large enough for dam and reservoir construction, developed hydropower in 1983; this provides half the country's electricity (and should increase to 70 per cent by the end of 2012). Dominica generates about 40 per cent of its energy from hydropower, but expansion even in quite mountainous SISI, like Dominica and Samoa, is restricted by topography and/or rainfall seasonality. For most SISI access to energy at reasonable cost, and some movement towards energy security, are critical problems especially with larger vehicles and rising demand for electrification.

TOURISM

Tourism is of obvious importance in most SISI, sometimes by default as a reluctant choice when other sectors have floundered. Most Caribbean countries, 'because of significant entry barriers to other markets, [and] given their natural resource endowments, gravitated towards tourism as a development option' (Hayle et al. 2010: 36, Payne 2006, Bishop 2010). Indian and Pacific Ocean states followed. It took time to be recognized that for SIS 'tourism offers the best chance for development in terms of creating growth and employment, generating foreign exchange and reducing poverty' (Croes 2006: 455). It is so important that some places, again mainly in the Caribbean, have been simply characterized as Small Island Tourist Economies (SITEs) where tourism provides the basis for resilience (McElroy 2006), while, somewhat grandly, 'tourism is by far the fastest-growing economic activity in all SIDS' (Nowak and Sahli 2007: 49). Indeed most countries where tourism is particularly significant for economic growth are small island states (Brau et al. 2011). Moreover tourism has grown substantially since the 1960s, with cheaper airlines, higher incomes, paid and longer vacations. Within decades the landscapes, and images, of destinations such as Bermuda, Sint Maarten and Tahiti, were utterly transformed.

Tourism is highly subject to global financial shifts and costs. Tourism receipts almost universally fell after the global financial crisis (GFC); only some more upmarket destinations contradicted this trend, with the Seychelles having a record number of visitors in 2010. As migration characterizes SISI, so visiting friends and relatives tourism is of growing importance, especially but not only in the Caribbean (Scheyvens 2007). Upmarket tourism destinations can attract more distant markets; some 75 per cent of tourists in both the Seychelles and the Maldives are from Europe (Yahya et al. 2005). Distance and intervening opportunities make access to metropolitan markets crucial, since domestic tourism and intra-regional tourism are almost non-existent. Likewise poor transport connections and expensive resorts limit backpacker tourism, except in Fiji, which has become a significant stopover for many Europeans, yielding as much as $55 million a year (Jarvis and Peel 2010). Although Cape Verde has established a successful tourism industry, São Tomé has been hampered by the high price and unreliability of flights (Frynas et al. 2003). Anguilla experienced a decline in numbers at the end of the 2000s because of reduced airline capacity; the few airlines and planes flying into Vanuatu are routinely full. The Seychelles suffered when Air Seychelles was absorbed into Etihad, removing direct flights from European markets. The Cayman Islands, a rare SISI with a successful domestic airline, benefits accordingly, hence a case has been made for Caribbean states that are dependent on tourism to have locally owned airlines (Holder 2010). Without access SISI are condemned to constantly seek tourists; isolated Niue, with only one flight a week, has had to rely on a few idiosyncratic tourists – mainly yachtspeople, those who seek to tick off another place or a handful combining business with pleasure (Connell 2007e). Yet even more inaccessible Pitcairn, by dint of its fame, has acquired a tiny but valuable industry (Amoamo 2011a): the ultimate exception to the rule.

Tourism is particularly vulnerable to perceptions of political unrest and violence, including piracy for the Seychelles, although the impacts are often short-lived (Connell 1987b, Narayan 2005), and to environmental degradation, where reefs and beaches are damaged, shops crowded and human densities too high. Within SISI, tourism is often highly centralized around the main urban area (as for example in Grenada and Vanuatu) or on offshore islands (as in St Vincent) and its impact localized. Remote places cannot easily benefit from tourism since tourists are less likely to visit outer islands, knowing little about them, preferring to stay put, or without enough time or money (Connell and Rugendyke 2008, Cassidy and Brown 2010). The economic benefits from tourism in SISI are less than in larger states because of disproportionate imports of foods (e.g. Narayan et al. 2010) and other materials, and overseas ownership of tourism infrastructure with development subject to international (dis)investment decisions. (Samoa, St Lucia and Grenada have been hit by the decisions of overseas airlines to cut flights on the grounds of profitability.) In some unusual cases, such as the Maldives, most tourism workers are migrants and their incomes are transferred overseas; overseas workers, mainly from Bangladesh, are more pliable, can be paid less and are said to be more committed, though isolation (on resort-only islands) and social factors are as significant (Scheyvens 2011, Shakeela et al. 2011). Local people may be disadvantaged by their lack of relevant skills and knowledge, the absence of training opportunities and external vested interests. Yet tourism still generates enormous incomes in several small economies while typifying their external orientation.

Various small political dependencies, especially in the Caribbean, such as the Virgin Islands, Saba and Sint Maarten, have long been 'small economically underdeveloped tourism-dependent countries of the world periphery' (Husbands 1986: 176), where tourism is vital and little else exists. It has been more important in politically dependent territories rather than SIS, with the exception of the French overseas territories most of which have increasingly priced themselves out of other than a national market (McElroy and Parry 2010, Gay 2012). Even the smallest islands, such as Anguilla, St Barts, the Cocos-Keeling Islands, Rapanui and Pitcairn, seemingly bereft of any other economic activity, have tourist economies (Carlsen 1999, McCall 2008, Amoamo 2011a) that prevent them being mere 'government islands' (primarily dependent on subsidy and where almost all employment is in the public sector, or dependent upon it). That is even more true within particular SIS, such as for the Mamanuca and Yasawa islands in Fiji and the Grenadines in St Vincent, and other SISI, such as Bora Bora (French Polynesia) and the Isle of Pines (New Caledonia). Several dependencies, such as Bermuda, Guam and Norfolk Island, are dependent on tourism, and once remote islands, like the Falklands, the Faroes and St Helena, have increasingly benefited from cruise tourism (see below). In much larger SIS such as Antigua, St

Lucia, Bahamas, the Maldives and the Seychelles, tourism has characteristically been the largest contributor to the GDP for several decades, and is substantially in excess of the value of exports. In St Lucia tourism generates almost 64 per cent of GDP and in the Cayman Islands it accounts for about 70 per cent of GDP and 75 per cent of foreign exchange earnings, though both states have other economic options (Wilkinson 2003, Weaver 2003). In Anguilla it generates 72 per cent of GDP and 80 per cent of employment (Forster et al. 2012). Intervening opportunities, inadequate transport, high costs and limited amenities (whether recreational or visual) mean that tourism is largely absent from such remote SISI as Kiribati and Tuvalu. Most of the few SISI where tourism is unimportant, such as São Tomé, the Comoros, Solomon Islands, Kiribati and Tuvalu, are among the poorest of all developing countries.

Since at least the 1970s tourism has both generated more income for SISI than commodity exports, and had a disproportionate economic influence than in other world regions. Virtually the only resources in which most SISI have a comparative advantage are clean beaches, unpolluted seas, warm weather and water and at least vestiges of distinct cultures (Connell 1988a). Such advantages are far from illusory. While tourism is highly elastic and more subject to the whims of fashion than other economic sectors, the lure of beaches and coastal amenities seems eternal. Tourism is labour-intensive, creating many jobs (though few are well-paid, or with promotion prospects, hence many are taken by migrant workers, and also women, where alternative wage labour may be scarce). It also provides foreign exchange and generates activity in other sectors of the economy (especially in services such as construction, stores and transport). Trends towards luxury resorts have increased the labour-intensive element though managerial staff are often expatriates (Britton 1991, Shakeela et al. 2011). In the Bahamas half the national workforce are directly employed in tourism and perhaps a further quarter are indirectly dependent upon it (Connell 1988a). In Antigua some 95 per cent of all employment is said to be related directly or indirectly to travel and tourism (Hayle et al. 2010), which would leave scope for little more than a small public service sector.

A significant proportion of the incomes generated from tourism leak from SISI or never reach them. Holidays are booked with overseas airlines and travel agents; airlines, cruise ships, hire cars and hotels are usually owned by transnational corporations, while construction materials, fuel and labour may be imported. Payments made within SISI may be repatriated when food and especially drinks are imported, alongside such 'tourism artefacts' as swimming costumes, surf boards, snorkels and scuba gear. In the Pacific even 'island nights' with 'authentic' local foods rarely tempt tourists to try them and actually require more imports (Dixon and Jamieson 2005). By the 1980s, even in the larger Caribbean SIS about two-thirds of the food consumed by tourists was imported, and in small islands like Carriacou (Grenada) it was greater still (Momsen 1998, Mills 2004). Integration of agriculture with the tourist industry has been disappointing, usually because of low volumes, uncertain supplies and inaccessibility. Large resorts are least likely to be linked to local producers (Scheyvens and Russell 2012). Handicrafts too are often imported. Estimates suggest that between 40 and 85 per cent of tourist expenditures leave the SISI with the lowest proportions retained in the smallest islands. So great has that leakage been, especially because of foreign ownership, that as in Fiji (Rao 2002) some have questioned the value of tourism, with its social costs, as a worthwhile development strategy. Yet at least a third of tourist expenditure remains in SISI, though little of that trickles down from urban areas.

The lack of diverse tourism experiences and opportunities, or physical heritage, especially in the smaller SISI, limits interest and confines expenditure to resorts. Diversifying away from 'sea, sun and sand' towards physical and cultural heritage has posed problems in many SISI like the Turks and Caicos (Cameron and Gatewood 2008). Ironically, the most luxurious resorts such as Wakaya (Fiji) and Mustique (St Vincent) are almost totally insulated from their surroundings (on islands purchased by foreign companies) and guests may never use money in the destinations. All-inclusive resorts owned by international chains – classic enclaves – have taken similar forms. As tourism grows resorts tend to replace locally owned accommodation, and land alienation displaces others. Ecotourism has been limited in most SISI, emphasizing the urban bias of the industry, and even in village-based ecotourism projects income distribution is usually highly skewed (Connell and Rugendyke 2008). Nonetheless, as Anguilla's national tourist organization recognized some decades ago, tourists keep that island going as they 'drop their dollars in little bundles all across the island making everybody happy: the resorts, restaurants, taxi-drivers, car rental companies, cruise boat operators, and the government' (quoted in Connell 1993b: 137). Income is more effectively diffused in such small islands.

Where once many SISI, from Tonga to the Seychelles (Connell 1988a, Wilson 1994), were fearful of the social costs of tourism, over-hasty modernization and the pernicious values of tourists, such fears largely subsided as the economic benefits became obvious and other facets of island economies struggled. Tourism has hastened modernity, and in many places the economic gains from tourism have been used in much the same way as remittances (Chapter 6), for example on improved consumption, housing and technology (e.g. Connell and Rugendyke 2008, Kerstetter and Bricker 2009, Scheyvens and Russell 2012). Social costs have ensued: uneven development, conflicts over land and incomes, abandonment of culture (and its role in social organization), while some relatively new phenomena have appeared, such as prostitution and, less obviously, crime and an increase in sex tourism and HIV/AIDS (La Grenade 2008, Connell and Negin 2012). In the Maldives 'luxury resorts have been contrasted with the high overcrowding, poverty, malnutrition and substance abuse of ordinary residents' (Shakeela and Weaver 2012: 1343), while in the Caribbean, tourists have been seen as 'an alien patronizing presence whose main concern was non-productive self-indulgence and hedonism' (Potter et al. 2004: 411). Yet some material forms of culture (e.g. music and dance) have survived and been enhanced. In Rapanui, where tourist numbers vastly exceed the local population, the outcome has been described as 'a run-away materialism [where] individual avarice and virulent consumerism are undermining social cohesion' (Fischer 2005: 257, 260). Social change is rarely so dramatic and usually merely emphasized and accelerated by tourism.

The absolute dominance of the tourism industry in several SISI has contributed to minor versions of 'Dutch disease', with distortions in price structures, the crowding out of other economic sectors, and perceptions of new forms of dependency. Ironically, in the heavily subsidized French overseas territories tourism itself has become a victim of 'Dutch disease' (Gay 2012). Some of the best coastal agricultural land has been lost to tourism, and agricultural production has declined through the withdrawal of labour, rather than being stimulated by tourist demand, hence, as in the Seychelles, food prices rose as tourism got underway (Kaplinsky 1983; cf. Nowak and Sahli 2007). Small-scale production and seasonality have weakened attempts to develop local linkages between tourism, agriculture and fisheries. Tourism infrastructure has often been geographically concentrated which both isolates tourists in resort enclaves, weakening the economic and social impact on wider society, and results in infrastructure provision being biased towards tourist areas. Tourism is therefore something of an enclave economic activity, but its economic contribution is enormous and fluctuations in tourist flows are less than those of commodity prices.

Cruise Ship Tourism

The growth of cruise ship tourism has extended the reach of tourism since the 1980s, and put increased pressure on SISI that are more accessible to major western tourism markets, particularly those of the Caribbean. Between 1989 and 1999 the numbers of cruise ship passengers rose from 4.0 million to 8.7 million per year, and industry forecasts – which appear to have been realized – were for some 21 million passengers by 2010. Larger ships offered new economies of scale, which led to reductions in the cost of cruises, in turn opening up new markets (Macpherson 2008). The geography of cruising changed over the same time period. Within the major destination of the Caribbean, there was a shift towards the western Caribbean at the cost of the southern and eastern Caribbean and the Bahamas, as cruise shipping companies sought new destinations to encourage repeat tourism. Since cruise liners bring their own accommodation, the effective capacity of the islands increased 20-fold. Between 1995 and 2004 the number of cruise ship visitor arrivals in the Cayman Islands more than doubled, increasing from 682 900 to 1 693 300, giving a ratio of 20:1 of tourists to residents, one of the highest in the world. By contrast, eastern islands, such as Guadeloupe, experienced a decline in numbers. In the Pacific those SISI closest to Australia were the main beneficiaries.

The busiest port in the SISI is the Bahamas, with around 1.8 million cruise ship arrivals in 1994 and 4 million by 2009, because the short distance from Florida is very convenient for both short and long cruises. Other significant destinations include the USVI (about 1.2 million in 1994 and 1.8 million in 2009) and Sint Maarten, the Cayman Islands, Aruba and Grand Turk (Turks and Caicos), all with more than a million passenger arrivals in 2009. Outside the Caribbean no similar significant destinations exist, other than Bermuda, with New Caledonia and Vanuatu of greatest significance in the Pacific. However several small Atlantic islands such as St Helena and the Falkland Islands have visitor numbers disproportionate to their populations. In 1982 about 500 tourists a year visited the Falklands, but by 2008–09 that number had grown to almost 69 000, with 62 600 arriving on cruise or expedition vessels to see the distinctive scenery and battlefields (Ingham and Summers 2002, Royle 2006). Even isolated Pitcairn Island has been able to sustain a small tourist economy through occasional cruise ship access (Amoamo 2011a), while cruise ships also visit remote, unoccupied and otherwise largely inaccessible islands such as South Georgia, Castaway Cay (Bahamas) and Mystery Island/Inyeug (Vanuatu) with significant income gains for the few with claims to such islands. By contrast, the extension of air transport had a less dramatic impact on the growth and expansion of tourism in most SISI, although the Galapagos and Rapanui have been prominent exceptions (Kerr 2005, McCall 2008) because of constraints on numbers.

In Caribbean ports especially the numbers of cruise vessels can contribute to overcrowding, especially if more than two cruise ships visit simultaneously, which is common in more popular cruise ports such as St Thomas (USVI), Sint Maarten and Nassau (Bahamas). Bermuda has long regulated cruise ship tourism (and the number of air arrivals), limiting them to six per week, so that only one would normally be there at any one time, resulting in an annual maximum of about 200 000 visitors. Congestion places pressure on environments and contributes to reduced satisfaction amongst residents, tourists and cruise ship visitors, which may account for stagnant tourist numbers in the Cayman Islands, St Thomas and Sint Maarten during the 2000s. In the BVI people began to protest over the increase in cruise ship numbers (from 122 000 in 1995 to 466 000 in 2005) when it was perceived that these numbers would have a negative impact on the BVI tourism image and thus on income. Similar protests have occurred in English Harbour (Antigua) when the government proposed to increase the number of cruise ship arrivals, in the belief that this would be at the cost of yachting, which then provided a greater contribution to GDP and employment than cruise ship tourism. That may be so, since in St Lucia cruise ships account for the majority of tourist arrivals yet generate less than 10 per cent of revenue (Moberg 2008). Countries have a weak regulating position against the oligopolistic cruise ship industry, wholly owned beyond the SISI and by giant global corporations such as Disney and Time-Warner, with cruise ships displacing the 'stay-over' industry and resulting in a 'descending spiral of competition' (Bresson and Logossah 2011: 154, Wilkinson 1999), as attempts by host countries to secure greater revenue by imposing taxes have usually 'led cruise operators to remove the offending islands from their itineraries' (Moberg 2008: 124). Consequently SISI have sought to achieve 'cruise conversion', encouraging visitors to return as stay-over tourists, but few do (Wood 2004). The newest cruise ships offer an ever-increasing range of on-board attractions and services, so that the cruises are the primary attractions and ports of call (especially for repeat tourists) ever less important, a move stimulated by the cruise companies to increase on-board revenues, consequently reducing local income generation.

While cruise ship visits do not generate the permanent full-time employment, and taxes, associated with land-based tourism, they offer a range of economic opportunities, for the local provision of goods (particularly café food and handicrafts) and services (especially transport). Ship-based tourism incurs fewer costs and lower financial risks, and creates fewer environmental pressures, than land-based hotel tourism for SISI governments. The costs of marketing, accommodation, electricity, water and waste disposal are largely external. However ships place pressure on water supplies, and generate waste streams including sewage, grey water and hazardous wastes, which can be a significant source of pathogens with the potential to threaten human health and damage aquatic life. Only in a number of heavily frequented regions and ports is this likely to be much more than a minor irritant rather than a significant threat to reefs and lagoons. Above all, cruise ship tourism has extended the range of the tourism industry.

Environmental Impacts

Tourism has been almost entirely coastal in SISI and has placed considerable pressure on reefs and coasts, through reconstruction (e.g. seawall and marina development), reclamation, waste and sewage disposal and recreational uses. The sheer numbers of visitors and watercraft are a concern in many hotspots, over the carrying capacities of dive sites, pollution, over-fishing and overall environmental and aesthetic degradation. Numerous studies have demonstrated the harmful effects of tourism on reefs, directly through trampling and anchoring boats, and indirectly through waste deposition (e.g. Allison 1996, Juhasz et al. 2010). Graffiti adorn heritage buildings and even the iconic *moai* statues of Rapanui (McCall 2008). In the Caribbean, especially, tourism has accentuated the processes of urbanization (Weaver 1993), while in Vanuatu a considerable proportion of coastal land has been sold off to foreign investors, not always legally, ensuring that the best land and the largest, most profitable tourist ventures are in foreign ownership, and local small-scale ventures are marginalized (Slatter 2006, Wittersheim 2011). Mangroves have been removed and golf courses constructed (the latter placing particular demands on water supplies, and creating localized pollution through fertilizer use). Wetlands have been dried out in the Caribbean and mangroves removed from national parks in the Seychelles to make way for tourism (de Albuquerque 1991, Clifton et al. 2012). Golf courses, spas and swimming pools are particularly demanding of fresh water (Ioannides and Holcomb 2003, Bishop 2010). Even where, as in the Maldives, a form of 'frontier tourism' has been practiced, where new atolls are opened to tourism as the tourism life cycle demands pristine beaches and seas, lagoon water quality has deteriorated (Brown et al. 1997). Tourism places considerable pressure on fresh water resources, often in islands where it is already scarce. In Barbados, for example, tourism absorbs 12 per cent of the country's water, at three times the per capita rate of the local population, and achieving reduced consumption has been difficult (Charara et al. 2011; Chapter 7). Comparable pressures on and uneven uses of energy, especially for air-conditioning in upmarket hotels, with tourists in Barbados using four times as much electricity as local people (Moberg 2008, Gössling and Schumacher 2010), are widespread.

Coastal erosion has followed the removal of mangroves and development of tourism infrastructure, as at Denarau (Fiji) where accelerated erosion followed attempts to stabilize beaches, using dykes and seawalls. Moving sand to enhance the beaches used by tourists in Vanuatu accelerated erosion elsewhere to the detriment of local people (Connell and Rugendyke 2008). Sand mining in Barbados and Dominica, in association with the construction of new tourism facilities, severely eroded some beaches (Archer 1985, Cambers and James 1994). In the Maldives reef blasting, to make channels for boats, changed the currents and led to erosion, so that much of the sand for tourist beaches was brought from other islands. Soil has even been imported from Sri Lanka (Scheyvens 2011). Construction of second homes, again especially in the Caribbean, and particularly in canal developments, has placed similar pressures on coastal environments (as after mangrove removal at Pacific Harbour, Fiji). Several countries, joined by the Maldives in 2011, enable non-citizens to purchase land, much of which is coastal, intensifying local use of other coastal land. The 'fixing' of coastal zones, by tourism and other infrastructure, accentuates vulnerability to hazards and climate change.

Early migration to the coasts of SISI occurred in colonial times when defences against human predators and malaria were less necessary; a second, much more imposing phase has resulted from a quest for economic opportunities through commercial agriculture, fisheries and tourism, accompanied by urbanization, sand mining and vegetation clearance, particularly the removal of protective mangroves: a coastal squeeze that has made coastal areas more vulnerable. Such pressures and consequent environmental degradation and deterioration have been particularly evident in the Caribbean, where tourist numbers are much greater and 'frontier tourism' impossible. Thus 'if sediment loss further continues Martinique is at risk of losing not only the majority of its famous beaches and its valuable mangrove habitats but also its prestige as a beach tourist destination' (Schleupner 2007: 100, Schleupner 2008). In rapidly urbanizing Noumea (New Caledonia), the locus of most tourist faculties, coastal change resulted in the loss of 29 per cent of mangrove areas between 1960 and 1989, with heavy subsequent lagoon sedimentation (David et al. 2010). Resort construction in Fiji, alongside a new causeway, prevented the flushing of the adjacent lagoon, where inappropriate sand mining and vegetation clearance had already caused accelerated coastal erosion and smothering sedimentation (Terry et al. 2006). Such degradation of lagoons and coral reefs and reduction of biodiversity may reduce the attraction of tourism destinations, particularly critical in smaller SISI, that are almost exclusively dependent on tourism, and where landscapes have been most transformed by it.

Tourist operators have often practiced conservation, to preserve clear seas, fishing and diving prospects, hence tourism is usually less directly environmentally damaging than most coastal developments. In Fiji several hotels are in advance of the government in stimulating conservation, though tourists may themselves disrupt conservation measures. Local people have challenged tourist development in particular forms. In Ulithi (Yap, FSM) tourist divers have been confronted by local fishermen, and in St Lucia fishermen have similarly been concerned at the tourist impact on marine degradation (Trist 1999, Rubinstein 2003). By contrast labour-intensive game fishing in the Seychelles has provided jobs for fishers. Similarly, shark-diving in Palau is not only labour-intensive, but provides more revenue than shark fishing and ensures ecological sustainability (Vianna et al. 2012). A challenge for many SISI is to make a transition from traditional tourism economies and infrastructure that can deplete and alter natural landscapes and resources, to low impact, nature-based ecotourism, which is more difficult to market and manage: that is from the mass resort and ship tourism which dominates most SISI to some form of 'slow tourism' that might generate equivalent revenues (Timms and Conway 2012).

Ecotourism opportunities tend to be in relatively remote places, tourists are few (from inadequate marketing hence lack of knowledge) or uninterested and spend relatively little; returns are therefore limited, numerous projects have closed down and owners have been bitterly disappointed by outcomes (Connell and Rugendyke 2008, Slinger-Friedman 2009, Wilson 2010). What might seem viable possibilities, such as whale-watching in Tonga and shark-diving in Palau, are disadvantaged by intervening opportunities, while megafauna tend to be more common where people are few (Moyle and Evans 2008, Vianna et al. 2012). Even where small-scale local resorts are reasonably accessible, as they have been in Samoa, marketing has proved difficult and they have fared relatively poorly compared with larger, often foreign-owned, establishments that are marketed through international brochures (Scheyvens 2008). In a further irony almost all of these local resorts, which were literally 'on the beach', were wiped out in the 2009 tsunami, although many have been restored. An unusual exception is Saba, which receives about 25 000 tourists a year, a number limited by the absence of cruise ships (through the lack of anchorages) and a tiny airport (reputed to have the shortest commercial runway in the world), focused on diving and hiking. Superior access might have meant otherwise.

Tourism has increased migration to the coast and urban centres in search of employment opportunities, intensifying pressure on the coasts and resulting in declining agricultural production inland (Connell and Rugendyke 2008). In many parts of the Caribbean, especially in the drier Leeward Islands and such SISI as Sint Maarten and Anguilla, and in the Galapagos, agriculture effectively disappeared because of tourism employment (Aldrich and Connell 1998, Miller et al. 2010). Climate change is assumed to pose problems for tourism in SISI, because of rising temperatures, possible degradation of coral reefs, coastal erosion and an increased risk of vector borne diseases (Mimura et al. 2007). Expectations, initially based on preference studies from Bonaire and Barbados, are that tourists will be driven away by negative environmental changes including temperature rise, but the impact will be slight, unless cyclones intensify (Uyarra et al. 2005, Gossling and Hall 2006, Moore 2010, Forster et al. 2012). While this may occur on smaller islands, on larger islands environmental (and other) changes have resulted in tourism simply moving to another part of the island and the tourism cycle taking on new forms, while tourists may shift to other destinations or travel during other periods of the year. In some destinations, such as Fiji (Becken 2005), tourism operators are seeking to manage the environment to reduce and mitigate the impact of climate-related changes. However, as in the Maldives, Barbados and the Seychelles, this is rare since the industry assumes that profits can be made in a shorter time period than climate change, and is more concerned over such issues as the GFC, skilled labour shortages and immediate environmental management (Payet 2008, Belle and Bramwell 2010, Becken et al. 2011, Sovacool 2012a). Yet, in a final irony, as management has responded to the 'litany of ecological impacts' on island environments, long-haul air travel to islands has been identified as one source of greenhouse gases contributing to climate change so that, in a small way, 'island tourism is contributing to its own demise' (Carlsen and Butler 2011: 3) – the 'Maldivian dilemma' – by enabling the paradox of 'last chance tourism' ('see it before it disappears') whether in Antarctica or Tuvalu (Eijgelaar et al. 2010, Farbotko 2010b, Prideaux and McNamara 2013). Although SISI are the most likely places to be negatively affected by climate change the impact on tourism is impossible to determine (Perch-Nielsen 2010). Future tourism is more likely to be affected by socio-economic influences of various kinds.

Tourism in Time

Tourism may be uneven and unpredictable, incomes may leak away and social costs be unwelcome, while small economies are dependent on decisions taken elsewhere (whether by potential tourists, airlines or cruise companies), but it often constitutes the core of national and local development and few SIS are without a tourism ministry. With more modern transport several destinations, like the Cook Islands, have been able to move away from being 'destinations on the edge' (Burns and Cleverdon 1995) to being dominated by tourism. Marketing is important in a complex and dynamic globally competitive industry. Any reduction in tourism in SISI would obviously lead to reduced national incomes, the consequences of which might also threaten ecosystems. Reduced tourism revenue would potentially fuel a 'downward spiral' where decreased environmental management would lead to negative impacts on biodiversity (and more general amenity), which might subsequently further reduce tourism receipts (Campling and Rosalie 2006). Only the USVI – the islands of St Croix and St Thomas – have been said to have experienced such a level of environmental degradation that they have descended into a 'tourism death spiral' (Potter et al. 2004: 430). However while it was observed many years ago that, for the Seychelles, 'paradoxically its future as a tourist paradise is most seriously threatened by the depredations of the tourists themselves and by the amenities provided to attract them' (Benedict and Benedict 1982: 109), new generations of tourists are simply unaware, up to a point, that the contemporary environment is no longer as pristine as once it was. In Fiji no significant differences in satisfaction were detected between tourists at pristine and degraded reefs; they were unable to distinguish them and enjoyed the social aspects of diving more than the ecological and environmental factors (Fitzsimmons 2009). Indeed, in numerous places, like the Maldives, tourist lack of awareness of the fragility of marine ecosystems results in environmental damage (Domroes 2001). Throughout the SISI governments are seeking new and expanded tourism markets, whether through cruise ship tourism, new niches (such as meetings and convention tourism, adventure tourism or bird and whale watching) or new sources (Russia, India and inevitably China). Bora Bora has sought to be a honeymoon destination, and the Cayman Islands, Fiji and the Cook Islands the places for weddings. Scale, however, precludes much diversity. The geography of tourism is linked to accessibility, and thus to oil and energy costs, which may rise rapidly in the future. Forecasting the future of tourism in SISI is inevitably problematic, but islands are likely to remain attractive destinations.

SERVICES

The absence of significant manufacturing has meant that the service sector makes a much greater contribution to GNP than in larger states, and a significant part comes from the public sector. After public service, the largest components of the service sector are tourism, followed by retail services, but some SISI have developed a distinct finance sector. Business development has been constrained by many of the same factors that have limited manufacturing: government attitudes, limited technical and management skills, small markets and lack of capital. In the rural sector especially, small businesses, notably stores, are often not oriented to profitability and have supported communities rather than contributed to a more extensive commercial development (Rakoto 1986). Where public service, finance and tourism dominate, and agriculture and fishing are trivial, as in the Bahamas, Bermuda and the Cayman Islands, employment in the service sector incorporates over 90 per cent of the workforce. In most SISI similar percentages prevail but often almost wholly dominated by the public service, as on 'government islands', since such employment is eagerly sought after as stable, relatively well-paid and in pleasant surroundings. On Niue, an island of about 1300 people, the only households without at least one government worker are of retirement age. Government employment enables small SISI and outer islands to stay afloat.

Finance

From the late 1960s, when international capital became more mobile and rapid global telecommunications were established, many SISI became offshore financial centres of various kinds: 'attempting to become a tax haven was a sensible livelihood strategy for many micro-territories faced with the end of colonial rule and the lack of other obvious sources of income or export earnings' (Moore 2011: 1770). That was even truer of dependent territories. Tax havens such as the Bahamas began in the 1930s, but most of those in British colonies like Bermuda and the Cayman Islands began in the 1960s, encouraged to do so by the British government. Some of the wealthiest SISI, notably such UKOTS as the Cayman Islands, BVI and Bermuda, successfully moved into the financial services sector, which generated more than a third of their income. The Cayman Islands was the world's fifth largest banking centre with assets exceeding $600 billion, Grenada had over 30 000 registered banks, about 1 to every 30 people in the country, while in 2007 the BVI hosted 800 000 registered international business corporations (Potter et al. 2004,

Palan et al. 2010). By contrast, Pacific tax havens have only ever made a modest contribution to national incomes (Sharman 2005). Tax havens offered low or nil taxation, secrecy and ease of incorporation. Many SISI were too small to draft effective legislation and regulation without the assistance of those who would most benefit from inadequate legislation, which opened possibilities for corruption, and latecomers to the offshore finance market were more likely to become engaged in less reputable areas. Tax havens experienced boom years in the 1970s when the electronic mobility of capital became particularly rapid, deregulated markets and privatization were the norm, international trade and investment were expanding and rapid increases in global oil prices created a surge in the volume of Eurodollars that required deposit. Most SIS have sought some involvement in contemporary financial services through tax havens, and by the end of the 1990s the Caribbean had become the core of an industry involving more than 20 SISI.

In this century there has been a global assault on tax havens from states, the EU and civil society through pressures to reduce tax evasion and associated illegal activities, hence their economic benefits to SISI (and elsewhere) have been greatly reduced. Terrorism brought new controls after 9/11. Companies like the notorious energy giant Enron used multiple overseas subsidiaries, and at one point had 600 subsidiaries in the Cayman Islands and another 100 in the Turks and Caicos (Potter et al. 2004). Russian oligarch and mafia money was reputedly laundered in Nauru and elsewhere. Seychelles was reputed to specifically target criminal money. The decline of tax havens was most severe in several UKOTS, since the British government had ultimate responsibility for the financial liabilities of dependencies. New regulation particularly affected the Cayman Islands (Roberts 1995, Clegg and Gold 2011, Hampton and Christensen 2011, Shaxson 2011). Moreover the GFC also struck the substantial global hedge fund industry, more than half of which had been domiciled in the Cayman Islands, BVI, Bermuda and the Bahamas (Palan et al. 2010). In Bermuda, where offshore finance was significant enough to have boosted local economies, but also to have resulted in spiralling housing prices, the decline was much less dramatic, since illegality was less evident. In Pacific SISI too the activities of offshore finance centres were significantly curtailed by external intervention, from the same countries that were key aid donors (Van Fossen 2003, Sharman 2005). Under pressure from Australia, Vanuatu opted to end its tax haven status in 2008. Niue, Nauru and the Marshall Islands also closed down their offshore financial centres. At its peak in the 1980s the Vanuatu tax haven employed over 300 people, a substantial component of the urban economy (Connell and Pritchard 1990), but those days have gone.

Although Vanuatu has retained more offshore insurers, banks, casinos, and tax-free real estate than other Pacific states, the number of offshore banks fell from 150 to 9.

Regulation was particularly unwelcome not merely because of economic losses but because SISI had 'become locked into their relationships with the offshore finance industry by their dependence upon the earnings potential of predominantly imported skills and expertise, [thus] any attempts at diversification into other sectors would be constrained by the need for wholesale reskilling and the acquisition of new knowledge bases' (Hampton and Christensen 2002: 168). Larger, more established centres have been able to retain significant capacity since their regulatory capacity and the veneer of 'good' regulation is more effective than that of smaller jurisdictions such as Anguilla, Montserrat and the Turks and Caicos Islands (Clegg and Gold 2011). SISI such as the Cook Islands, Vanuatu and Nauru (Connell 2006a), and especially the UKOTS, several of which depended particularly heavily on the offshore finance sector, have lost employment and revenue, weakening their economies. Further significant decline in the UKOTS might lead to increased direct control from the UK (Clegg and Gold 2011, Hampton and Christensen 2011), as has occurred for the Turks and Caicos Islands.

Illegal financial behaviour has sometimes 'spilled over' into other spheres, such as tourism (Block and Krausner 1987), while SISI such as the Bahamas, Bermuda and Cayman Islands have experienced social costs that were partly a function of unusual inequality: 'poverty in the midst of plenty' (Palan et al. 2010). Despite employment and income benefits, tax havens attracted highly skilled workers 'in chronically short supply' from other sectors of the economy where national benefits were potentially greater (Sharman 2011: 59). Global clampdowns, emphasized in the wake of the GFC, in which tax havens were implicated (Palan et al. 2010), have ensured that fewer remain particularly successful. Reaching minimal acceptable standards in a climate of more intense global regulation and monitoring has become too difficult for most. Nonetheless the BVI has remained the world's largest provider of offshore entities, with a million incorporated companies (Leigh et al. 2012), while larger more established centres, such as Bermuda and the Bahamas, where illegality was rare, have been able to withstand both the GFC and intensified regulation without substantial hardship. By contrast, for Barbados and Vanuatu, the costs of externally imposed compliance were simply too high (Vlcek 2010). The IMF recognized 23 offshore financial centres in 2011, including eight SIS and eight other SISI, mostly UKOTS in the Caribbean. Tax havens and corporations have continued to avoid regulation, global investment continues, secrecy is welcome and transparency is not, rules can be bypassed and most havens, at least in the Caribbean, will survive and probably prosper.

Service Diversity

Other service industries have had some success. The Marshall Islands is the third largest shipping nation in the world, through providing flags of convenience. Bahamas, Vanuatu and São Tomé are some way behind. Tuvalu 'acquired' a fleet of 12 Iranian oil tankers in 2012. Several Caribbean SISI have sought to develop medical tourism and Barbados has established IVF facilities. The Bahamas and Cayman Islands, and also Jamaica, are seeking entry, based on their proximity to North America, and their existing tourism infrastructure, but face strong competition from Mexico and Costa Rica (Corbin 2011, Connell 2013b). Montserrat had a famous recording studio, where Paul McCartney, Elton John, Eric Clapton and others recorded albums, before its destruction by Hurricane Hugo. Deregulation and electronic technology have, however, largely been unable to transform the economies of SISI. Even in services, where electronic communication is central, locational disadvantages, a shortage of skilled labour, a different trade regime and greater scrutiny have restricted the service sector, beyond small-scale retailing and transport services. Services such as banking and insurance are often foreign-owned, even regionally; Trinidadian and Jamaican companies own most such services in the eastern Caribbean. Various SISI, especially in the Caribbean, where islands have good air connections, even including the land-short Seychelles, have enabled the growth of a 'second home' industry, with expatriates purchasing homes – so providing immediate revenue – and purchasing subsequent related services and stimulating some local employment.

Smaller SISI especially have had unusual strategies for generating revenue. Aruba, Grenada and several other Caribbean SISI have hosted offshore gambling sites; Antigua battled with the United States to establish an offshore gambling industry, generating some foreign exchange and job creation but in 'a highly tenuous offshore services sector … if the USA were to legalise Internet gambling' (Potter et al. 2004: 367, Cooper 2011). Vanuatu, São Tomé, Tonga and several other SIS have sold diplomatic passports, mainly to Chinese entrepreneurs (Van Fossen 2007). In 2011 Dominica and St Kitts came under pressure for selling passports under an Economic Citizenship Program to Iranian nationals. Hosting 'reality' television shows has brought in occasional wealth (and tourist publicity). Fiji earned about $20 million from staging

Celebrity Love Island in 2005, and the Cook Islands, Vanuatu and Samoa have all hosted versions of *Survivor* and other series. For many years the sale of stamps, later 'telephone sex' and the renting of its internet domain (.tv) have been valuable economic strategies for Tuvalu. Throughout this century .tv has generated $4 million a year for Tuvalu. Rather earlier, in 1981, stamps represented 99 per cent of the value of Tuvalu's exports. São Tomé too has benefited from 'dial-a-porn', though the rise of the internet has reduced such possibilities, and earned some income with stamps portraying such celebrities as Marilyn Monroe and the Beatles. Some of the smallest dependent territories, like Pitcairn and Wallis and Futuna, have had significant overseas markets (often for stamps whose designs had nothing to do with those territories, and which never actually emanated from there) while multi-island states such as the Cook Islands, Tonga and St Vincent brought out stamps for particular islands. Antigua even produced stamps for the unpopulated Redonda. Pitcairn, which markets both stamps and its domain name, has been described as a 'postage stamp republic' (Steinberg and McDowell 2003). At least in the 1980s stamps were valuable and successful forms of economic activity (Bertram 1986), but maintaining that has proved impossible, because of the propensity of several states to bring out too many stamps of limited aesthetic value or local relevance and the global decline of philately. Likewise, coins have been minted: Niue minted SpongeBobSquarePants and Star Wars coins in 2011. Such gimmicks usually earned little and were quite unsustainable. Stamp, coin and passport sales, being a tax haven or providing flags of convenience are disproportionately more attractive to the smallest states (Slemrod 2008), despite some social costs, simply because the small revenue that they generate has some value.

Strategic Services

Global superpower rivalry has enabled several SISI to benefit from strategic land rentals. The Marshall Islands gains enormous incomes through the rental of Kwajalein atoll for a missile range, and some islanders received massive compensation payments for the post-war use of their atolls for nuclear testing, but the long-term social costs of such economic benefits were immeasurable. While the inhabitants of the BIOT never chose to lease out their territory to military interests, and gained no financial benefit from it, other territories have chosen (or been 'encouraged') to do so. Densely populated Bermuda had space for an American base until 1995. Antigua received annual rental from an American base used for training regional security forces (Connell 1988a). Ascension is primarily a British airforce base, but also a United States airforce base and a European Space Agency rocket tracking station. Satellite tracking stations are even more widespread. The United States has used the Seychelles as a base for drones, while in 2011 the Seychelles agreed to the establishment of a Chinese naval base there, and a year later Russia was similarly seeking a base there, shifting the apparent balance of power in the Indian Ocean, with implications for strategic development in the Maldives. A year later the Cocos Islands were touted as a base for American drones operating in Asia. Guam has been primarily dependent on the American military for its economic growth (and bases occupy as much as a third of the island). Its future political economy, and that of the adjoining Northern Marianas, depends on its perceived strategic role as part of United States' 'forward defence'; thousands of military personnel are locally seen to have had a significant positive impact on the economy but a harmful one on local culture (Owen 2010). In several SISI the military also provide meteorological services, infrastructure construction and emergency relief.

More recently some states have provided a service to metropolitan nations by leasing islands (or parts of them) as detention centres, contemporary versions of placing political exiles and lepers on remote islands, or using them as quarantine stations. Nauru earned a substantial income in the mid-2000s, when little else was possible, with the refugee processing centre contributing some 20 per cent of national GDP and employing 100 people (Schandl et al. 2011: 48). Christmas Island and Manus (PNG) were also destinations for Australia's 'Pacific solution', a scheme that was revived in 2012, where asylum seekers were located beyond national territory. Guam and Tinian (Northern Marianas) have played similar roles. In the Australian case detention offered an unusual and unanticipated income generation strategy for Christmas Island, where the decline of phosphate mining threatened viability and the arrival and accommodation of asylum seekers boosted the population and the economy, albeit at considerable social and political cost (Mountz 2011, Needham 2011). Such uses of islands indicate both the particular challenges of achieving development in small SISI and how unusual strategic shifts can make isolation advantageous. Such distinctive 'development strategies', whether military bases or tax havens, offer little sustainability. Two-thirds of the island of Tinian (Northern Marianas) was alienated for American military uses for over 35 years but never used, hence no revenue was generated and the land could not be used for alternative purposes.

As the case of Diego Garcia indicates, and as the Marshall Islands demonstrated half a century earlier, SISI have not easily been able, or even willing, to defend the interests of their inhabitants in the face of much larger states. Strategic services have not always been chosen but rather thrust upon dependent territories at some social and health costs. Like trade negotiations, strategic relationships are even more asymmetrical and SISI have few negotiating opportunities and limited room to manoeuvre. As Henry Kissinger infamously said in the 1970s: 'There are only 90 000 Micronesians; who gives a damn?' Nonetheless, in less extreme circumstances, 'strategic issues linked to global cold war politics can be used by small states to lever aid and other types of support from richer patrons' (Minogue 1987: 136). And, unremarkably, this has extended beyond the cold war into a different, equally suspicious global age.

In such varied ways rental incomes are particularly significant in smaller dependent territories, where autonomous sources of income are few, a context that distinguishes them from larger states. Yet even service incomes are not easily won and fluctuations in morality (that ended Tuvalu's 'phone sex' income), global tastes (the virtual demise of philately) and especially strategic interests alter their significance. But economic benefits may be small, social costs can be great, and sometimes devastating (Vine 2009) and moral quandaries are raised. That they remain of significance is indicative of the limited status of conventional development strategies.

EMPLOYMENT AND INCOMES

Creating economic growth, productive employment and income-generating activity is necessary, yet is particularly difficult where population growth is often faster than economic growth, where the number of school leavers is greater than the rate of employment creation (and where most school leavers seek formal employment). Public service employment is widely favoured because of its usually superior wage levels, conditions of employment (prestigious, clean and relatively secure) and incidental fringe benefits, that do not preclude simultaneous involvement in the private sector, to the extent that some are even 'phantom workers' – often unproductive or absent except on pay days (Connell 2009c). In most SISI public service employment has stabilized or been cut back in recent years but is unmatched by growth in the private sector.

Unemployment is an urban phenomenon (and unmeasurable in rural areas where subsistence is significant). Rising urban unemployment has occurred in most SISI in recent years, despite few adequate measures of its extent. That has not always been so, as in Grenada where measured unemployment fell from 27 per cent in 1994 to 12 per cent in 2001, as construction, manufacturing and tourism expanded, but was concentrated amongst poor, unskilled and rural workers (Barrientos 2004). Nonetheless, as globally published unemployment rates indicate, data for SIS are quite unreliable, being little more than guesstimates. The World Bank records a figure for only six SIS, and these range from an improbable 1 per cent in Tonga to 17 per cent in São Tomé (World Bank 2012). Where reliable measures exist unemployment is never below 10 per cent and is often much higher. Thus St Lucia has had an unemployment rate of 25 per cent, higher than in most other Caribbean SIS (Henry-Lee et al. 2010), but probably similar to that in many urban areas.

Unemployment is particularly high amongst urban youths, with growing numbers of unemployed and marginalized youths in most urban centres, such as Port Vila (Mitchell 2004, 2011). Even two decades ago, inadequate access to employment, land and credit had led to increased levels of unemployment in the Tongan towns of Neiafu and Nuku'alofa (Gailey 1992b). In 2001 half those unemployed in St Lucia were aged between 15 and 24 (Henry-Lee et al. 2010), though some older people had probably given up seeking work. In 2005–06 the unemployment rate for 15–19 year olds in Honiara was estimated at 75 per cent and for 20–24 year olds was 49 per cent (Russell 2009). Cape Verde had a 2009 unemployment rate of 13 per cent (with 17 per cent in the capital) but youth employment was 31 per cent. Broadly comparable conditions occur in many SIS, but are less evident in smaller SISI (Abbott and Pollard 2004). Consequently urban settlements especially can be seedbeds of discontent and incentives to migration (Chapter 5), but amongst those least able to achieve it, or gain from it.

Until quite recently there was a widespread belief that poverty, however defined, did not exist in most island states, especially in the Pacific, since urban and rural 'safety nets' existed, where extended families could and would support those who for whatever reason experienced temporary problems. Islands seemed idyllic; how could poverty exist? However, safety nets are breaking down and it is no longer possible, if it once were, for urban people simply to return and be supported by rural kin. Increasingly visible urban poverty bears some relationship to internal migration since the urban poor include those who have migrated from impoverished rural areas, where income-earning opportunities are few, towards possible jobs and services.

Poverty undeniably exists in Pacific SISI. By the mid-2000s half of the Kiribati population was below the poverty line, Fiji had 34 per cent below the poverty line and only Vanuatu of the nine other Pacific SIS had less that 20 per cent of its population below the poverty line. Samoa, which a decade earlier had rejected any notion of poverty, had 20 per cent below its own poverty line. Poverty was well entrenched although only in Fiji, FSM and Tonga did it appear to be significantly increasing (AusAID 2009). In three Fijian villages within 25 km of the capital, incomes came mainly from selling produce (fish, vegetables and brooms) and some local employment (at low wages) but very few households could cover living costs that centred on food, water, school fees, church tithes and village functions. Some households sold almost all their fish, and consequently had an inadequate diet based on rice and cassava, could not afford school fees for some secondary schoolchildren and lived in bush material houses without electricity or water. In two of the villages the average household lived in 'extreme poverty' on less than $1 per capita per day (Zann and Zann 2008). Similar situations are true of many other villages in many SISI where inadequate access to land (and cash crops) is not compensated by access to employment. Not surprisingly, within these villages, and villages in Roviana (Solomon Islands) (Furusawa and Ohtsuka 2006), there were substantial differences in incomes with those who were better off having access to off-farm employment within or beyond the village.

Outside the Pacific, the extent of poverty (however defined) was earlier and more widely recognized, with the IMF and World Bank producing Poverty Reduction Strategies for many SIS including Cape Verde, Comoros, Dominica, Grenada and São Tomé. In Comoros in 2004 some 37 per cent of households were classified as being poor and in Cape Verde 30 per cent of the population were poor (and some 14 per cent 'ultra-poor') in 1989, but that proportion had risen to 36 per cent in 2002. Grenada, regarded as having the highest proportion of people in extreme poverty in the eastern Caribbean, had 32 per cent in poverty and 5 per cent in 'extreme poverty' in 2002, while Dominica, St Kitts, St Lucia and St Vincent all had more than 25 per cent below the poverty line. In St Lucia that had risen from 19 per cent in 1996 (Moberg 2008). Similar figures exist in the Turks and Caicos and Anguilla and, of the SIS, only the Bahamas had a rate less than 10 per cent (Bourne 2009). Poverty may be associated with multiple deprivation; in Dominica 23 per cent of poor households included a person with disabilities, 10 per cent had someone with a long-term illness and 27 per cent had people who were diabetic or hypertensive (Perch and Roy 2010). Despite the former existence of Poverty Reduction Strategies, poverty appears to be generally increasing, though formal data are inadequate.

While income poverty is more easily measurable, a non-income component of poverty consists of people's capabilities, vulnerabilities and participation: broadly, a poverty of opportunity, alongside inadequate access to services. Some households are poor in the sense that they do not have enough food, clean water or access to adequate housing or basic education (Bryant 1993, Abbott 2007). In Vanuatu in 1998 only 39 per cent of the population had access to clean water and 50 per cent to adequate sanitation, while comparable figures for Kiribati from 2000 were 62 per cent and 50 per cent, respectively. In Fiji some 37 per cent of rural people had access to reliable metered water supplies, compared with 95 per cent in urban areas (*Fiji Times*, 30 April 2012). Poverty and inequality are most visible in urban settlements. In Honiara as much as one-third of the urban population are below the national poverty line and perhaps half of these are in urban settlements, so that 'many families are mired in poverty [leading to] family breakdown, street children, sleeping out, sexual exploitation, alcohol and drug abuse, delinquency, violence and lawlessness' (Russell 2009: 16, Maebuta and Maebuta 2009). Social disruption has followed poverty. Domestic violence has become a growing problem and women are more likely to be the victims of other violent crimes (Amnesty International 2011, Rasanathan and Bhushan 2011, UNDP 2012). A similar situation is evident in Suva and other large urban centres. In the Caribbean, crime can be linked to poverty and income inequality (World Bank 2007, UNDP 2012). Guadeloupe has the highest crime rate in France with Pointe-à-Pitre said to have been 'taken over by drug dealing and prostitution' after 6 p.m. (Borredon 2011).

Social problems have increased. Experimentation with drugs is more common. An epidemic of cheap heroin has swept through the Maldives with the UN estimating that 40 per cent of the country's youth use hard drugs (Doherty 2012), typifying the shift in the burden of drug problems from developed to developing countries. Crystal methamphetamines transit Micronesia. Youth are especially vulnerable to risky sexual activity, early pregnancy, behavioural problems and a difficult transition from school to work, even where work exists (Barrientos 2010). Various urban centres have long been sites for 'weekend warriors' fuelled by alcohol. Named gangs are gradually emerging. Despite being relatively new in many SISI, alcohol consumption has contributed to enormous discord (Marshall 1979, 2004). Violence can be the outcome of poverty and unemployment, especially amongst urban youth, as in Port Vila (Mitchell 2000); in Apia (Samoa) this has increased as rural communities 'banish' troubled village youths to the urban area where social control is absent (*Pacific Beat*, 29 June 2011). In Vila a resurgence of sorcery has followed urban anomie (Rio 2011). Begging has also become visible in many large urban centres. Such visible evidence of inequality has created tensions and frustrations.

Much poverty is also hidden in outer islands and remote areas, emphasized by a lack of 'voice' and limited access to educational and health facilities, and formal employment. In several SISI government services in remote areas have tended to contract (and were usually least developed in the first place), skilled workers are unwilling to be posted there and material supplies are unavailable (Chapter 5). In Antigua and Barbuda contraction and concentration of services in Antigua has caused Barbudans to face higher costs of goods and services, including food, while Nevis was even less able than St Kitts to cope with the GFC (Perch and Roy 2010). Even in the small island of Grenada, rural communities were overrepresented amongst the poor, as were large families (Barrientos 2010). In most SISI, a poverty of opportunity pervades social and economic life.

Access to education is reasonable in most SISI, and universal in many, while literacy rates are high, but the quality of education is often poorer in rural areas and truancy and drop-out rates are higher. In Pacific SIS remote schools often experience problems in recruiting and retaining experienced teachers, maintaining communication with central agencies, infrequent visits from school authorities to ensure standards and delays in receiving supplies and salaries (Coxon and Munce 2008). In Fiji around one-third of all primary schools, mostly rural, have fewer than 100 pupils, increasing the difficulty and the cost of providing good quality education. In some villages there are no primary schools and parents send their children to boarding facilities, but the accommodation and food provided are 'poor', depressing school performance (Bacchus 2008) and contributing to a vicious cycle of inadequate opportunity. High drop-out rates even from primary schools reflect inadequate incomes (though, when education has become free, as recently in Vanuatu, attendance rates significantly increase). In St Lucia over a third of children leave school without any qualifications, there are not enough places in secondary schools and tertiary education is a luxury for a small percentage (Henry-Lee et al. 2010). In various SISI, therefore, education is far from adequate to equip school leavers for local wage employment. In Vanuatu most children in secondary education come from medium and high income households, a situation broadly similar even for primary school children, which also enabled the rather better off to lobby more effectively for improved access to education. Across the Pacific, fewer than 3 per cent of students attain tertiary education. Rural areas are relatively disadvantaged.

Most SISI have limited data on income distribution, especially over time, again partly because of the task of measuring incomes in semi-subsistence economies. Nonetheless, there are indications that inequalities may often be significant and not diminishing. In Tonga the 2009 Household Income and Expenditure Survey showed that absolute poverty had risen from 2.8 per cent in 2001 to 3.1 per cent in 2009. Meanwhile, the number of people living below the basic needs poverty line jumped from 16.2 per cent in 2001 to more than a fifth of the population (22.5 per cent) in 2009. While this group were not living in absolute poverty, they struggled to meet extra costs such as education, transport and utility bills, an evolving situation blamed on 'geographic isolation, exposure to natural disasters and climate change, low economic growth and high risk to external shocks' notably a decline in tourism and remittances. While outer islands experienced the most rapid rise in poverty, doubling from 12 per cent of the population in 2001 to 23 per cent in 2009, the largest numbers in poverty were on the urban fringes of Nuku'alofa, many of whom were migrants from outer islands (Tonga Ministry of Finance 2011). In Fiji, where the most finely grained data on poverty exist, the major pockets of poverty are in urban squatter settlements and in a few rural areas, where the sugar industry has declined (Pabon et al. 2012). In Samoa and Fiji the rise in poverty was also most evident in outer islands and rural areas. In Samoa in 2008 the lowest income earning households (10 per cent of the total) received less than 2 per cent of total income whereas the top 10 per cent of households received 34 per cent; inequality had increased during a period of economic growth that had not therefore favoured the most disadvantaged (Government of Samoa 2012). In Fiji, over a longer time period from 1975 to 1991, the extent of both poverty and inequality grew steadily, when the Fijian economy was growing relatively rapidly, creating obvious concern (evident in the statement by a Director of one NGO at the start of this chapter) and the growing realization of relative deprivation. Moreover over that period Fiji had achieved greater levels of economic growth than many SISI, hence the extent of poverty may well be greater elsewhere. The two trends towards greater absolute poverty and increased inequality, and their negative social consequences, are certainly more widespread than in Fiji.

In some SISI poverty and income inequality have resulted in a downward spiral – an urban 'poverty trap' from which there appears scant hope of escape. Indeed, in some places there is growing evidence of 'chronic intergenerational poverty' (Brown 2002). A parallel downward spiral may also exist in rural areas where youths choose not to enter the agricultural sector but, despite high urban unemployment, move in search of urban jobs (which their limited education often denies them), resulting in uncultivated land, low productivity and loss of traditional knowledge and techniques, but no compensating urban economic growth. Single-parent households and those with disabilities are commonly relatively poor (though both in the Caribbean and the Pacific many single parent households actually depend relatively successfully on the remittances of distant males), as are those where there is functional or real illiteracy. Such households bear the primary burdens of poverty. In what can seem idyllic but are beguiling contexts, poverty and unemployment pose problems.

DISCONTENTS OF MODERNITY

Generating modern economic activity let alone economic growth has proved difficult for a range of reasons stemming from limited human and physical resources, inadequate infrastructure, high energy costs and high wages but, above all, location and the lack of economies of scale. Weak economies are tied to global political economies. Agriculture and fisheries have remained valuable as the once anticipated conventional shift towards manufacturing has proved disappointing, with little more than narrow import substitution and the decline of subsidized manufacturing trade zones and protected exports (such as textiles). The loss of established preferences (for agricultural and manufactured goods) and the decline of agriculture, even abandonment in parts of the Caribbean, have placed greater pressures on tourism (and remittances) to sustain economies. This has been the most obvious success story of the modern economy, where SISI have comparative environmental advantages, and have turned these into employment and incomes. A small number of islands, such as Bermuda, Tahiti and Rapanui, however inaccurately, are defined to the outside world through tourism. Advances in transport technology have enabled more places to benefit yet, even here, some islands are simply too remote, and without obvious attractions, to compete. The tyranny of distance may be weakest for tourism but the once vaunted 'end of geography' is yet to arrive.

Inability to compete successfully in the more obvious components of the modern economy has resulted in various SISI turning to less conventional strategies, whether exporting stamps or internet domains, or seeking niche products like kava and black pearls. The possible significance of seabed and marine resources, hitherto difficult to access and assess, has resulted in states attaching greater significance to maritime boundaries and seeking to extend them. One consequence has been conflicts between such adjoining states as Fiji and Tonga, over areas around Minerva Reef, where oil reserves may exist, and between Vanuatu and New Caledonia over the Matthew and Hunter Islands (mini versions of disputes in the South China Sea). Strategic location has proved both advantageous and disadvantageous. Relatively successful finance industries of various kinds, some legal and some rather less so, have declined following greater external scrutiny. Few exceptional strategies, notably those involving the marketing of strategic location, have enabled sustained economic development.

In many SISI, especially where populations are still growing quickly, unemployment and poverty have become more evident markers of contemporary economic status, especially as agricultural stagnation and new aspirations foster rural–urban migration. The quest for modernity has brought challenges. The outcomes of poverty are particularly evident in the growing urban centres: income inequality, inadequate housing, limited access to clean water, education (especially post-primary education) and health services, thus tending to perpetuate the structural context of poverty. The following chapter therefore examines the factors that have contributed to more obvious urbanization and its social and environmental consequences.

5. Urban futures?

> The nation's poor have a number of common problems. They live in unhealthy homes with little or no sanitation; they are increasingly exposed to crime; they feel ignored by the state, service institutions and their own communities; their children are more likely to be sexually abused by adults; they are deeply cynical about politicians and politics in general.
>
> *Fiji Times*, 21 April 2003

As the previous chapter has shown, with or without the creation of a modern economy, one distinctive and visible outcome of change throughout SISI is urbanization. Until the second half of the twentieth century most towns – there were no cities – were tiny, administrative centres: colonial outposts rather than centres of national development. That changed on the eve of independence, as towns took on new functions – a nascent industrialization, that never progressed far, alongside secondary and tertiary education, national hospitals, parliaments, broadcasting and so on. Towns underwent particularly rapid growth, the outcome of which raised difficult management problems rather than stimulating economic development. Urban populations have grown on all but the tiniest of islands (with Pitcairn's 'capital' called Adamstown), but alongside that growth have come social and environmental problems, visibly evident in the emergence of squatter settlements, an informal sector and urban sprawl. Urbanization is a combination of inmigration, usually the outcome of rural–urban migration and less frequently of immigration, and to a lesser extent of urban fertility. To account for urbanization, and its causes and consequences, it is first necessary to trace basic components of population change.

POPULATIONS

Only in parts of the Pacific are any SISI still going through the demographic transition, from high to low fertility and mortality, yet population growth rates remain rapid in some places. However, even basic data on population are often inadequate to assess trends, especially in the Caribbean and, perhaps surprisingly, in some of the smallest SISI, hence population change estimates – like all quantitative data – must be treated with caution (Table 1.1). In the population context, as elsewhere, there are great differences between SISI. While the average population growth rate in Pacific SISI is about 2.2 per cent, in Vanuatu and Solomon Islands it is around 2.6 per cent but falling. In Kiribati the population growth rate is about 2.3 per cent, resulting in extremely high population densities, especially in the capital city (see below). In several states, such as Fiji, Tonga and Samoa, and most Caribbean SISI, growth rates are less than 1 per cent, mainly because of high levels of outmigration. In Dominica, despite one of the highest fertility rates in the region, the population fell by 10 per cent between 1990 and 2000 in the wake of the banana wars (Moberg 2008). In some SISI, such as the Seychelles, a combination of migration and steadily falling fertility rates (partly as a result of that migration) has slowed the population growth rate to below 0.5 per cent, so that an aging population threatens the financial stability of the state, hence a major challenge is to sustain fertility at the replacement level (Campling and Rosalie 2006). That is also true of St Kitts and some relatively large SISI, including Barbados and St Vincent, where populations are not growing. Between 2000 and 2010 the FSM went from slow growth to population decline. Populations are also falling in several small and remote dependent territories such as Montserrat, St Helena, Tokelau, Niue, Pitcairn, Wallis and Futuna and Tristan da Cunha. Montserrat's population peaked in 1946 and has been declining ever since. St Helena's population collapsed after its inhabitants acquired full British citizenship in 2002 and migration ensued (Hogenstijn and van Middelkoop 2005). The Marshall Islands and Cape Verde are unusual in having a high rate of population growth even alongside significant emigration, in contrast to Samoa and Tonga, and more recently Tuvalu, where migration has long been something of a 'safety valve' for high fertility rates. Island populations are growing in UKOTS such as the BVI, Turks and Caicos and Cayman Islands, where subsidies and economic growth prospects, centred on tourism, are better. In the few remote territories where economic growth has occurred, notably the Falkland Islands, population growth has followed.

In many remote Pacific areas, such as south-east Tanna (Vanuatu), where children are of productive value and access to contemporary family planning techniques is largely absent, population growth rates remain relatively high (Winthorpe 2004). Similarly, where adoption is common, as in the Marshall Islands, fertility rates are also relatively high. Total fertility rates (TFRs: the average number of children born to a woman) remain very high in some of the poorer SIS, being 5.1 in São Tomé and 4.7 in the Comoros. High rates are also evident in Melanesia (with Solomon Islands on 4.6 and Vanuatu on 4.4) but appear to be falling. Similar rates are also evident in Polynesia – with Tonga and Samoa at 4.2 and Tuvalu on 3.7 – emphasizing the significance of international migration. By contrast no Caribbean SIS have TFRs greater than 2. Throughout the Pacific, life expectancies have risen over the past quarter of a century, but remain lowest in the Melanesian states. Infant mortality rates (IMRs) are also highest in Melanesia, with Solomon Islands (24 per 1000), Vanuatu (21), and also Kiribati (52), Nauru (48) and FSM (38), being the greatest. In Solomon Islands the IMR may actually be increasing largely because of inadequate, and sometimes declining, provision of health services to rural areas, and overall life expectancy barely increased between 1985 and 2010 (Table 1.1). Some of the worst health and mortality problems are experienced in the growing urban settlements, especially in urban areas established on coral atolls (see below). In the Caribbean, IMRs are substantially lower, often in the teens. The lowest mortality rates, including IMRs, are in the dependent territories. Several UKOTS, such as Bermuda (2) and Anguilla (10), have very low IMRs, while Mayotte has a rate of 16 compared with the neighbouring independent Comoros on 75. In the Pacific the IMR in Samoa (23) is double that of adjoining American Samoa (10). Political dependencies are usually better financed, and often 'more economically developed, socially advanced and demographically progressive than the independent island countries' (McElroy and Pearce 2006: 536, Dunn 2011). In several SIS not only are low absolute sums spent on healthcare but they also represent small proportions of national budgets, hence access to effective services is limited (Taylor et al. 2005, Connell 2009c). Disparities between life expectancies and mortality rates in dependent and independent SISI remain significant.

Although prolonged rural–urban migration, increasing pressure on both rural and urban services, rising youth unemployment, social discontent and significant maternal mortality all contribute to a favourable climate for establishing population policies, developing and implementing policies has proved difficult, where educational gains are few and economic growth weak. The factors that are most conducive to successful population policies – integration of population and development policies, improved rural development and communications to spread new values and reduce the economic significance of children, formal sector employment opportunities for women, increasing age of marriage – have been most evident in Caribbean and Indian Ocean SISI. In much of the Pacific a small, nuclear family is not a preferable alternative to extended kinship and the migration opportunities this may enable. In urban Vanuatu the age of women at first childbirth is gradually falling. The outcome of high population growth rates has produced a 'youth bulge': a preponderance of young adults in the population, placing strains on land resources, employment, education and social organization (Mitchell 2011, Noble et al. 2011). Lack of economic growth has meant not only that many SISI have been unable to reap a 'demographic dividend', where the share of working-age population rises relative to that of children and young people, but that the 'youth bulge' has meant an excess of youths who cannot find employment or adequate roles. Such youths may engage in anti-social and disruptive behaviour, as has occurred in the capitals of Tonga and Solomon Islands, although perceptions of violent and troubled youths are usually greater than reality (Mitchell 2011). In fewer SISI, where migration is considerable, such as St Kitts, Palau, Niue and Wallis and Futuna, a different problem is apparent in an aging population and a high dependency rate, so that the demographic dividend is even more elusive. For many SISI, a crucial development issue is that of maintaining, let alone improving, present standards of living in the face of continued population increase and urban migration.

RURAL–URBAN MIGRATION

Population mobility is inherent to SISI. Without migration life could be particularly difficult. In extreme form, in small, remote high islands such as Tikopia (Solomon Islands), achieving sustainability had a high human cost, involving celibacy, infanticide and other extreme practices to stabilize population size and growth rates; even in contemporary times island chiefs regulated the population to ensure that it did not pass the number assumed to be sustainable (Kirch 1997). On Marshallese atolls populations were once restricted to two-child families; land was so scarce that chiefs were buried on land, and commoners buried at sea (Spennemann 2007, Rudiak-Gould 2009). Many coral atolls in pre- and early contact times had complex regional local and reciprocal socio-economic–political linkages to ensure sustainability (Alkire 1978); some that involved warfare and violent conflicts over land, resources and fishing grounds (D'Arcy 2006, 2009). In the more modern era such practices gave way to dependence on distant governments and/or migration, and a reduction in sustainability.

Long before the twentieth century, men from various SISI travelled in search of wage employment, in response to non-existent local opportunities, far beyond the confines of current island states (Chapter 6). In what are now Kiribati and Tuvalu, extremes of population pressure on ecosystems were recognized as early as the 1880s (Munro and Bedford

1980), while islanders were being resettled in the 1940s. Only in the late nineteenth century did tiny urban centres begin to emerge with some attraction for migrants, most obviously in the Caribbean. By contrast, in Melanesian towns migrants were actively discouraged until the eve of independence through a form of urban apartheid (Connell and Lea 2002). Only around independence, did rural–urban migration begin there in earnest.

Major influences on migration involve rising expectations over what constituted a satisfactory standard of living, a desirable occupation and a suitable mix of accessible services and amenities. Changing aspirations, the increased necessity and desire to earn cash, a preference for white-collar and pink-collar employment, a taste for respectability and distaste for agricultural work have given migration a primarily economic rationale: a movement in search of wage employment, at best scarce in most rural areas, and the security of the 'fast money' of wages rather than the 'slow money' of crop sales (Finney 1967). An economic rationale, real or latent, underlies most migration moves. In the outlier atolls of Solomon Islands, migration is 'a household strategy, by which migrants and their families (including those staying behind) diversify sources of incomes in order to minimize risks, such as loss of income and crop failures' (Birk and Rasmussen 2013). In Vila, the most important reason was simply '*long winim smal vatu from no gat rod long winim vatu long aelan* [to earn a little money since there's no way to earn money on the home island]' (Mitchell 2000: 192). For many that was reason enough. While migration has an economic basis it has also been shaped by the desire for better services, whether health, education or housing, which might stimulate household economies, especially as rural provision was lacking and relative deprivation more obvious. Nonetheless migration was rarely straightforward. Politics could intervene; the intense violence around Honiara at the end of the century caused 'unusually large numbers' of people to return to the island of Bellona (and probably elsewhere) and resume a more 'traditional life' (Reenberg et al. 2008: 816).

Where fishing is a mainstay of the economy, significant changes in fish stocks for whatever reason can also be catalysts to migration, especially where there is little prospect for developing alternative income sources (Hamilton et al. 2004). The close relationship between access to fish, alternative income-generating possibilities and a propensity to migrate (however reluctantly) is evident for Rodrigues and the Galapagos. Slower terrestrial changes, however inexorable, have less significance for migration, since agricultural systems are typically more flexible and adaptable. Necessity stimulates invention and modification, as much as migration. Thus in Siwai, Bougainville (PNG) and Ha'apai (Tonga) people responded to reduced soil fertility by switching from sweet potato to cassava (though perceived as less palatable and nutritionally valuable) rather than contemplate migration (van der Grijp 1993, Connell 2007f). Such options for change and potential migration responses depend on multiple factors, including the existence of alternative occupations, the capacity to modify ecosystems in useful ways and known and realistic migration options.

Urban kin influence and support migration. They may provide demonstrations or create images of modern lifestyles, send remittances (visible symbols of success) and offer fares and accommodation for new migrants to the city. By the 1960s, as in Tonga, 'unless adult males can find some local employment outside agriculture such as fishing or storekeeping, they *must* migrate' (Marcus 1974: 91; my italics). A decade later, migration was often seen 'as an almost inevitable decision that they [villagers] will have to make sooner or later and once this view is accepted a sort of migration momentum develops' (Walsh 1982: 7). The spreading taste for commodities influenced work habits and towns, and metropolitan countries (Chapter 6), exercised a powerful allure and a sense of a future. Migration became characterized by movement away from remote places to more accessible coastal locations, particularly urban areas and capital (usually primate) cities. Even on small islands like Rendova (Solomon Islands) people moved away from the least accessible villages to places where income-generating opportunities were superior (Grant 2012). Livelihoods became attuned to the rhythms of urban and international worlds; bright lights versus boredom had some influence in the smallest islands. National populations increasingly concentrated on more central urbanized islands.

A recent neologism but now common phrase, especially in the Pacific and the Maldives, is 'outer islands'. The Bahamas have 'out islands', though these tend to be perceived as tourist destinations (as in the Seychelles). Mauritius has an Outer Islands Development Corporation responsible for the remote islands of Agalega and unpopulated St Brandon. Vanuatu, Tuvalu and the Cook Islands unquestionably have outer islands. Such islands are perceived as different, sometimes culturally, like the outer islands of Yap and Chuuk in FSM and the Polynesian outliers of Solomon Islands, but ubiquitously because they lack the development options, service provision and trappings of modernity of centres. Such islands have tended to go through phases of change. The first such phase – labour migration – affected many atolls, critically in Tokelau, where many people were taken against their will and perished (Maude 1981). Small islands had little to offer colonialists other than labour, hence a second phase varied from benign neglect to lack of external interest, other than from missionaries (and anthropologists). In the Cook Islands, relatively isolated from colonial influence, the expanding capitalism of the world system meant little, as the islands became recipients of welfare rather than arenas of exploitation (Simpson 1990; see Knapman 1976). Such welfare gains, in the late colonial and early postcolonial years, marked a third phase where some outer islands could even be seen as a 'pampered periphery' undergoing a transition from subsistence to administrative subsidy, perceived by some as a 'moral decline' (Bayliss-Smith et al. 1988, Rudiak-Gould 2010, 2013). A limited local economy had shifted dramatically towards a better funded public sector with islanders entering 'the material life of the nation more frequently by way of the backdoor of bureaucracy than through the front door of the market mechanism' (Rutz 1982: 51). Woleai, an outer island of Yap State (FSM) with a wage labour force of about 50 in 1990, all but one of whom were in the public sector as teachers or health workers, Ujae (Marshall Islands) and Fakaofo (Tokelau) typified this transition (Connell 1992, Hooper 1993, Rudiak-Gould 2009). That was where and when colonial and post-colonial benevolence largely ended.

In a final post-independence phase isolation has been less benign and more the product of urban bias (and mismanagement), contributing to urban migration and in numerous cases – from Bougainville (PNG) to Rotuma (Fiji) and from the Western Province to Bellona in Solomon Islands – to aspirations or even attempts at secession. Tourism enabled a handful of outer islands to flourish. In Solomon Islands it has been said that 'the state is mainly only relevant and visible in Honiara' (Richmond 2011: 129). Small but nearby islands such as Nevis, Union Island and Barbuda, seemingly ignored and bereft of support and attention, have similarly threatened to break away from larger neighbours. More isolated islands like Anuta and Rodrigues went unwillingly to independence as part of larger nations (Feinberg 1986, Paul 1987). Tuvaluans stressed cultural identity and economic autonomy (Macdonald 1975) to emerge from the periphery of the Gilbert and Ellice Islands into nationhood. Mayotte succeeded in separating from the Comoros, and Anguilla from its unwieldy link with St Kitts, while what are effectively clusters of outer islands, such as Wallis and Futuna and Tokelau, have resisted any movement towards even tentative decolonization, in favour of continued colonial ties (Connell 1991a, 2009a). Such political dependence reduced the extent of migration otherwise apparent in outer islands.

Nowhere more than in outer islands was it more apparent that 'incorporation in a wider polity and the *escape* this offers through outmigration, counters the limits of small size and narrow resources' (Ward 1982: 182; my italics). In the Reef Islands (Solomon Islands) islanders had become 'painfully aware of their isolation' (Davenport 1972: 59). On Moch (FSM) islanders increasingly recognized the need to have an established community on the central high island on which to depend not merely for income (Pam and Henry 2012). Any private sector has disappeared from most outer islands, almost before it was established, and public services have suffered because of the unwillingness of providers to live in remote areas and the breakdown of transport infrastructure. The rise of container shipping and bulk carriers, and the decline of the copra trade, have resulted in ships simply bypassing many islands, a process well in place in the Pacific by the 1970s (Brookfield 1978, Ward 1982). Even large islands such as Vanua Lava (Vanuatu) receive a cargo ship only every 2 or 3 months, so that exporting goods is difficult and relief supplies are inaccessible. Air services have likewise bypassed small islands (Ward 1998, Foster 2005). Only exceptionally, and in dependent SISI, such as French Polynesia, have local air services been extended in this century (Malogne-Fer 2004). Since the 1980s outer islands in Tonga have gradually become more distant in terms of time from the capital, while Auckland has become closer (Ward 1998). Even after the arrival of the mobile phone, electronic communications similarly favoured international destinations rather than national peripheries.

Populations of outer islands are ubiquitously contracting compared with the centre. Some 94 per cent of Fiji's population is on the two main islands with the remaining 6 per cent on over 90 smaller populated islands, and the balance is still shifting to the centre. At independence, the population of Tuvalu was more or less evenly divided between the eight main islands; a quarter of a century later half the population was on the urbanized island of Funafuti. Within other archipelagic states most outer islands have lost population in recent years, evident for Aniwa (Wilson 2010) and Lofanga (Tonga) (Kronen and Bender 2007). Much the same pattern of growing central populations and absolutely declining populations in remote islands is also true of the Cook Islands, the Bahamas, BVI, American Samoa, the Northern Marianas, FSM, Kiribati, Marshall Islands, Palau and even the Cayman Islands and the Turks and Caicos. Even on much larger outer islands such as Rodrigues, the majority of the population now live in Mauritius proper (Bunce et al. 2010, Gemenne and Magnan 2010). Population decline also occurs even where people explicitly state that they wish to remain, as on Aniwa or in Tuvalu, as households spread their children to other areas to enhance livelihood diversity (Chevalier 2010, Wilson 2010). Decline has been most rapid on atolls where both development prospects and provision of services are limited, as in Mokil (FSM) (Oles 2007a), Sikaiana (Solomon Islands) (Donner 1993, 2002), the south-west islands (Palau) (McKnight

1977), Ha'ano (Tonga) (Evans 1999) and Nanumea (Tuvalu) (Chambers and Chambers 2001), and certainly also elsewhere. For most of these islands urban residents are more numerous than islanders, but data are inadequate to assess whether that process is still accelerating.

While very small islands have some advantages for resilience because of the greater per capita availability of marine resources to complement land resources (Keegan et al. 2008, Jones 2009), resources are still limited, and social worlds truncated. Yet only a handful of islands have experienced actual depopulation in recent times, and most, such as Manam, Bikini and the Chagos archipelago, are special cases for environmental or political reasons. The Phoenix Islands (Kiribati) were depopulated in the 1960s after a settlement scheme from the 1930s failed, due to the scarcity of fresh water, the declining market for island copra and the isolation which made service provision exceptionally difficult (Maude 1968). The tiny northern islands of the Northern Marianas have been intermittently populated in the twentieth century with populations resettled on Saipan after cyclones and volcanic eruptions. The outer island of Naqalevu (Fiji) was depopulated in the past half century, as apparently were both Sorol (FSM), which had never had a population of more than 13 during the nineteenth century, and Merir (Palau), where in its final phases: 'The island is dying, at least as far as the present generation are concerned ... The women are too old to cultivate taro in any quantity, and the men cannot keep the taro groves cleared' (Osborne 1966: 49; cf. Bayliss-Smith et al. 1988). Other tiny island populations, such as those of the South-West Islands of Palau, Ngulu (Yap, FSM) and Pitcairn, have survived long after their demise was predicted (Connell 1988b), and in the cases of the atolls seemingly with quite unsustainable populations, emphasizing the powerful desire of islanders (and others) to remain amidst familiar landscapes and lifestyles, despite social costs (Lutz 1988, Marks 2008), in unusually challenging circumstances.

Rare efforts to decentralize urban populations and establish regional growth centres, as in Kiribati and Solomon Islands, have been short-lived failures. Only the most exceptional integrated efforts, where economic growth was combined with social benefits, have stimulated return migration, as in Takapoto (Pollock 1978) but, even more frequently, in French Polynesia as elsewhere, few have returned and outer islands have continued to lose population (Lockwood 1990, Malogne-Fer 2004). National perception of the difficulty of providing adequate services to outer islands in the Maldives is such that the government has sought to provide subsidies for people to move from small 'primary islands' to 85 'focus islands' to enable more effective service provision, concentrating people on many fewer than the country's 250 populated islands, and avoid the overcrowded capital city, Male, being the only migration destination (Bertaud 2002). While that has proved unattractive, and risks households that do move becoming too dependent on government, it recognizes the challenges in providing adequate services throughout archipelagic states.

Population decline in small islands (and the few remote upland areas in SISI, including several Caribbean hillsides) is widespread. Employment opportunities and services are concentrated in urban centres and, in small states where human resources and capital are often limited, some centralization is inevitable. As physical infrastructure decays, youths emigrate, numbers fall and services are inadequately provided, so incentives to remain decline, and the downward spiral is enhanced. Ironically the contemporary migration patterns that in many places began in the 1960s, and alleviated stresses arising from ecological problems resulting from population pressure, gave way several decades later to situations where the loss of labour was so great that subsistence activities were threatened by 'radical depopulation' (Evans 1999: 160). The more educated have tended to migrate first taking advantage of superior urban educational and employment opportunities. In many small islands roles for youths, especially girls, are limited, and social (and marital) opportunities scarce. As others move away, the dependency ratio increases and more is expected of those who stay, hence they too seek to migrate. Contemporary patterns of migration have diversified, and become both more selective and skilled, demographic structures have changed, as the restructuring of global and island economic landscapes present different development contexts, but urbanization is a universal outcome.

RESETTLEMENT

While environmental change is only exceptionally a factor in migration, occasionally it has been the only factor where natural hazards have effectively forced migration, while political and economic factors, from nuclear testing to mining, have also contributed to some parallel collective and permanent population shifts. Volcanic eruptions have brought the abrupt depopulation of particular islands, such as Tristan da Cunha, Manam (PNG) and Agrihan (Northern Marianas), although not usually permanently, and the partial depopulation of others including Ambrym (Vanuatu) and Montserrat. Floods and tsunamis have contributed to migration over much shorter distances. In the post-war years resettlement has also occurred in response to human-induced changes, from the overcrowded Gilbert and Ellice Islands and later Tokelau. Much less benevolent resettlement depopulated islands such as Bikini (Marshall Islands), to make way for nuclear testing, and the Chagos archipelago (BIOT), to make way for a United States military base, and from some substantial mine sites. These various migrations were very different in causes and consequences, and especially in the willingness of local people to participate, but outcomes were remarkably similar.

Catastrophic environmental events have resulted in very necessary migration, but usually anticipated to be short term. Only volcanic activity has led to the abrupt and definitive depopulation of any islands. In Vanuatu Lopevi was depopulated in 1960, and some of the population of Ambrym were resettled on the central island of Efate after a major eruption in 1950 (Tonkinson 1977, 1979). Volcanic eruptions on Niuafo'ou (Tonga) in 1946 brought the resettlement of the population on larger islands to the south, though many later returned (Rogers 1981). Half the population of Montserrat left, mainly for Antigua, after the eruption of Mount Soufrière in the mid-1990s, and others were resettled in the north of the island (Philpott 1999, Skelton 2000: Chapter 7). The entire population of Tristan da Cunha was evacuated to Britain after the 1961 volcanic eruption, but most returned two years later (Munch 1964, 1971). More recently about 9000 Manam islanders – the entire population – were resettled on the nearby New Guinea mainland after devastating eruptions in 2004 in which five people died.

Other environmental hazards have had short-term physical impacts. In the western Solomon Islands, Upolu (Samoa) and Niuatoputapu (Tonga) many survivors of the 2007 and 2009 tsunamis chose to move to higher ground inland. Other hazard events, including cyclones, have rarely resulted in relocation. Numerous examples exist of migration following cyclone and flooding events, for example on Tikopia, where Cyclone Zoe prompted migration to Honiara, thus diversifying livelihoods and income-generating opportunities, but some family members remained to safeguard houses and gardens (Rasmussen et al. 2009). By contrast, after Cyclone Bebe struck Funafuti (Tuvalu) in 1972, the entire population of the southern islet, Funafala, about 100 people, moved to the urban centre of Funafuti; most never returned. Cyclone Heta, that devastated Niue in 2004 destroying almost a quarter of the houses, prompted consideration of the permanent abandonment of the island, as had Cyclone Ofa two decades earlier (Barker 2000). After Cyclone André in 1983, as many as 75 per cent of the population of the outlying island of Agalega were reported to have migrated to Mauritius some 1000 km to the south (Gemenne and Magnan 2010: 52) though many later returned. Similarly, after Cyclone Martin in 1997, which killed 19 people and destroyed 95 per cent of the houses on Rakahanga atoll, most residents were resettled on Rarotonga, 1000 km away, but some went overseas, never to return (Ingram 2006). In Barbuda cyclones resulted in people simply moving inland (Potter and Sluyter 2010: 141). In atolls that is impossible. In Kiribati, where houses have been built away from the coast, the President has observed: 'We keep moving back from the shoreline. In a country like Kiribati, with very narrow islands, the room to move back is very limited' (quoted in Green 2010: 19; cf. Connell and Lea 1992). Local resettlement is almost impossible. In each case, intended temporary mobility often became more permanent.

Resettlement has also occurred in response to diverse human-induced changes. The instigator of the first Gilbertese resettlement in the Phoenix Islands in the 1930s argued that the optimum population density of the Gilbert Islands (the central island chain in Kiribati) had actually been reached by the year 1840 (Maude 1968: 319), while in 1865 one missionary described them as 'genuine Malthusians' (Munro and Bedford 1980: 3). Nearly a century later population densities were much higher, food shortages had occurred, poverty was common and the colonial administration proceeded with resettlement. Two decades later in the 1950s, the Phoenix Islands settlers and others from the Gilbert Islands were relocated in Solomon Islands, after recurrent droughts. Vaitupu (Tuvalu) islanders were resettled on Kioa Island in Fiji around the same time. Carteret and Takuu islanders (Bougainville, PNG) have experienced similar pressures on scarce atoll resources, and some resettlement occurred (Chapter 7). In Kiribati, in the year of independence, the settlers in Solomon Islands, once pitied were now envied: 'Settling overseas, beyond the oceans of our islands, is something to be sought after. Why? Because our population is still growing … So now many consider them, the resettled ones, as the fortunate ones and they consider us to be the unfortunate ones' (Schutz and Tenten 1979: 127). That statement by two i-Kiribati was prophetic in terms of a difficult future for Kiribati. A similar movement, explicitly justified as a response to 'future problems of overcrowding' and initially envisaged to include the whole population, was instigated by New Zealand for the Tokelau atolls. Government-assisted migration to New Zealand began in 1963. When the scheme ended in 1976 some 528 people had been resettled with many more migrating freely (Wessen et al. 1992, Huntsman and Kalolo 2007). The atoll population has stabilized and slowly declined as the Tokelauan population has grown in New Zealand.

Resettlement also occurred where islanders were moved away from land more or less forcibly acquired for military purposes. In the 1940s islanders were moved from Bikini and Enewetak, and later Rongelap, to other Marshallese islands, notably Kili and Majuro (Kiste 1974). About

2000 Chagos islanders were moved to Mauritius, from where some have resettled in the UK (Vine 2009, Jeffery 2010). Short-distance migrations have followed the development of large mining projects. On large islands, notably Bougainville, Guadalcanal and Lihir, islanders were moved over quite short distances, eventually leading to disputes over access to adequate land, compensation, housing, employment and environmental degradation (Connell 2012c). Gold Ridge (Solomon Islands) is following the same trajectory. Banaba (Kiribati) was depopulated to make way for phosphate mining, with islanders resettled on Rabi island (Fiji), an island already alienated for plantation use (Tabucanon 2012). In every case migrants left reluctantly, with considerable trepidation. Most sought to remain close to their land and return to it when they could. While such forced migrations had nothing to do with either environmental changes or human pressures on the environment, the outcomes and consequences were very similar (Connell 2012c). Partial urbanization was a ubiquitous outcome.

Resettlement within SISI has become particularly difficult, mainly because of land disputes and constraints; it also has implications and resonance for sovereignty, identity, employment and access to services. Resettlement is unwelcome, even in response to hazards, by settlers and 'recipients'. Local opposition and resentment, centred on land tenure, has prevented effective resettlement. Landowners were increasingly reluctant to cede land to others, however worthy their claims and needs might be, even with shared kinship ties. In Montserrat landowners refused to sell land and resettled islanders were forced to rent houses in a barren location, resulting in feelings of shame and loss of self-esteem where house ownership was the norm (Skelton 2008). In post-colonial times land boundaries have become frozen and land too valuable for most islanders to be willing to enable displaced people to use, lease or purchase it, even where land had previously been alienated (Campbell 2011, Connell 2012c). Resettlement cannot be achieved effectively without substantial financial resources and management ability and neither are readily available, especially in the post-independence era.

Finding places within island realms for people displaced by hazard will become more difficult. Mismanagement, cultural tensions and local opposition have posed problems. Many involuntary migrants have sought to return home or move to cities. Resettlement after hazard has been a partial catalyst for urbanization. Unless local circumstances change, resettlement moves are likely to require an international focus: a colonial model revisited. Regional moves are no longer politically feasible. Significantly, in 2009 the President of the Maldives stated that it would set up a sovereign wealth fund based on revenues from tourism, that would be used to acquire land in other countries, while in 2012 Kiribati purchased a Fijian island (though not officially for resettlement). Resettlement may become an international strategy rather than a form of local migration.

URBANIZATION

Urban areas are invariably growing faster than overall population growth within SISI, and in many SISI urban populations exceed rural populations. It is no longer possible to regard islanders as primarily rural people. Migration is an important component of urban growth in much of the Pacific, but in the Caribbean natural increase is more significant (Hope 1986). Informal settlements are growing particularly quickly in several Pacific SISI, but especially in the larger Pacific SIS, as the supply of land and formal housing is inadequate to meet needs. Informality occurs where incomes are low and irregular, public housing policies fail to meet demand for low-cost housing, and urban planning and management remain weak (Connell 2011b). In Suva (Fiji), for example, the expansion of settlements is a result of rapid rural–urban migration and 'over-urbanization' where formal housing is too scarce (or too expensive) and migrants especially are without employment and adequate incomes. That has been accentuated by the demand for services (especially education), the expiry of land leases and the breakdown of extended family support. Almost a third of the population of Suva is now in over 100 settlements, where housing is sub-standard, not hazard resistant and often in marginal locations that are particularly risk prone. In Honiara about 20 000 of the 70 000 residents are in settlements, and Vila has a similar proportion, most on the fringes of the city. Many settlers have few sources of income, no legal residential status and no security of tenure, limited access to services (such as water, electricity and sewerage) and live at high densities, the combination of which poses social and health risks.

Many urban poor live in low-lying settlements, with the better off on higher ground: socio-economic inequalities are geographical. The poorest have little support from the rural economy and no opportunity to move away from town when poverty, unemployment, old age or social disorder make life difficult. In settlements like Blacksands (Vila) migrants often contemplated return migration, because of insecure land tenure and unemployment, but usually chose to remain in town for their children's sake. Most households had incomes below the national average and at least a quarter had problems meeting school fees, paying rents and providing food (Mecartney 2001). Although urban food gardening remains important almost everywhere (Thaman 1995), urban growth and pressure on peri-urban land increases the distance to food gardens, depletes firewood and contributes to poor people's food and fuel crises. Low incomes, lack of social support and intermittent unemployment give a sense of biding time, waiting for unforeseen and uncertain opportunities, seeking multiple jobs, maintaining strict budgets and abandoning some 'traditional' obligations, simply to get by. Further complications arise where footholds in the city are tenuous, for example where landowners who have negotiated leases with urban settlers die and new arrangements are negotiated. This has sometimes resulted in landowners attaching more stringent regulations on land, for example banning some economic strategies such as fishing, using mangrove swamps for collecting crabs, shells and wood, and restricting gardening, numbers of houses (and ethnic groups) and the establishment of trade stores (Numbasa and Koczberski 2012). Such changing practices attest to increased competition for land and other urban resources. Many urban residents survive rather than prosper in the city. Holding several jobs, termed 'occupational multiplicity' in the Caribbean (Rubenstein 1987, Gmelch and Gmelch 1997), where they can be found, and being in the unregulated informal sector, increasingly characterize urban life.

A consequence of urban population growth being in excess of job creation is the growing informal sector, characterized by diverse forms of marketing and recycling. It plays an important role in employment creation, labour absorption and the marketing of rural produce, while urban residents, many self-employed, purchase a substantial proportion of their food (and other goods) in the markets (Maebuta and Maebuta 2009, Russell 2009, Connell 2011b). Only in the market sector of the economy has the informal sector been of universal importance, though, as in Tonga, the success of informal markets brought conflict with the formal sector (Brown and Connell 1993, Besnier 2004). Even in relatively developed SISI, like Martinique, where wage and salary employment are more accessible, the informal economy is extremely important, and not solely an urban phenomenon though it flourishes best there (Browne 2004). The widespread rise of crime, prostitution and begging are symptomatic of the problems of limited economic growth, unemployment, the social costs of urbanization, and the decline of traditional social control mechanisms. Occasionally it enables high earnings; incomes from sex work in Honiara were higher than anything possible in the formal or informal sector (Russell 2009: 58). In Suva and Honiara especially the rise of urban poverty and the informal sector has been marked by new repressions of the poor and marginalized in anti-urban policies, where residents are forced out of urban areas, most dramatically by the bulldozing of settlements and by attempts to devolve responsibilities from the state, rather than efforts to devise welfare and employment policies to reduce urban problems (Connell 2003b, 2011b, Russell 2009, Mitchell 2011, Birk and Rasmussen 2013). Even urban markets and market vendors have been opposed by urban and national governments, despite their ability to provide food and substantial employment, albeit insecure, for both youths and women. In Honiara governments were bulldozing markets simultaneously with reports recognizing them as invaluable for employment generation (Maebuta and Maebuta 2009, Russell 2009). Similar opposition has been widespread and long-standing (Bennell and Oxenham 1983). Anti-urbanism is not however matched by pro-ruralism.

Urban planning, even land-use planning, is rare in SISI, and unregulated urban expansion has increased the costs of urbanization, especially for infrastructure provision. Public space and recreational space are scarce. Many settlements are entirely unregulated. Indeed, at least in the Pacific region, 'urban growth across the region is proceeding so fast that governments simply cannot keep pace with facilities and services' (UNEP 1999: 21), producing 'anarchic housing and land markets' (Potter et al. 2004: 264). Policies that once had some ephemeral success have been overwhelmed by rapid changes, especially urban population increases, bureaucratic inefficiency and land shortages (Jones and Lea 2007, Storey and Hunter 2010). Management has worsened rather than improved. Few effective housing policies exist, and state housing is undeveloped and beyond the reach of the poor, so that people have had little option but to provide for themselves. Even 'site and service' schemes have become rare. A widespread 'evasive or benign neglect of the housing sector by public institutions' is typified in Windward Island SISI where the state 'intervened in the housing arena only when the demand for land for commercial purposes has suggested the efficacy of clearing low-income groups from particular residential sites' (Potter et al. 2004: 249). Basic infrastructure is rarely the result of public initiatives. Governments have demolished as much as constructed, so that many people are in effect urbanizing the towns themselves, in the face of formal intransigence and neglect.

Earning power is increasingly concentrated amongst urban bureaucracies while the absence of developed state mechanisms (such as progressive taxation, unemployment benefits and pension schemes) for affecting transfers of income minimizes redistribution towards rural areas other than through personal remittances. Growing inequalities, coupled with rising expectations, are the concomitants of urbanization. At the same time as expectations rise, employment crises and growing populations have emphasized inequalities in access to land, housing and other services. Urban permanency and high unemployment have put considerable pressure on already inadequate urban services, and emerging differences have cut across traditional social divisions. Hints of class formation have emerged in Pacific towns, but differ from the more obvious class and racial distinctions in Caribbean towns.

URBAN ENVIRONMENTS

In some urban areas population densities are exceptional – notably in Ebeye and Majuro (Marshall Islands) and Tarawa (Kiribati) – with serious overcrowding and environmental degradation, manifest in disease (respiratory illness, tuberculosis and gastro-intestinal problems) and nutrition problems. Localized densities in Majuro, Tarawa and Male are even higher; a decade ago, in parts of Tarawa, densities were as high as 15 000 people per sq km (White et al. 1999, ANU Enterprise Pty Ltd 2009). In Ebeye, albeit with around 12 000 people, it was then over 100 000 per sq km while much larger Majuro also had a population density of over 70 000 persons per sq km (Chutaro 2005). Male may have as many as 100 000 at a density of 18 000 per sq km. Moreover in each of these towns densities appear to be still increasing steadily, as they are in Funafuti. In Funafuti, Tarawa, Majuro and Male, and in Victoria (Seychelles), land is so scarce that it has been reclaimed for urban development, both by the state and on an individual piecemeal basis. The Maldives has an ambitious scheme to reclaim land adjoining the international airport that would triple the existing capital city population by 2020, though the island, surrounded by a 3 metre seawall, is already one of the most densely populated in the world.

Land shortages are an urban phenomenon, alongside land tenure disputes and conflicts, resulting in increasing numbers of informal settlements on unsuitable land, illegal dumping and other harmful practices. At least since the 1980s, many recent migrants have settled in marginal, usually low-lying urban areas prone to flooding, rats, insects and disease problems, as in Nuku'alofa (Tonga), Nadi and Suva (Fiji), Funafuti (Tuvalu), Apia (Samoa) and Castries (St Lucia). Such areas are particularly vulnerable, having been partly reclaimed from swamps, the outcome of better sites being occupied or beyond economic reach. Various reclaimed areas are prone to sewage spills and coastal hazards, including cyclones. By contrast in Castries (St Lucia), where some 60 per cent of urban housing is informal, settlements are built on steep slopes surrounding the capital, and are prone to landslides (Anderson et al. 2011). In such marginal areas, occupied by the relatively poor, management is absent, services and water security are scarce and continued unregulated growth worsens such problems.

The socio-economic impact of hazards is greatest in urban areas since many low-income residents live in more marginal areas, and have restricted ability (whether savings, insurance, or social ties) to cope with disaster. In the Caribbean, exemplified in Grenada, limited resources, need for access and skewed land ownership can 'drive the poor to settle on vulnerable land' and engage in unsustainable resource use, so that floods, landslides and storms have a disproportionate effect on the poor (Barrientos 2004: 13). Even in Fiji, where urban sewerage is relatively extensive, only a third of residents in the Suva-Nausori capital area are connected to a centralized system; the rest are served at best by septic tanks. Since the topography of the metropolitan area is rugged and soils impermeable, much effluent flows directly into streams and coastal waters. All Fiji's four sugar mills are located near waterways, and in the town of Labasa have turned the river into a 'deadzone' without marine life (Kumar 2010; Chapter 7). In Navua township (Fiji) the families most affected by floods in 2004 were former agricultural workers, displaced from rural areas because trade liberalization and cheap imports had reduced their incomes from sugar cultivation, or through the non-renewal of agricultural leases, so that the costs of the floods forced many further below the poverty line (Lal et al. 2009b: 33). Indirect costs include households' reduced ability to pay school fees, resulting in children missing school, additional costs for health care from an increased incidence of water and vectorborne diseases, but reduced ability to purchase food and medicine.

Access to adequate fresh water supplies is a growing urban problem, especially on atolls, while good sanitation practices are often lacking (either individually through inadequate knowledge of good practice, or collectively through the poor management of garbage and various wastes, etc.) thus compounding this problem. A rise in prevalence rates for both gastroenteritis and salmonellosis in eastern Caribbean SISI may be attributed to contaminated water supplies, and inadequate sewage disposal, particularly in poor areas. The burden of disease in SISI tends to fall most on those who live in inadequate urban conditions (Chapter 7). In the highly urbanized island of Ebeye, where a 2001 cholera outbreak killed six people, sanitation practices remain poor, and although the entire coast was unsuitable for either fishing or swimming, it continued to be used for both and for defecation (Bower and Overmars 2001, Lui 2009). Tarawa particularly experiences several critical environmental issues including groundwater depletion alongside increased salinity and pollution from sewage and animal excreta (hence reduced access to fresh water); marine life and seawater contamination from human and solid waste (with around one-third of South Tarawa's population using beaches as toilets); over-fishing of reefs and lagoons; non-degradable waste disposal; coastal erosion, beach mining and deforestation eliminating sources of food, medicinal plants and habitat and increasing the vulnerability of coastlines; and breakdown of traditional subsistence production, resulting in poor nutrition and health-related problems (ADB 2002: 28). Access to fresh water is a severe challenge, compounded by problems of garbage and sewage disposal. Kiribati has no legislation to deal with solid waste management or pollution of the lagoon (Chapter 7). Lagoon pollution, exacerbated by the closing of the lagoon for causeways, threatens public health. Somewhat similarly in Funafuti (Tuvalu) newer urban settlements have been particularly prone to flooding associated with cyclones, and damage has been exacerbated by changes including causeway construction, airstrip sealing, deforestation (for firewood), groyne construction, 'one-off' reclamation and sand and aggregate mining (Connell 2003a, Lazrus 2009). While urban Tarawa and Funafuti experience extreme difficulties other atoll urban centres face similar circumstances.

Inadequate access to water is not solely a problem in atoll states. In Honiara nearly 30 per cent of the poorest households depend on unsafe water sources and have no satisfactory sanitation system, while 49 per cent of settler households cook on open fires, hence firewood, sawdust and kerosene (for cooking and lighting) are important consumer goods (Russell 2009: 15). Colonial infrastructure has often decayed into obsolescence resulting in significant water losses. Even when settlements have formal access to water, facilities are poorly maintained, supplies are not guaranteed and settlements are usually the first affected when chronic water shortages occur (Amnesty International 2011). In Fiji, reticulated water supply reaches less than 50 per cent of urban households, and water losses from leaking pipes are massive, approaching 70 per cent of all water supplied (Kumar 2010). Informal settlements often depend on shared pit toilets, sometimes very close to water courses subject to flooding. In Pacific SIS no sanitation master plans exist even for cities as large as Vila with 40 000 people. A lack of infrastructure provision in peri-urban areas has made households dependent upon rainwater and wells, which are increasingly susceptible to contamination and contribute to gastroenteritis and dengue fever (Government of Vanuatu 2002: 25). Environmental conditions in many informal settlements threaten the health of residents and ultimately of the wider urban area.

The physical spread of cities, and commuting, into agricultural hinterlands and along coasts has created a wider urban environmental footprint. Thus the Tagabe catchment area, a key source of water for Vila, has come under pressure from urban expansion, rural runoff, industrial wastewater and informal settlements, but authority for the river is divided between urban and rural councils and several customary landowners (ADB 2004). Water catchments in Vila, Tarawa and elsewhere, have been built over with housing, at much greater risk of pollution. Pressures on peripheral areas also result from urban food gardens and even cemeteries (with burials increasingly randomly located or close to houses). Growing urban (and therefore coastal) populations have resulted in stressed coastal and lagoon environments, from dredging, coral and sand mining, reclamation, construction of seawalls, jetties and causeways, and removal of mangroves and other coastal vegetation, for construction and firewood. More generally, construction, whether for housing, tourism facilities, roads or airports, has placed pressure on coastal resources.

Sand Mining

In many urban areas sand mining has exacerbated coastal erosion. Beach sand mining is a traditional practice on many islands, but usually on a small scale. With accelerated urbanization, coastal migration and construction of more permanent houses and other buildings, much larger quantities of sand (and gravel) have been removed. Where sand has been taken slowly, a bucket or a sack at a time, it has been difficult to prevent, control or discourage. On Montserrat, where beach sand mining was long practiced, a construction boom in the 1960s, associated with tourism, combined with a transition to cement houses, led to a huge demand for sand and aggregate and resulted in serious beach erosion. Legislation to control sand mining, introduced in 1970, was inadequately implemented, and Hurricanes David in 1979 and Hugo in 1989 compounded the problem, with serious beach erosion, taking more sand from already eroded beaches (Cambers 1999). On Kosrae (FSM) local people remove sand for house construction or landscaping. With a growing population this has exacerbated coastal erosion problems, but no inexpensive alternative to beach sand exists for poor communities, and enforcement of environmental laws has proved difficult (Ramsay 1999).

Within the last four decades, the Maldives has experienced a massive boom in cement houses and high-rise buildings and thus of sand mining. Until the late 1980s, almost all cement houses and buildings were built with local coral sand and coral aggregates. Subsequently imported sand and aggregates from India started to replace local sand. The post-tsunami reconstruction boom, associated with further resort development, created a sand shortage, while increased oil prices raised transportation costs and prices of Indian sand more than doubled. Costs were even higher in the relatively poor outer islands, hence outer islanders had little choice but to mine their own beaches, despite it being prohibited. Ironically, sand mining for post-tsunami reconstruction made those islands more vulnerable to erosion, SLR and storm surges. On Male, coastal degradation caused by sand mining has been countered through protective measures, such as concrete coastal defence structures, that cost millions of dollars (http://sondhu.blogspot.com/2011/01/sand-mining-might-erase-some-islands.html). Sand and aggregate 'mining' has been particularly damaging in such coral atoll contexts. Removal of sand and gravel from lagoons, notably in Funafuti, Tarawa and Majuro, has scoured lagoon floors, reduced marine productivity and contributed to coastal erosion (Smith and Collen 2004, McKenzie et al. 2006, Ambroz 2009a). Alternative sources of aggregate are offshore hence lagoon exploitation is likely to continue, as populations grow, with the resilience of atoll structures being further reduced, and where government efforts to ban mining in sensitive areas have been thwarted by illegal night-time mining (Webb 2005, Ambroz 2009a). Pressure on coastal resources is universal.

Waste Disposal

A characteristic of all islands is that non-biodegradable waste has increased with greater consumerism and consumption of imported and packaged goods, while suitable disposal sites have become scarcer, especially close to urban areas, where much waste is produced and land disputes are frequent. SISI face struggles to find and manage dump sites and the disposal of toxic and hazardous waste is a particular issue, especially in atoll states where sites are particularly hard to locate and protect, and landowners in tropical countries resent garbage dumps more than almost any other urban land use. Few sites enable disposal without leakage into lagoons and water lenses. Household waste streams contain biodegradable material (vegetable wastes) but with growing proportions and volumes of inorganic paper, plastics and metals (UNEP 1999, Carden 2003), making disposal more problematic, and putting coastal ecosystems under pressure.

Waste disposal is a particular problem on atolls, both because of land shortages (and land tenure) and because dumped waste may pollute lagoons and water lenses. In the exceptionally urbanized atoll of Male (Maldives), waste has been transported to other atolls and used for lagoon infilling (and a base for the international airport). That is expensive and rarely an option. On Majuro, batteries and waste oils are frequently mixed with household waste, resulting in soil and water contamination, since dump sites are poorly constructed and maintained (Morrell 2008). Management of electronic waste is virtually non-existent, at least in Pacific SISI, though the toxic chemicals in electronic goods pose environmental health risks. Plastic waste pervades urban watercourses. Solid and liquid wastes deposited in coastal lagoons, and other inshore waters, increase the levels of toxicity (Chapter 7), making them hazardous for recreation and fishing, damaging habitats and reducing biodiversity (Morrison and Kaly 2010). Waste (and unmanaged surface water) also provides breeding grounds for mosquitoes, rats and flies.

The capacity to collect, sort and dispose of solid waste is stretched thin in many urban areas. Garbage collection is often irregular and may only serve formal housing areas. Much waste is thrown on open dumpsites, including the sea, in the streets or burnt in open areas. In informal settlements in Nasinu (Suva, Fiji), a particularly poorly serviced area, only 19 per cent of households had their rubbish collected while 52 per cent either burned or buried it. Fully one-fifth of households reportedly threw their rubbish into a nearby river or dumped it on nearby land (UNESCAP/POC 2002). Similar patterns occur in urban Vanuatu. The Port Vila Municipal Council does not collect any solid waste from informal settlements, the largest and fastest growing housing areas around Vila. Poor management of municipal waste collection (and dumps and vehicles) results from lack of adequate funding and management (simply 'good housekeeping') and other factors (including access to appropriate land). In the Marshall Islands the Majuro Atoll Waste Company declared a State of Emergency in 2008 due to a shortage of waste bins and the lack of financial resources to service urban centres and safely dispose of waste. Limited knowledge (and capital) and few realistic alternatives result in dumping waste in rivers, roadsides and coasts. Atmospheric pollution is rarely perceived to be a critical problem in SISI, but in such urban areas as Castries, Suva and Vila, vehicle exhaust and fires have led to localized pollution. In Suva (Fiji) burning municipal waste is a major health problem with some toxins in the smoke being carcinogens and persisting in the environment. Similar circumstances occur elsewhere.

Recycling of cans and bottles rarely occurs, since their sources are often very distant from island destinations, the informal sector is weak (and discouraged) and it is not cost-effective. This is particularly problematic in more remote atoll states, such as Kiribati, where 'the impacts of unmanaged urbanization, a continued crisis of inadequate sanitation, a lack of solid waste disposal controls and ineffective freshwater management offer equal threats to sustainability' as much touted issues of climate change and SLR (Storey and Hunter 2010: 167; Chapter 7). Waste materials continue to increase without obvious signs of improved management.

Sewage

Inadequate sewerage systems in some urban areas have resulted in high coliform contamination in surface water and groundwater, especially near informal settlements (where dependence on groundwater is often greatest). Sewage pollution in lagoons has resulted in the decline of some marine species, such as benthic foraminifers, which in turn has reduced the rate of sediment supply to reef islands, raising problems for their sustainability (Osawa et al. 2010). Adequate sewerage systems require substantial amounts of water, yet in some contexts, such as urban Tarawa, this is not readily available, and other 'sanitation methods', including traditional defecation on beaches, in the bush or in pit toilets, remain the most frequent practices (Storey and Hunter 2010). Most sewage and solid waste continues to be disposed of along the waterfront; settlement residents are more likely to dump solid waste, use the beach as a toilet and use dirty water for drinking as a result of being cut off from infrastructure and services (Teiwaki et al. 2005). The majority of households cannot or choose not to pay for services and are also dependent on wells and rainwater. Related problems occur in Male, and other islands in the Maldives. Growing populations have meant that houses usually had larger land areas in the past and wells (the main source of water) and toilet areas were separated. With house areas becoming smaller, septic tanks and sites for open defecation are closer to wells. Defecation also occurs on the beach and in the bush. Such traditional practices are problematic on islands where 50 per cent of people do not have toilets (Live and Learn Environmental Education 2006). Infrastructure provision and maintenance, and changing attitudes, both hampered by poverty, have not kept pace with growing populations and increased demand. While atoll states experience unusually severe problems in sewage management, because of a combination of water shortages, land tenure constraints and limited resources, sewerage systems are inadequate in many SISI.

Health and Disease

Urbanization and environmental changes have influenced health issues in urban centres, notably the rise of NCDs following lifestyle changes, including greater dependence on purchased food. Most SISI have experienced an epidemiological transition from infectious diseases to NCDs – a transition most advanced in the Caribbean but of enormous significance in the Pacific. Changing diets and more sedentary lifestyles have resulted in NCDs reaching epidemic proportions in several states, especially in Micronesia and Polynesia, where they are the leading cause of death (Inaoka et al. 2007, Connell 2009c). In the Marshall Islands a third of all women and a quarter of men have diabetes, which has been called 'the most expensive crisis ever to hit the Marshalls with the growing numbers out of control' and accounts for nine out of ten hospital admissions (*Radio Australia*, 8 September 2011, 13 February 2012). In Samoa almost 53 per cent of adults are obese, 23 per cent have diabetes and 21 per cent high blood pressure; the cost of sending patients overseas for treatment was estimated at about $4.5 million in 2011, while costs within Samoa were significantly greater (*Radio New Zealand International*, 15 November 2011). In the Caribbean, too, hypertension and diabetes are a substantial economic burden in states such as the Bahamas and Barbados (Abdulkadri et al. 2009). Suicide rates are high in several SIS, especially the Marshall Islands and Fiji where in 2011 incidence reached an 'epidemic level' – the second highest rate in the world – especially amongst Indo-Fijians in the sugar cane belts of the west and north, where poverty was increasing (*Radio Australia*, 7 November 2011).

HIV/AIDS is slowly increasing in most SISI, has some link to poverty, urbanization, tourism, greater mobility and migration, and renegotiation of sexual identities, and is not easily amenable to policy formation (Barrow 2008, Henry-Lee et al. 2010, Connell and Negin 2012). The rise of NCDs and HIV/AIDS poses a challenge to health services in most SISI, since many health services are usually weak, under-resourced and under-staffed, with high staff turnover, and NCDs are costly and labour-intensive to treat. Health services are particularly weak in smaller SISI (Connell 2007a), but telemedicine (telehealth) is very slowly emerging as a possible solution to some of the problems of limited finance, inadequate human resources, isolation and fragmentation, though it is costly and has usually required aid subsidies and skilled human resources to implement and maintain (Adam 2000, Hedson 2000, Kupu 2000, Smith 2004). A continued high incidence of infectious diseases in some countries, notably in Melanesian states but also in Kiribati, has meant that several Pacific SISI especially confront the double burden of fighting communicable diseases such as malaria, dengue, leprosy and waterborne diseases, alongside the growing epidemic of chronic NCDs including heart disease, diabetes, obesity and cancers. The Secretary-General of the SPC, Dr Jimmie Rogers, has declared that NCDs are a greater threat to the Pacific region than sea-level rise (pers. comm., 2012). Rates of smoking and alcohol consumption are not declining. These trends cannot easily be reversed, as tobacco companies focus on developing countries, while such environmental changes as the rise of impoverished informal settlements and inadequate waste management, and climate change (Chapter 7), have further contributed to health problems.

BEYOND THE CITY

Population growth and migration have accelerated an irreversible coastal urbanization at high densities. In many respects urban life represents the best of times and the worst of times. For an elite, mainly entrenched in politics and bureaucracy, and often with entrepreneurial interests, urban life has proved successful. For those without incomes, or the education to acquire formal employment or appropriate connections, resident in overcrowded informal houses without services and the income to obtain them, some of whom grew up in the towns (and thus would find it difficult to 'go back' to rural areas, even if they wished to do so), urban life has been a disappointment. Youth, bulging, bored and disenfranchised, have occasionally rebelled against such circumstances. The capitals of most SIS are far from the models of economic growth and development, centres of culture and crucibles of nationhood that were once anticipated for them. Informal housing, employment, services and livelihoods – and competition for all of these – have become more dominant. Nonetheless, especially in countries where international migration opportunities are scarce, such as Solomon Islands, Tuvalu, the Maldives and Kiribati, these primate cities remain extremely attractive destinations for migrants from rural areas, especially outer islands, where income and employment opportunities are even more bleak.

Given the problems of employment generation, service provision and environmental management that exist in most SISI, and that have become obvious in urban areas, it is unsurprising both that towns have become stepping stones for migration overseas (although that is relatively rare) and, more generally, that in recent decades there has been a widespread resort to overseas migration, in the quest for better livelihoods and superior opportunities. Experiences and perceptions of the wider world, its values, and its material rewards further underlie the migratory experience. Indeed despite national urban growth, a lack of facilities, international migration and transport routes have meant that cities like Miami, Auckland and Honolulu dominate national capitals, even for seemingly mundane activities such as shopping, but also as centres of social change and headquarters of transnational companies. That has taken place in the context of an expanding ‘culture of migration’ where migration has gradually become expected and normal, but tied to the aspirations and needs of those who stay at home and thus the needs of extended families that have themselves become transnational. This global diffusion of island populations in various continents and forming multiple diasporas is examined in the following chapter.

6. Leaving the islands: international migration

> The Caribbean person is truly the first transnational figure who has emerged from colonial people. Their achievements have been immense to their home countries and their homeland rocks to which they are inextricably bound.
>
> Ralph Gonsalves, Prime Minister of St Vincent and the Grenadines, 2007; quoted in Palmer, 2009: 51

Islands have never been apart from migration. Mobility was and is at the core of island dynamics. Many islands were settled in waves, most dramatically through the middle passage that took slaves from Africa to the Caribbean, succeeding Arawaks and Caribs, the equally complex migration in the Indian and Atlantic Oceans that brought Africans, Indians and European colonialists to empty islands, and the colonization by the Lapita people and their successors who became the Polynesian and Micronesian settlers of much of the Pacific. While Melanesian islands were settled thousands of years ago, the Seychelles, Bermuda and Barbados were unpopulated until the seventeenth century and later. Until the late nineteenth century the history of the Caribbean was largely defined by migration. The year 1834 marked both the end of the abolition of slavery in British colonies, and signalled the beginning of large scale emigration from the smaller, already ecologically stressed Leeward Islands – St Kitts, Nevis and Antigua – to rather larger islands (Potter et al. 2004). Cape Verdeans similarly began to migrate once slavery ended (Carling and Akesson 2009). Polynesian and Micronesian islands have been characterized by the repeated mobility of fragments of their populations in great canoe voyages, hence one Tongan anthropologist has emphasized that Pacific islanders inhabit a 'sea of islands' where population ties across the ocean are at least as important as those within the islands (Hau'ofa 1994; see D'Arcy 2006). By the mid-nineteenth century Pacific islanders had been forcibly taken to South America. Later many left voluntarily for plantations in Queensland and elsewhere. This chapter examines their legacy in contemporary migration streams and their impact on island economies and societies.

From the start of the twentieth century contract labour took Gilbertese (i-Kiribati) and Tuvaluans to Nauru and Banaba (for employment in the phosphate mines) and later to work in the international shipping industry (for which both countries now have training schools). Cook Islanders were digging phosphate in French Polynesia, while Montserrat migrants were mining it in Redonda. Small Caribbean islands, such as Saba, where agriculture was particularly difficult, produced migrant sailors: such seafaring traditions enabled many to seek better employment in the United States Navy. Migration was so significant in the more impoverished Caribbean islands, like the Bahamas and the USVI, that by the twentieth century it had become 'a livelihood option for all classes of people … embedded in the social fabric of the islands' (Potter et al. 2004: 59) which rather later became a 'visa mentality' and 'obsessive desire' to migrate (Sunshine 1985). Many other SISI experienced a long history of, usually circular, migration often to plantations; by the start of the twentieth century as many as 20 000 migrants from various Caribbean islands had gone to construct the Panama Canal or work on central American banana plantations (Richardson 1986, McElroy 2011) and migration from Cape Verde to the United States was already massive. By the 1930s migration from the Caribbean to the United States was becoming more common. The need for an 'escape valve' had persisted and expanded, since local economies could not create enough jobs and migration was one means of stifling labour unrest (Palmer 2009).

Island realms have experienced a long history of international migration despite relative quiescence in the mid-twentieth century. That history was partly the outcome and growing awareness of uneven economic opportunities between home islands and distant destinations. When international migration revived in the second half of the century it returned initially in a similar form, with men travelling long distances for employment, but evolved to embrace the movement of women and families, increasingly as permanent migrants to large metropolitan states (notably the United States, the UK, Australia, New Zealand, France and Canada; all but the last being colonial powers). A final phase incorporated the independent migration of women, often as skilled workers, and extended to newer and more distant destinations in the wake of global economic shifts. Mobility is widespread, long established and with a household and government rationale that has barely changed.

CONTEMPORARY MIGRATION

Thousands of Caribbean islanders from French, Dutch and British possessions migrated to the metropoles from the 1950s onwards. West Indians were recruited and migrated in huge numbers to Britain, although migration streams gradually shifted towards the United States. International migration came of age in most SISI in the 1960s, with the global 'long boom', rising demand for labour in metropolitan states, jet air travel, cheaper fares, rising expectations and demand for access to services and employment in metropolitan states. In the Pacific, by the end of the decade, Samoans had not just migrated to Australia and New Zealand but lived in every state and territory of Canada and the United States, not least Alaska. In the same decade the smaller states, such as Niue, Dominica, St Kitts, Montserrat, Tokelau and Pitcairn, were already experiencing absolute population decline and moving towards a 'culture of migration' where migration was normative and central to island life (Connell 2008), a condition earlier 'embedded in the social fabric' of the Caribbean and as 'migration ideology' in Nevis and Cape Verde (Byron 1999, Akesson 2008, Carling and Akesson 2009). In Tonga islanders have 'developed a migratory disposition, that is, a logic of life strategies and organized action in which migration is desirable, possible and inevitable' and which 'colors everyday life' (Besnier 2011: 40, 2012: 494). Exemplified in the outer islands of Tuvalu, children hoped and expected to work away from the islands while parents welcomed this and assumed they would be better suited to this than themselves (Chambers and Chambers 2001). Even as migration gradually became expected and accepted, it was no rupture or discontinuity with island experiences, but an integral strand in more diversified and extended household and national concepts of livelihood and development. For small states especially it was a semi-permanent safety valve, not primarily for population growth but as an appropriate and legitimate means to economic and social wellbeing for weak local and national economies.

Just as Caribbean islanders have moved north to America, many Polynesians from Niue, the Cook Islands, American Samoa, Tonga and Samoa have gone south to New Zealand (whence some have gone on to Australia) or, as the New Zealand economy has stagnated and immigration restrictions become tighter, to the United States, legally or illegally. In the past 20 years emigration has accelerated from the FSM and the Marshall Islands, as the signing of the Compacts of Free Association guaranteed migration rights in the United States and its territories. Between 1990 and 2004 more than 13 000 people left the

Marshall Islands for the United States, especially after government jobs were lost in public sector reforms (a situation also true of the Cook Islands). FSM and Palau are following a similar course, as migration in Micronesia increasingly resembles that in much of the South Pacific: a steady outflow, growth of relatively permanent urban communities overseas (beyond student groups), the return flow of remittances and growing interest in migration. Universally the demographic balance is shifting overseas, and for the smallest states, including the Cook Islands, Niue, Tokelau, Anguilla and Montserrat, substantial migration has resulted in a majority of the ethnic population living overseas. For some remote parts of several states, such as the northern atolls of the Cook Islands and Barbuda (Antigua and Barbuda) that is even more true; three times as many Barbudans are estimated to live in New York city alone than live in Barbuda (Rubenstein 1983, Potter and Sluyter 2010). Niue has a resident population of less than 1500, but over 5500 Niue-born live in New Zealand alone, alongside their many dependents; in Niue relatives are invariably elsewhere (Connell 2008). Highly dependent on New Zealand aid, its long-term viability became more uncertain after Cyclone Heta in 2004 when the cost of rehabilitation prompted New Zealand to re-examine depopulation.

A few dependent territories, notably the USVI, Curaçao and the Turks and Caicos in the Caribbean, Guam, the Northern Marianas and American Samoa in the Pacific and Mayotte in the Indian Ocean, have become stepping stones or 'transhipment stations' (Maingot 1983), sometimes illegally, for onward migration to metropolitan states, usually the United States. In each of these SISI population growth has been rapid, as a consequence of economic and political stability and significant well-paid government employment. Extensive subsidies have turned these and other dependent islands, such as Niue, Norfolk Island, Christmas Island, Britain's south Atlantic territories, Sint Eustatius and St Pierre and Miquelon in the north Atlantic into 'government islands' (and anomalies of empire) where public service employment enables a prosperity and population stability that would otherwise be impossible.

In several dependent territories, like the USVI, both legal and illegal immigration have been substantial (Olwig 2007, Roopnarine 2008). This is particularly true of Mayotte where French subsidies have stimulated massive migration from the independent Comoros to the extent that there were an estimated 60 000 undocumented migrants in 2010, around a third of the population, yet Mayotte has still been particularly successful at 'expelling undocumented foreigners' (Jolivet et al. 2010, Hopquin and Canavate 2011). The Bahamas has long been a destination for many illegal Haitian migrants (Marshall 1981), as has the Turks and Caicos, where they constitute something of an underclass of both legal and illegal workers. Martinique is second only to the United States as the preferred destination for St Lucians (Moberg 2008). Many Comorians, Haitians and Cubans have drowned in desperate migration attempts. Where economic growth has been significant, such as the USVI and the Cayman Islands, in the heyday of international finance, the locally born population make up less than and just over half the population, respectively; in the Cayman Islands migrants came from as many as 117 countries, including various other offshore finance centres and the entire wider Caribbean, an extraordinary diversity for a small territory (Connell 1994). Successful tourist development has also stimulated immigration, as in the USVI, Sint Maarten and Aruba – ranked seventh in the world in terms of its immigration rate (Croes and Hooimeijer 2010). In Guam (and the BIOT) migration has followed military expansion; Christmas Island's population quickly doubled after it became the key destination for 'boat people' refugees seeking entry to Australia. Even small islands are not necessarily isolated. In these places, especially, migration is the principal regulator of demographic change.

In the present century, migration has continued to respond to global fluctuations. The rate of migration from most SISI to major destinations, namely the UK, New Zealand, Australia and the United States, has tended to decline in recent years due to economic recession in destinations and the restructuring of migration controls, with policies preferring skilled migrants, rather than enabling family reunion. Certain sectors in metropolitan countries, notably health and IT, are particularly short of skilled workers, hence new patterns have emerged. Fijian nurses and teachers have gone to various destinations, including the Marshall Islands, Palau, New Zealand, Bermuda and the United Arab Emirates (Voigt-Graf 2003, Rokoduru 2008, Connell 2009c), while rugby players have gone beyond the 'traditional' destinations of New Zealand and Australia to Japan, France and the UK. In the Caribbean, Aruban nurses go to Curaçao, Guyanese nurses and doctors go to several states, with many eventually going on to Canada, the United States and the Netherlands (Connell 2007c, Lewis 2011). Active recruitment of nurses and contract migration stimulated such processes, ultimately resulting in the formulation of a WHO Global Code of Practice, partly designed to meet SIS needs, to discourage excessive migration and recruitment (Connell and Buchan 2011). Numbers are substantial; the World Bank reported in 2009 that three times the number of nurses trained in the overall Caribbean region were employed outside it, and emigration accounted for 70 per cent of attrition – a situation 'without parallel in the world' (quoted in Lewis 2011: 67). Yet recruitment of nurses and teachers has continued, as in St Lucia, to the detriment of local services, and 'the flight of human capital may intensify' within the evolving CSME region (Henry-Lee et al. 2010: 76). New occupational niches have also shaped migration; Fijians have moved overseas as shipping crew, soldiers, tourism workers, police, pilots and private security workers (Voigt-Graf and Connell 2006). Recent migration from Fiji and Tonga to the Middle East (Iraq, Kuwait, Afghanistan and also Sudan) emphasizes how new and highly paid overseas employment opportunities are welcomed, even in an unappealing and violent security and social context. Such new migration flows have quickly turned Fiji into a major recipient of remittances. There are, however, real disadvantages, both in terms of the loss of skills to Fiji and circumstances where more than a dozen workers have been killed. Palau, the FSM, the USVI, Guam, the Northern Marianas and the Marshall Islands have similarly lost numerous citizens in active service in the Middle East.

In earlier decades male migration usually preceded female or family migration (especially for regional migration in the plantation era) but the demand for service sector workers, in areas from domestic service and care-giving to tourism, IT and health, has shifted the balance in favour of women, particularly in the Caribbean, even though women migrants are most likely to be exploited (Ferguson 2003: 16). In the Caribbean both parents may migrate, leaving 'barrel children' behind for whom the social consequences can be damaging (Jokhan 2008, Moberg 2008, Lewis 2011), though in the Pacific and Cape Verde similar separations, and fostering, have been less traumatic (Akesson et al. 2012). Even so, in many cases, families migrate as units.

Whereas early migration tended to follow colonial, post-colonial and linguistic ties, skilled migration especially has become more global. Missions sometimes paved the way through education and scholarships. Many of the earliest Tongan migrants to the United States were Mormons; the Church provided recruits with plane tickets and access to high school, a situation that encouraged conversion. Sport, IT, health and warfare have gradually globalized their labour markets. The commodification of labour has taken new forms, with new pressures from many SIS to negotiate migration ties with metropolitan states, as part of an expanding 'outward urge'. The economic future of several states partly hinges on the continued flow of remittances, which make up exceptionally high proportions of national income, and hence on the continuity of migration. Interest in migration has long been such that in Samoa, when prospects for emigration seemed particularly poor at the start of the 1980s, threatening the culture of migration, the 'broken dreams' of potential but unsuccessful migrants even contributed to a rise in youth suicide (Macpherson and Macpherson 1987). A quarter of a century later migration has never been more important for island futures.

MIGRATION RATIONALES

Migration decisions are usually shaped within a family context, as migrants leave to meet both family and personal expectations; the key one usually being material support for kin. In Nevis young people are eager to migrate because it enables them to fulfil contradictory obligations: to help their families and establish a degree of adult independence (Olwig 2003). Yet migration has rarely been an individual decision to meet individual goals, nor has it usually been dictated by national interests. Within and beyond SISI, economic factors underpin most migration, whether for the immediate benefit of migrants (and their families) or the long-term benefits of their children. Migrants invariably seek better employment and income-generating opportunities elsewhere.

Migration is directed at improving both the living standards of those who remain at home and the lifestyle and income of the migrants. Bluntly 'there are few opportunities for socioeconomic advancement in Tonga and migration is perceived as the only solution' (Lee 2004: 135). In Samoa migration was simply 'to seek wealth for all' (Muliaina 2001: 25). Yet that was embedded in society: 'the most overt reason for leaving is the impoverished labor market in the [Tongan] islands … the rising cost of living and the importance of money as a claim to dignity in an otherwise rigidly ranked social order' (Besnier 2012: 495). By the 1970s in Samoa, 'migration was a far more lucrative investment than anything available in the village' (Shankman 1976: 71). Consequently, 'families deliberate carefully about which members would be most likely to do well overseas and be reliable in sending remittances' (Gailey 1992b: 465). In Nevis parents likewise decide who will migrate, and make appropriate arrangements (Olwig 1993). Through such processes, extended households, as in Tonga, have transformed themselves into what have been described as 'transnational corporations of kin' which strategically allocate family labour to local and overseas destinations to maximize income opportunities, minimize risk, and benefit from resultant remittance flows (Marcus 1981, Munro 1990, Flinn 1994). In numerous cases this means the dispersion of extended households across several continents, well evident in the Caribbean and Polynesia (Byron 1999, Olwig 2007, Connell 2012a). To an even greater extent therefore than for internal migration (where health, education and social reasons explain some part of migration), international migration is more evidently an economic phenomenon.

The primacy of economic factors, necessarily shaped by other factors, was true a century ago in the Caribbean, and long before the great waves of Pacific migration, when Samoans, for example, travelled for 'economic motives, to visit family, for education, to follow the heart … adventure' (Salesa 2003: 179). Droughts and famines early prompted migration from Cape Verde. Even in contexts where environmental degradation has been highly politicized, as in Tuvalu, the key objective remains superior access to employment opportunities and the social and economic benefits derived from that, such as access to housing, health and education services (Mortreux and Barnett 2009, Chevalier 2010, Paton and Fairbairn-Dunlop 2010, Shen and Gemenne 2011, Shen and Binns 2012). Economic success enables new social respectability and status (Perminow 1993, Olwig 2003, 2007). Changes in values, following increased educational opportunities and the end of the expansionary phase of bureaucratic (largely urban) employment, have further oriented migration streams outwards, as local employment opportunities have not kept pace with population growth.

Migration is not simply the outcome of individual and household calculations and decisions. Politics intervene at every level, both negatively and positively. In certain circumstances political issues are absolutely dominant as in the case of the BIOT, where the indigenous people were forcibly removed and resettled on Mauritius, to make way for a military base (Vine 2009, Sidaway 2010): an extreme example of the particular vulnerability of dependent territories and their peoples. More frequently instability, nepotism, corruption, violence and coups produce 'difficult' political circumstances in several states. In Fiji and the Comoros coups and the resultant economic disruption created distinct and selective waves of migration. In Fiji that resulted in the relative decline of the once majority Indo-Fijian population, which further accelerated after the 1987 and 2000 coups, emphasizing how migration was 'a barometer of fear' (Narayan and Smyth 2003). Indo-Fijian newspapers and magazines in Australia were full of matrimonial advertisements and their number and content offered a good indication of the desperation and determination of some Indo-Fijians to migrate (Voigt-Graf 2004). Violence and coups led to persistent migration from the Comoros (Boina 2011), and skilled migration from Bougainville (PNG) and Solomon Islands during the late 1990s political crises. Migration from Tonga took place in the context of a stagnant economy and an inflexible hierarchical political system in the 1990s, that inhibited socio-economic mobility. For whatever reason migration remains, in different forms, a strategy of moving from a poorer area to a richer one in search of social and economic mobility.

Migrants not only help to maintain family and communal networks but even enlarge their social fields of interaction, incorporating them into multilocal networks of support and empowerment. Migrant extended households can therefore be characterized 'by remittance transfers among various component parts of the "transnational corporations of kin" which direct the allocation of each island's family labour around the regional economy' (Bertram 1986: 820). For households in Samoa

> having young wage earners abroad diversified families' earnings streams and reduced their dependence on high-risk activities. Having family members in several locations abroad diversified earning sources and reduced risk levels still further. Families, using intelligence from migrants abroad, periodically surveyed risks and returns in various enclaves and encouraged others abroad to relocate in places in which returns were found to be higher and risks lower. (Macpherson 2004: 168)

In this way Samoans were, for example, encouraged to join the United States military because jobs were assured, wages were higher and education could be obtained without loss of earnings. 'If this analysis depicts Samoans as calculative and instrumental, it is because in relation to risk and return they are necessarily so … [as] risks and returns available in various places were formally canvassed and modeled by families' (Macpherson 2004: 168). Moreover, while the transnational corporation of kin has usually been seen geographically, with migrants scattered between countries, households have also sought to position children strategically in key areas of employment, such as banking, education or the police force, combining spatial and professional deployment: occupational multiplicity writ large. In several Pacific SIS nurses have joined the profession primarily because it provided migration opportunities (Connell 2009c). Such strategies disadvantaged sending countries, especially where a significant technological gap between source and destination countries stimulated demand (Di Maria and Lazarova 2012), as skilled workers became scarce.

While extended household strategies and consensus certainly occur, and demonstrate the significance of access to the migration–remittances nexus, at least in Tonga and Nevis families are not necessarily in agreement about their economic aims and functions, while migration may be precipitated by social crisis rather than economic calculus (Lee 2004, Olwig 2007). In Kiribati and Tuvalu, where seafarers are away from their wives and parents and send remittances to both, there are frequent disagreements about their allocation and use (Dennis 2003: 35). Disputes over outcomes occur in various places. Moreover, over a decade ago, in many Tongan villages, and in Tokelau, remittances were becoming individualized so that the idea of a collaborative transnational community of kin was thought to be declining (Hooper 1993, James 1993b). Yet simultaneously the reduced dependence on local kin that remittances bring increases the reliance on overseas kin. The extent of greater individualism is impossible to determine, but conflicts over use emphasize, rather than downplay, the role and significance of remittances.

REMITTANCES

Remittances play an increasingly important role for households and states, especially in smaller islands, having long been crucial in smaller Caribbean SISI (Philpott 1968, Rubenstein 1983). Early success in generating incomes and stimulating remittances resulted in Tongan migrants being described as 'worth their weight in gold' (de Bres and Campbell 1975). So crucial are remittances that in some SISI, such as the Cook Islands, islanders embark on frequent tours to the diaspora, often in church or dance groups, with one objective being fundraising by stimulating donations and remittance flows (Alexeyeff 2009). In this century SISI have become some of the largest recipients of remittances in terms of their contribution to GDP. Where SISI have the same currency as the principal source of remittances they cannot easily be calculated, but from New Zealand to the Cook Islands may be as high as 60 per cent of GDP (Brown, Leeves et al. 2013). The independent states of Samoa and Tonga, with 22 per cent and 20 per cent respectively of their GDP coming from remittances, are officially surpassed in percentage terms only by Tajikistan. Cape Verde and Grenada are also among the top 10 global recipients (Melde and Schicklinski 2011). In 2003 remittances to the Comoros represented 12 per cent of GDP (da Cruz et al. 2004). However most data on remittances to SIS are inadequate, the World Bank publishes no data for the Comoros, Tuvalu and Kiribati, amongst others, and remittances to Cape Verde, and other SIS, are known to be underreported (Pop 2011). Each of these four SIS, and probably also St Kitts, Dominica and St Vincent, are likely to receive remittances that make up more than 10 per cent of their GDP.

The vast majority of households in SIS receive at least some remittances. As many as 60 per cent of all households in Tonga have at least one overseas migrant who remits and 90 per cent of households received remittances (Brown et al. 2013). About two-thirds of Cape Verde households receive remittances (Pop 2011). In one typical St Lucia rural area only 28 (21 per cent) of 133 households had no household members abroad and most received remittances (Moberg 2008). In Kiribati the 2000 census recorded that overseas remittances, primarily from seamen, were the primary source of income for as many as 30 per cent of households even in urban Tarawa. In 2002 at least 35 per cent of all households in Tuvalu received remittances. In both countries these were usually the main source of income in the outer islands (Abbott and Pollard 2004, Borovnik 2006). If remittance receipts could be disaggregated by island, smaller, more remote islands would have even higher proportions of their incomes coming from remittances, as occurs in relatively poor Savai'i in Samoa and the remote Niuas group in Tonga.

Remittances are a significant part of national income in many SISI, hence more than 20 years ago four smaller Pacific island states, two independent and two dependent (Kiribati, Tokelau, the Cook Islands and Tuvalu), were conceptualized as MIRAB states, where MIgration, Remittances, Aid and the resultant largely urban Bureaucracy were central to the socio-economic–political system (Bertram and Watters 1985). That notion of MIRAB is also applicable to larger states such as Samoa, Tonga and Cape Verde, and several Caribbean SISI, where remittances constitute high proportions of national income (Connell and Conway 2000). While this acronym is disliked in the Pacific for cultural reasons and because of its implication of a 'handout mentality', even 'Dutch disease', it nonetheless suggests the ongoing centrality of migration and remittances in SISI (Bertram 1999, 2006). It has generated other acronyms such as MURAB, which placed extra and appropriate emphasis on the attendant urbanization that a growing bureaucracy engendered in island states such as Tuvalu (Munro 1990), and even MIAB, where migration did not initially stimulate a significant flow of remittances, as in some parts of Micronesia (Ogden 1994, Friberg et al. 2006) and also Niue and French Polynesia, though there is now a shift in this direction (Grieco 2003, Connell and Brown 2005). While metropolitan aid has reduced the relative significance of remittances in most political dependencies, such as Niue, Pitcairn, Wallis and Futuna, the Austral Islands (French Polynesia) and much of Micronesia (where the MIAB acronym emerged), they are reported to be still of significance in Tristan da Cunha and St Helena (Sear et al. 2001: 19). Other acronyms have emerged, including MIRTAB where tourism plays a parallel role in directing external income flows; in some Caribbean SISI the combination of remittances and tourism have produced a 'travel economy' (Palmer 2009). It is scarcely surprising that one further acronym was MIRAGE, where GE stands for government employment, with the clear implication that this was scarcely a sustainable, legitimate or viable form of economic development (Harold Brookfield, quoted in Bertram and Watters 1985).

Just as whole countries have been conceptualized as MIRAB economies so the acronym has also been applied to particular islands, including Ware (PNG) (Hayes 1993), Fakaofo (Tokelau) (Hooper 1993) and Lofanga (Tonga) (Kronen and Bender 2007). Remittances from migrants constitute a significant and usually the primary source of income in such outlying islands as Carriacou (Grenada), Ponam (Manus, PNG), Kabara (Fiji), Mauke (Cook Islands), Nanumea (Tuvalu), Falahola (Vava'u, Tonga) and Barbuda, becoming even more crucial after the decline and virtual collapse of the copra trade and largely ensuring the economic survival of such islands (Richardson 1975, Carrier 1984, James 1991, Hayes 1993, Felgentreff 1996, Chambers and Chambers 2001, Marsters et al. 2006, Potter and Sluyter 2010). In Nanumea remittances grew from being about half the island income in the 1970s and 1980s to some 75 per cent in the 1990s, as world copra prices slumped (Chambers and Chambers 2001: 156). In islands like Ponam and Nanumea, having large families in the hope that one or more children might eventually provide remittances was a conscious economic strategy (Connell 1998a). For the tiny coral atoll of Manihiki (Cook Islands), migration and the resultant remittances have been seen as so crucial that they constitute nothing less than a socio-economic strategy for collective survival (Underhill 1989). Even so, while remittances crucially maintain livelihoods in remote areas, there is an overall urban bias in their distribution (da Cruz et al. 2004, Pop 2011, Brown et al. 2012).

Migrants send remittances for many reasons – but above all out of a sense of duty. In Tonga remittances are motivated by 'a combination of *fatongia* (obligation), *fetokoni'aki* (helping) and *fe'ofa'aki* (kindness) (van der Grijp 2004: 188). Similarly remittances to Rotuma are not simply cash transactions but 'represent the investment of time and effort and care which Rotumans call *hinasi* – love in action' (Rensel, quoted in van der Grijp 2004). On Eauripik (FSM), while everyone on the island was 'useful', people could became 'worthless' if they migrated and sent nothing back (Scourse and Wilkins 2009). Migrants from Cape Verde who 'forgot' their relatives were frowned on while those who sent remittances were 'praised as exemplary migrants' (Carling 2004: 125). However, migrants are ultimately motivated by factors other than altruistic family support, such as asset accumulation and investment at 'home', as intergenerational flows of remittances take on more individualistic elements (Brown 1998). Rationales change and evolve.

Within the Melanesian states especially, where international migration is limited, remittances from urban to rural areas continue to play an important role, especially where migration is from small islands, such as Ponam or Ware (PNG), or Aniwa and Paama in Vanuatu (Wilson 2010, D'Arcy 2011). In the outlying Reef Islands (Solomon Islands) remittances were 'the only means' of generating 'sufficient income for necessary household expenditure' (Birk and Rasmussen 2013). Otherwise, because of lower earning power, domestic remittances are of lesser economic significance though of considerable social significance. However, even for remote parts of such large islands as Tanna (Vanuatu), remittances from urban migrants are the single largest source of income in several villages (Winthorpe 2004). Only in a very few areas such as Kadavu (Fiji), where there is an unusual and valuable agricultural crop, have remittances declined in significance in past decades (Sofer 2009). In many such places demand for access to international migration opportunities and superior incomes is considerable.

As the collapse of the Pacific copra trade in the 1990s unfortunately demonstrated, remittance flows are usually more stable over time than income from agricultural exports or tourism. Indeed the essence of MIRAB was that remittances and aid insulated small island economies from shocks and reduced vulnerability through the continuous operation of 'stabilising negative feedback loops' (Bertram 2006). In Tonga, and elsewhere, 'regular remittances from a dependable relative with a reasonable job overseas are often better than relying on a locally employed family member, whose salary is considerably lower and whose responsibilities to many are greater and more immediate' (Besnier 2011: 43). The financial situation of particular households is partly determined by the number of relatives who have migrated from there (Gailey 1992a, Kronen and Bender 2007). Where data can be disaggregated, women are more frequent and more generous remitters, and consequently in some cases women were specifically selected by households as migrants (Muliaina 2001, Connell and Brown 2005, Bautista 2010). Since remittances are countercyclical, notably in response to disasters, they contribute to stability by smoothing consumption. Characteristically, after several Samoan cyclones, remittances have been used to purchase food, rebuild houses, replant plantations and supplement lost income, simply enabling villages and countries to return towards normalcy (e.g. Sutherland et al. 2005, Naik et al. 2007). Unusually, on Grand Comore, the frequency of volcanic eruption has discouraged some remittance flows through concern that houses may eventually be destroyed and investment is foolish (Morin and Lavigne 2009). More generally, however, the remittance economy is so flexible that where circumstances change, and islanders prosper, as on Manihiki (Cook Islands), where a black pearl economy developed, and when the squash industry boomed in Tonga, remittances may reverse direction and move from peripheral islands to the metropolitan core (van der Grijp 2004, Marsters et al. 2006). Such circumstances are however rare, and 'normal service' has resumed in both places.

Remittances are vulnerable to global or national financial crises that affect the earning power of migrants. In the wake of the GFC, over the period 2009–11, remittances from metropolitan states to many SISI declined, though not to the anticipated extent (Melde and Schicklinski 2011). In neighbouring Samoa and Tonga remittances in the former stayed more or less stable, while in Tonga they fell by as much as 20 per cent (placing some recipient households in difficult financial positions). Tonga received the majority of remittances from the United States (notably California), more drastically hit by the recession, while Samoa received more of its remittances from New Zealand (and Australia), less affected by the recession and where social security payments (the 'dole') partly offset reduced earnings. Moreover specific remittance response to the 2009 tsunami ensured that Samoan receipts did not collapse. Where there is need, in some Australian contexts at least, Pacific island migrants have remitted so much that social costs are imposed on their own households.

A proportion of remittances do not reach their intended destinations since transaction costs, which vary according to the mode of transfer, are considerable. In the mid-2000s costs were higher than average in Cape Verde (Pop 2011), but particularly high in Pacific SISI. Using bank drafts or money transfer organizations, such as Western Union, to send money from metropolitan states to Tonga (as over 80 per cent of remitters did) resulted in transaction costs ranging around 15–20 per cent. Households lost substantial sums and the GDP of remittance-dependent countries, such as Tonga and Samoa, was reduced by about 4 per cent (Gibson et al. 2007). World Bank and donor government pressure on banks resulted in the instigation of a dual ATM system for money transfer, while independently mobile phone transfers were introduced in 2011. Both systems reduce transaction costs to about 4–5 per cent. With the greater use and more widespread availability of ATMs, an increasing proportion of people with bank accounts, the remarkable rise of mobile phones (now owned by almost all households in countries like Fiji, Samoa and Tonga, despite their absence a decade earlier) and social networking, remittances may effectively become more valuable. Within a year a substantial number of Samoan households were receiving telephone transfers, and women especially benefited from improved access to money. While transaction costs are much less in the Caribbean, similar changes are imminent there.

Because of its significance, the sustainability of remittance-dependent development is particularly important, but uncertain, especially if the flow of migrants dwindles. With family reunification overseas and greater integration of migrants into host communities, willingness to remit might be expected to decline over time. Only recently have second and third generations of islanders grown up beyond their island 'homes', hence the extent to which they will remit, and to whom (and even whether they can be described as 'islanders' or 'migrants'), are unknown. Second generations are more likely to act as individuals, rather than perceive themselves as members of wider transnational social groupings, with a reduced likelihood of sending remittances to their 'home' countries especially if they take up host country citizenship (part of acquiring local rather than amorphous transnational identities). Links between second-generation Samoans in New Zealand and Tongans in Australia with kin in Samoa and Tonga, respectively, are declining, as social and language ties weaken, people 'marry out', any sense of obligation and belonging declines, and overseas born 'islanders' become a majority (Muliaina 2001, Lee 2003, 2004, 2009). Most respond only indirectly through their parents and contribute limited remittances. Half (46 per cent) of all second-generation Tongans in Australia sent no remittances and 29 per cent claimed no contact with Tonga, while the individual level of remittances fell sharply between generations (Lee 2011). Some Polynesian migrants have moved to regional New Zealand and Australia, partly to escape the constant pressures from visitors to the main cities requesting remittances (Brown et al. 2012). 'Remittance fatigue' has been identified amongst second-generation Comorians and Cape Verdeans in France, but remittances nonetheless continue as long as close relatives remain at home (da Cruz et al. 2004, Carling and Batalha 2008). For Cape Verdeans and Nevisians, a parallel decline in social connectivity is yet to result in a decline in remittances (Carling 2004, Olwig 2007), perhaps also reflecting discrimination in destinations and a resultant desire to maintain island ties.

Both unskilled and skilled migrants, such as nurses, sustain remittance levels at high levels, over long time periods, often more than 20 years (Connell and Brown 2004, Potter et al. 2004, Brown et al. 2006), despite early assumptions that remittance decline would be rapid, and especially rapid for skilled workers, as they would be less likely to need to insure family at home (Gibson and McKenzie 2011b). As long as there are kin in the islands some remittances reach them, and the most probable brake on remittances will be a slowing of migration, for whatever reason, and the death of potential extended family recipients. While in the long term remittances may only be sustained by continued migration, cautious optimism suggests that remittance flows will not decline significantly in the near future, but neither will remittances continue indefinitely, especially from non-migrant generations.

The Use and Impact of Remittances

Conventional wisdom suggests that remittances are overwhelmingly used for consumption with smaller amounts for investment. After debt repayments remittances have invariably contributed to more modern forms of consumption, notably of food, vehicles and houses (the last of which provides welfare benefits, especially in association with superior toilets, better access to clean water and solar panels). By the 1970s, when Cape Verde was one of the highest per capita recipients of remittances in the world, house building was so important that it also imported more cement per capita than any other African country (Connell 1988a). In the Comoros 'remittances spent on household consumption often represent much-needed expenditure on poverty-reducing items: nutrition, shelter, education, and health, with islands receiving more remittances experiencing less poverty that others' (da Cruz et al. 2004: 9). By the end of the nineteenth century remittances to Barbados had already contributed to both housing and shop construction (Marshall 1982). In the early migration era a modern house defined and distinguished a household receiving remittances, and in the Caribbean was the definitive marker of both 'bettering one's condition' and gaining respect (Olwig 2007). 'In-kind' remittances, including clothing, food, electrical and electronic goods and furniture, support consumption. Cars and other vehicles have also been sent as remittances, especially in Samoa, after the changeover to driving on the left in 2009. The use of remittances for consumption and basic needs has reduced the extent of poverty in recipient countries.

A significant proportion of remittances support 'traditional' customs and obligations, providing respect, autonomy and continuity (Evans 2001; cf. Besnier 2004). It is more appropriate to engage in exchange and gift giving rather than sales and purchases, hence commercial practices are downplayed in favour of social obligations, especially where these enhance status, as in the *grande mariage* of the Comoros (Morin and Lavigne 2009, Boina 2011). In the Pacific a significant element of remittances are communal social objectives, particularly support for churches. In the Comoros numerous village projects, ranging from mosques and schools, to roads and health posts, are funded from remittances sent by diaspora organizations (da Cruz et al. 2004). Remittances thus reinforce traditional values centred on personal, family and village obligations, enabling cultural and even economic continuity, so maintaining 'the social integrity of … society despite diasporic fragmentation' (Besnier 2004: 19). The social and the economic cannot be disentangled.

It has often been argued that remittances (and aid) are not conducive to private sector development and discourage agricultural work, while the MIRAB model implies a strong bureaucratic outcome, which may stifle productivity. With additional sources of income from remittances (and indirectly from aid) people 'can be expected to evaluate the return on [agricultural] investments relative to alternatives … [I]t would be expected that as the alternatives to commercially oriented agriculture would improve, so a reallocation of household effort away from agriculture would take place' (Bertram and Watters 1985: 511). Sturton stated that 'The Tongan economy displays all the characteristic markings of the "Dutch disease" where a dominant export activity attracts a disproportionate command over resources, pushes up domestic production costs, and reduces international competitiveness. In the Tongan case the "booming" sector has become development assistance and migrants' remittances' (1992: 3). A similar analysis of Cape Verde found only a minor Dutch disease effect (Bourdet and Falck 2006). Faeamani has argued that in Tonga, through the combination of the loss of young adults and an inflow of cash in the form of remittances and goods, 'there is a consequent reduction in garden size and production' (1995: 140), but this is more likely to be a function of labour shortages (Evans 1996, 2001). In Tarawa too, dependence on remittances has discouraged the development of urban agriculture (East and Dawes 2009) but land shortages probably played a part. On Kosrae (FSM) remittances released pressure on local marketable resources (notably mangrove crabs), by enabling the purchase of imported foods, fuel (kerosene and electricity) and construction timber (and concrete) so that overall pressure on mangrove ecosystems was reduced (Naylor et al. 2002). Similar processes elsewhere in Micronesia have resulted in the decline and abandonment of even the former staple of taro (Marshall 2004a, Oles 2007b), with consequent social, economic and environmental costs. In Montserrat, by the 1950s, agricultural production was declining rapidly while remittances were contributing to improved health, housing, education and nutrition (Lowenthal and Comitas 1962). Similarly, in nearby Nevis, remittances contributed to the demise of both sugar cane and sea island cotton, ironically the crops that had financed the migration of many Nevisians, while the labour force had shrunk (Byron 1999). In each case various factors were involved and the direction of causality was unclear.

It is implausible that remittances do not have some disincentive effects, although there is limited incontrovertible evidence of this. Indeed, as James has noted: 'It has been argued that remittances take away the motivation of locals to produce, but the facts of local entrepreneurship seem to contradict this since large consignments of local products … are sent to relatives overseas for sale among the Polynesian population' (1991: 2). Agricultural decline has nonetheless been widespread, but for many reasons (Chapter 3). To some extent, this is 'what is required for efficient economic behaviour: that the family allocates its resources to the highest productive use, even if it happens that this particular use is not "productive" in the "domestic" economy, but rather in a "foreign" economy' (Poirine 1998: 77). After consumption, housing and some community goals (such as water tanks, sports clubs and churches), air fares and education (an investment in social capital), remittances are used for various forms of investment, sometimes in the agricultural sector, and for land purchase where possible (Connell 1988a), but mainly in the service sector, and especially stores. Even on small outer islands, like Falahola (Tonga), remittances have been used for economic ventures, ranging from agriculture to tourism, though remoteness has limited their success (James 1991). In the most difficult circumstances remittance recipients make efforts to invest where they can. In the outer islands of Kiribati, where most remittances provide basic needs, and the custom of *bubuti* (a request that cannot be refused) makes savings and business more difficult, all recipients still sought to retain some income to invest in land purchase, doughnut bakeries, stores or even in sewing material for blouses to be sold (Borovnik 2006).

Commerce has been a major beneficiary. In Samoa remittances have constituted the start-up money for many shopkeepers and small entrepreneurs. Half of all market vendors, all of whom received remittances, claimed that some had been used as capital for the purchase of seeds, fertilizer and tools to engage in food production for sale (Muliaina 2001). Remittances have also contributed to urban investment. The expansion of the Nuku'alofa flea market in Tonga since the mid-1980s demonstrated how many households used remittances as an investment in market stalls, enabling subsequent economic diversification (Brown and Connell 1993). Throughout Tonga the increased use of remittances for investment in fishing, agriculture, stores and transport businesses, attests to the shift from consumption to investment that has also occurred in several eastern Caribbean and Indian Ocean SISI (Connell and Conway 2000, Morin and Lavigne 2009, Melde and Schicklinski 2011). No part of the economy is necessarily abandoned but remittances enable some diversification out of agriculture. Thus in Cape Verde earlier remittances in the 1980s 'were driven mainly by altruism, but after the mid-1990s … became more investment driven' (Pop 2011: 93). Ultimately remittances are directed wherever agricultural or other productive opportunities appear attractive, such as into the 1990s boom in squash cultivation in Tonga (van der Grijp 1999), emphasizing that they have complex and important social and economic dimensions, and may both stimulate and depress certain sections of the economy. Where there are opportunities, and where consumption goals have been satisfied, remittances are used for investment, stimulate entrepreneurial and trading activity, increase the extent of formal sector employment and produce multiplier effects. When guest worker migration from countries like Tonga began in the 2000s (see below), those who had never worked overseas before tended to use remittances for immediate consumption objectives, while those with a previous migration history who had achieved basic consumption goals sought to establish or develop small businesses; the same situation that had occurred a generation earlier in eastern Caribbean SIS (Wood and McCoy 1985, Blanco 2009).

Remittances have become ever more valuable. Two decades ago, 'the capacity to call on international resources ha[d] become a crucial factor in influencing a family's local economic conditions. The lowest stratum in contemporary Tonga are [sic] those totally dependent on the nation-state framework, and the limited resources it embodies, without any overseas options at all' (Marcus, 1993: 29–30). Indeed it is increasingly argued, as in Tonga, that 'every family needs to have someone overseas. Otherwise the family is to be pitied' (quoted in Small 1997: 152) hence, in contrast to western societies, it is often single-female-headed households that survive most effectively (Gailey 1992a). More recent studies of migration and remittances in Tonga and Fiji have shown that households with migrants were more likely to have a higher income, independent of remittances, but the direction of causality was unclear (Brown et al. 2006, 2013). A consensus in very diverse circumstances is improbable.

A consistently substantial and growing volume of remittances has characterized smaller SISI, notably in Polynesia (making up a significant part of national income, in excess of the value of exports and aid). Remittances have been particularly important in the most remote islands, where development needs are less well met, and have gone through a partial transition from consumption to investment. Remittances contribute to valuable objectives, such as human resource development, maintain social networks and create social capital (Grieco 2003). Their reciprocal contribution exemplifies how remittances are constituted through a process that is simultaneously cultural continuity, social exchange and trade. Especially in smaller islands, exemplified in Ponam (Carrier 1984),

Rotuma (Bryant 1990), Mandok (New Britain, PNG) (Pomponio 1993), the Lau Islands (Turner et al. 2007), Aniwa (Wilson 2010) and Malaita (Solomon Islands) (Gegeo 2001), education is highly valued, both in a general sense and for the development of specific skills (for example, in health provision), in order to create human capital for potential migration. With imperfect knowledge, households are consciously making decisions in favour of the quantity and quality of education of children that boost their opportunities for migration and thus remittances (Brown and Connell 2004). Remittances have substantially contributed to welfare in most SISI, especially improved housing, and to superior levels of consumption, education, health and wellbeing. Despite concerns that remittances constitute a 'moral hazard' by reducing the incentive of recipients to work (da Cruz et al. 2004), limited evidence supports this, even when remittances are substantial (Connell and Brown 2005, Pop 2011). Nonetheless migrants are wont to feel that a lazy and pampered society has emerged for which they are partly responsible (Byron 1999), hence second-generation resistance. Other concerns exist that remittances emphasize intra-village (and country) economic inequalities and hamper social development (Small 1997), but at least in Tonga and Fiji, as a 'mature remittance economy' develops, where most households have regular access to remittances, so both poverty and inequality appear to be reduced (Brown et al. 2013). The limited evidence on the role of remittances in contributing to inequality is, however, ambiguous.

Remittances support what migrants and their kin want, have the greatest multiplier effects in poorer and more remote areas, and are countercyclical. Only indirectly do they support the national economy. Remittances have contributed to employment (especially in the service and construction sectors) and eased balance of payments problems, despite contributing to inflation. While remittances are certainly vulnerable to external shocks, their fluctuations and the shocks have been less than those caused by changes in commodity prices, aid receipts or tourist flows. Moreover remittances in SISI, at least in the Pacific, have been sustained at a high level and over long time periods, providing social protection – where governments may not – and, through a steady and reliable source of income, have enabled some degree of poverty alleviation.

BRAIN AND SKILL DRAIN

The proportion of skilled islanders as migrants has increased, partly through shortages in receiving countries which, as with health services in several countries, led to private sector recruitment (Connell and Stilwell 2006). Yet even in SIS, such as FSM, where there are no effective constraints to migration, skilled migrants predominate (Akee 2010). They move for much the same reasons as others: access to superior incomes, employment and services for themselves and their families. Additionally skilled workers are often frustrated with limited promotion opportunities at home (inevitably restricted in SISI), inadequate access to technology, materials and further training, and a perceived lack of recognition. Limited economic diversification, high tax rates, low remuneration, poor working and living conditions, particularly in remote regions, are further influences, or simply the ability to develop and use skills, as in the case of a Niuean nurse whose increasingly specialized career trajectory took her to New Zealand, the Gulf, Australia and finally Canada (Voigt-Graf 2003, Connell 2009c). Skilled wage and salary differentials are much greater, hence income opportunities are superior, the job market is better understood and opportunities exist to develop skills further. Ironically many migrants become part of a 'brain loss' or 'brain-waste' because their qualifications, despite gaining them entry, are deemed inadequate for employment in the destination: a lost investment in human capital, and prevalent in SISI where the technological gap with developed countries is greater. While skilled workers are more likely to move to metropolitan states, growing shortages within SISI have stimulated increased regional migration, to countries offering better work conditions and salaries, such as Fijian teachers and nurses going to the Marshall Islands, and Fijian tourism workers to the Cook Islands. Migrants from many Caribbean states move to thriving tourism economies nearby, alongside nurses moving from poorer to richer states, for example, from St Vincent and Grenada to Barbados, the Cayman Islands and the Bahamas.

Small states lose disproportionately more skills than larger states; they are more successful at producing skilled workers but less successful in retaining them (Beine et al. 2008). Fifteen SIS are among the 30 countries with the highest skilled migration rates to OECD countries alone; the top six are all SIS – Grenada, St Vincent, St Kitts, Samoa, Tonga and St Lucia. Emigration rates are especially high amongst women (Docquier and Marfouk 2006). Separately distinguishing the migration rates of doctors, a skill of particular value in most SISI, shows that the top six countries of emigration are again all SIS – Grenada, Dominica, St Lucia, Cape Verde, Fiji and São Tomé (World Bank 2011). However these data are based on doctors 'trained in country'; many SISI are too small to have medical schools, doctors are trained overseas and their migration rates are higher (Connell 2009c). Grenada has lost some 85 per cent of its tertiary graduates and 98 per cent of its doctors; St Vincent, St Kitts, Dominica and St Lucia have somewhat comparable figures. The Caribbean has some of the highest skilled migration rates in the world, extending to larger states such as Jamaica, Haiti and the Dominican Republic (Castellani 2007, Bristol 2010). Cape Verde is proportionately the greatest loser of educated emigrants, including doctors and nurses, in sub-Saharan Africa (Pop 2011). Data are unavailable for most politically dependent SISI where migration rates of skilled workers may be greater.

Losses of skilled labour have become serious in several SISI, but perhaps especially for some of the smallest, despite their needing few skilled workers. By the 1980s the brain drain was argued to be hampering human resource development in the Caribbean generally, and in Fiji and the Comoros (Connell 1988a). Lack of domestic production of skilled workers also played some part in deficits. The two occupational groups usually singled out as the main categories of skilled workers in deficit in SIS have been health workers and teachers, especially since both stocks and migratory numbers are greater and their contribution to human resource development is substantial. The rate of migration can be considerable. In Tonga, for example, of 344 nurses who graduated between 1987–2006 some 115 had migrated by 2009 (95 to New Zealand) and 86 others had been lost to attrition. In Samoa and Tonga the main losses of teachers have been science and mathematics teachers from secondary schools, partly through migration and partly from attrition (since science and mathematics teachers are most easily able to obtain better paid jobs within the public sector). Football and rugby players make up growing proportions of skilled migrants: a substantial brawn drain. No Tongan, Samoan or Fijian international rugby player lives in those islands and most have chosen to be internationals for other countries (Grainger 2011, Besneier 2012). Again, where required national numbers are few, such as of airline pilots and air traffic controllers or meteorologists – all scarce in Pacific SISI – losses can be more disadvantageous. But, much more generally, as in Dominica, a critical development problem is that 'many middle class Dominicans, the sort of people with ambition and education, who might be expected to offer leadership as "critical citizens" in a small-scale society, have long since left the island' (Payne 2006: 38). Experience and management expertise, alongside various specific skills, have departed.

Most SISI have also lost a considerable number of public servants to regional organizations – a more localized brain drain (Liki 2001). Even aid agencies, including international NGOs, provide many good jobs and create 'perverse incentives' that 'result in a flow of administrative and technical talent from the public sector' (Moore 2011: 1771). Such individuals remain in the region but are no longer specifically involved in national development. (A parallel loss can be that of skilled workers employed by transnational corporations, or in tax havens, whose interests are not necessarily those of island states.) Ironically, such workers are highly visible locally and provide a model of successful mobility.

The impact of skill losses has been particularly evident in the health sector, where there is a shortage of skilled health practitioners in almost all Pacific and Caribbean island states (Thomas and Hosein 2006, Connell 2007c, 2009c). Even in larger SIS like Fiji, the loss is significant and problematic. Doctors are twice as likely to migrate as nurses because wage differentials are greater, and because most nurses are women and men are often the primary migration decision makers (Brown and Connell 2004). However female Fijian nurses frequently take the decision to migrate autonomously, often leaving their husbands and children behind, mainly for higher wages, but also to escape marriage problems and customary obligations (Rokoduru 2008). Because of smaller numbers and greater training costs, the loss of doctors (and specialists) is most critical. An acute shortage of doctors and nurses in Fiji was exacerbated in the wake of the 2000 coups, and by active recruitment in some receiving countries, with the Ministry of Health having to recruit replacement doctors from the Philippines and the Indian subcontinent (Chandra 2003). Cuban doctors have also filled Caribbean and Pacific vacancies, and been invaluable in East Timor. In some smaller states the brain drain has been equally worrying; the Cook Islands, for example, lost more than half its vocationally qualified population in the single decade 1966–76 and much the same happened again in the mid-1990s when the national economy collapsed (Connell 2005). Many SISI experience shortages of accountants, engineers, IT workers and other skilled groups. In 2011 the Permanent Secretary for the Fiji Public Service Commission observed that: 'Work classifications that Government considers as scarce skills are in the areas of Doctors, Engineers, Meteorologists, Veterinary Officers, Town Planners, Marine Scientists, Agricultural Scientists and Geologists' (*FijiLive*, 18 June 2011). Even tiny Niue has imported refrigeration and other engineers from the Philippines because of its migration losses (Connell 2008). Changing aspirations and the migration of the more educated young contribute to the brain and skill drain from both national peripheries and entire SISI, perhaps ultimately worsening the welfare and bargaining position of those places. It is unusually difficult to replace skilled migrants, both because of the duration of training that is required and the limited need for rare skills. The outcome in the health sector at least is that basic needs are less well satisfied, especially in more remote areas, and there is a loss of morale amongst those who have remained, as working conditions deteriorate. Wards are closed, waiting lists and times lengthen, examinations are more cursory and large proportions of budgets are directed to referrals to distant places (Connell 2009c). The shift to skilled migration has been at some cost to service provision: national capacity to deliver services has been weakened. Skilled migration is unlikely to decrease, given worsening metropolitan skilled worker shortages in the 'standard' destinations, and increased shortages in newer, more distant markets.

Return migration has not solved labour shortage problems. What has long been true of the Comoros is more generally true: 'there has been a *natural tendency* for qualified Comorians to seek employment abroad and not return to the poverty and lack of prospects in the islands' (Newitt 1984: 91; my italics). Various Pacific SISI have turned towards Asian labour markets for replacements (Connell 2005, 2008), just as small Caribbean states have turned to Cuba and southern Africa. In the case of migration of Tongans and Samoans to the United States alone: 'Emigration results in the permanent loss of young educated skilled labour from the Pacific island nations. Skilled labour is in short supply and emigration probably hinders development' (Ahlburg and Levin 1990: 84). This is certainly true more generally in the health sector, where more costly (and sometimes less skilled or culturally appropriate) replacements have been required, and in the loss of sportsmen.

After global concerns, dating from the 1960s, about the negative impacts of the brain drain, and despite growing numbers, skill losses in the 1990s became regarded from outside the SIS with less concern, since it was argued that the prospects of working abroad stimulated more people to acquire valuable skills, not all of whom subsequently migrated, while skilled migrants were as likely to send remittances. While the overall economic benefits from migration are exceptional for skilled migrants from SIS (Gibson and McKenzie 2011b, 2012, Batista et al. 2012), outcomes for the countries are more ambiguous. Remittances from health workers at least have been substantially more than training costs (Brown and Connell 2004), and that is likely to be true of other skilled workers. The economic costs of skilled migration are not as great as feared, even if training is in the public sector and remittances are private, but it is impossible to accurately cost, in this case, the health (and other) disadvantages of high levels of emigration. Only where relatively low migration rates exist can human capital development benefit – the converse of the situation in SISI – as useful development skills are lost and losses are not mitigated by remittances or return. In several Caribbean SIS rates are excessive and benefits are fewer. In Fiji where, as expected, demand for tertiary education has increased for over a decade in response to migration opportunities, the migration rate has been such that wide-ranging skill shortages remain, especially in the health sector, since investment in education has been inadequate to offset declining stocks of human capital (Connell 2010c). Cape Verde seems to have been more successful in retaining more of the highly educated (Batista et al. 2012), which may be a function of superior local development opportunities and/or salaries. Despite its disadvantages, policies to discourage skilled migration are now less likely, and small states are training individuals for migration, based on long-established precedents in Kiribati and Tuvalu where seamen are trained for the international labour market (see below).

RETURN MIGRATION

At no time during the past quarter of a century has there been substantial return migration to most SISI, despite the overwhelming 'return ideology' that characterizes migration (Philpott 1968, Olwig 2007), with Comorians, for example, 'haunted by an obsession with the return home' (Boina 2010:163). Some SISI, like Wallis and Futuna, have seemed in danger of becoming mere 'nurseries and nursing homes'. Limited return migration is at least partly due to the great differences in income levels between the SISI and the metropolitan periphery, but also to social factors. Unsurprisingly, return migration is greatest where employment and economic opportunities exist. Few returned to Cape Verde, for example, until quite recent years when the national economy began to develop. Migrants who have returned have often been those with capital and skills, and where one member of the household was able to develop a business, rather than work in the public sector (Connell 2009a, 2009b). Once derided as merely the return of failures and the retired, at least in Cape Verde, Samoa and Tonga (and probably elsewhere), it encompasses a mix of retirees, unskilled workers and skilled workers, of different durations overseas, genders and age groups, who return despite the discrepancy in wages and working conditions (Potter et al. 2004, Maron and Connell 2008, Lee 2009, Brandao and Zoomers 2010, Gibson and McKenzie 2011b). Successful development of small businesses provides incentives for those who have sent remittances to return home and manage them. In overseas Polynesian households that included nurses the greatest propensity to return came from those with business investments at home (Brown and Connell 2004). In Cape Verde those without capital are hesitant to return; as one migrant put it: 'A failed return is not socially accepted. Someone who returns empty handed – no material wealth, no diploma – is ruined' (quoted in Brandao and Zoomers 2010: 278). Nonetheless, there as elsewhere, returnees sometimes came back empty handed and return migration covered a spectrum of socio-economic groups (Carling 2004). Return migration has nonetheless increasingly been regarded in a more positive light.

Interest at home, at least in an institutional sense, in the development potential of those people who have left is relatively new in SISI. Until quite recently, for governments at least, they were out of sight and out of mind, and their skills and experience had gone. Potential returnees were frustrated by local employment structures that rewarded tenure rather than productivity (Gibson and McKenzie 2011b). Encouragement for migrants to participate in the small societies they have left is never easy and sometimes actively discouraged. Migrants are sometimes said to have 'voted with their feet' so no special efforts should be made for them, especially if they threaten the positions of those who remain. Those who have left Rodrigues, for example, are perceived with a mixture of both resentment and envy (Gemenne and Magnan 2010: 50). Many return migrants experience resentment, jealousy and resistance to their new ideas (Gmelch and Gmelch 1997: 184, Connell 2009c). Indeed new ideas, norms and practices may not necessarily be either beneficial or welcomed, if they threaten an established social, economic and political order or are inappropriate for a different scale, direction and resource-base of development. Change is not always welcome in hierarchical societies where an established order prevails and limited opportunities exist. Returnees may experience a lost sense of identity and belonging (Conway and Potter 2007, Olwig 2007). Many remain in urban centres, in part for employment and in part to escape family obligations. In the eastern Caribbean some even return to countries other than their own to avoid social pressures (Duval 2004). Nonetheless there is new recognition in many SISI of the positive role of diasporas, which goes beyond remittances, to include the transfer of skills, knowledge, capital and technology acquired elsewhere.

Utilizing the development potential of the diaspora (as not merely a market for 'nostalgic' island produce – useful though this is) works best when migrants intend to return and have something to gain. Engaging diasporas has become more important as a form of 'do-it-yourself development' in an age of aid fatigue. The significance of return migration is that returnees share the same language and culture – particularly valuable in education and health care. In Samoa a renal unit was established and operated in significant part by short-term Samoan returnees. In Barbados fully half of those who returned were able to introduce changes in the workplace based on knowledge acquired abroad (Gmelch and Gmelch 1997, Conway and Potter 2007). Yet, more generally, little evidence exists of new knowledge transfers to home governments and businesses (Gibson and McKenzie 2012). In the Cook Islands, many skilled, qualified and experienced people have returned and used their skills in a range of occupations (Hooker and Varcoe 1999), though the Cook Islands are unusual since wages and salaries are more comparable with those in the main destination. In Barbados return migrants were often among the elite; as one returned nurse observed: 'All that I have gotten in life since I left school – property, experience, motivation, the will to get ahead I got in England' (quoted in Gmelch 1992). As in other parts of the world, the longer people stay away the greater skills and capital they may accumulate but the task of return becomes more difficult (Boccagni 2011, Jones 2011a). Using human capital and specialized skills effectively is difficult in very small labour markets. Even in larger island states such as Jamaica, matching the skills of returnees to job opportunities has been problematic (Mullings 2011). Skills acquired overseas are not easily used or effectively absorbed where workers are involved in activities that are absent at home, and with a work organization and discipline that is not easily transferable.

Some returnees have failed to settle overseas, experienced prejudice and returned in dismay, without capital or skills. An unfortunate recent development has been the enforced return of migrants, especially in the Caribbean but also in Tonga, Samoa, the Marshall Islands and Cape Verde (Carling 2004, Pereira 2011), who have been deported for criminal activities (often involving drugs in the United States) and whose return migration has contributed nothing positive. No comparable data are available for most SISI but between 2005 and 2009 over 15 600 Jamaicans were deported to Jamaica compared with just 6315 who returned voluntarily. Amongst SIS only the Bahamas had more than 1000 deportees over the same period (Hunte 2011, Lewis 2011). Few SISI have to confront such challenging circumstances.

Limited return is the outcome of widespread perceptions that fewer opportunities exist at home (which was often the rationale for initial migration) but is also the outcome of children of migrants being educated in the destination and losing some degree of contact with 'home' societies, including critical linguistic and social skills. Return migration is often deferred ('until children leave school', 'until enough money is saved', 'until retirement', etc.) until the point where it becomes implausible. This is also linked to a gradual shift in the demographic balance, towards metropolitan states; relatives are more likely to be found in destinations and little incentive remains to return to what is less likely to be seen as 'home'. After the death of key family members in Nevis, migrants often found themselves at 'a point of no return' where material and emotional ties had gone (Olwig 2007: 212). In every SISI, in every region, outmigration remains well in excess of return migration.

NATIONAL INTERVENTIONS

No SISI governments have sought to discourage international migration in recent years, though several have expressed concerns at the loss of skilled labour (including nurses and army personnel), high training costs, the breaching of bonds by scholarship holders or, in the case of Niue, particularly extensive migration. While island states have signed regional and global codes of practice that discourage the migration of skilled health workers, rhetorical concerns have not been translated into practical policies on retention or return, and new migration and employment programmes have gained considerable support.

SISI have played a deliberate role in stimulating certain forms of migration. Jamaica and Barbados began lobbying for migrant employment as early as 1947 (before the great migrations to the UK and later North America of the 1960s). Formal temporary migration schemes followed. Canada's Seasonal Agricultural Workers Program (SAWP) began taking workers from the Caribbean in 1966 to meet labour shortages on Ontario farms. During the nickel boom in New Caledonia in the 1970s leaders from Tonga, the Gilbert and Ellice Islands (Kiribati and Tuvalu) and Fiji all requested opportunities for their citizens to work there but all were rejected, with the French territory preferring to bring workers from the two other French Pacific territories, less because of language issues than because they would be 'docile labour and reliable voters' (quoted in Connell 1987a: 217). A contemporary echo of this was New Caledonia's decision to import as many as 4000 Filipino workers for the Goro nickel mine in 2005. In the 1980s Cape Verde sought new migration destinations in Canada and Australia following high unemployment in Europe (Connell 1988a). In 1977 the Fiji Minister of Labour argued that the provision of overseas employment was 'the greatest form of overseas aid any government can offer' (*Fiji Times*, 19 December 1977). At various times and in various contexts SISI have sought greater migration opportunities elsewhere, usually as unskilled and semi-skilled workers.

Several countries have specifically trained workers for overseas employment, notably in the Marine Training Schools of Tuvalu and Kiribati where men and some women are trained to work on the overseas cargo shipping line; their remittances are greater than the cost of the training, hence there are perceived net benefits (Borovnik 2006). Fiji likewise has a School of Maritime Studies, established in 1975, and around 600 overseas seafarers who contribute around $8.5 million a year in remittances (*Fiji Times*, 27 June 2011). Similar but smaller schools exist in Tonga, Samoa and the Marshall Islands. In 1989 Kiribati began a new programme for supplying crew to Japanese distant water shipping vessels, and by 2005 about 500 i-Kiribati were employed on Japanese, Korean and Taiwanese vessels (Barclay 2010). A multinational Australia-Pacific Technical College (APTC), serving most of the independent Pacific states, was established in 2007 through Australia's overseas development agency, AusAID, as 'a centre of training excellence … to assist Pacific island students to gain Australian skills and qualifications in a variety of trade areas', ranging from painting and decorating, to hospitality and automotive engineering. While ostensibly the programme upgrades local skills, an international qualification enables graduates to migrate more easily.

The 'Philippines model' of training an excess of health workers for delivery into the 'global care chain' provides possibilities for SIS, and has been taken up as 'managed migration' in the Caribbean, notably in St Vincent (Connell 2007c, Blouin and Debnath 2011, Lewis 2011). Such policies are based on the assumption that skilled workers would gain better paid, long-term jobs and remit at a higher level. Under the Kiribati Australia Nursing Initiative, funded by AusAID, i-Kiribati nurses are trained in Australia, principally to remain there and generate remittances rather than transfer superior skills to Kiribati. (An unexpected outcome was an increasing disdain by i-Kiribati men for the long period of training required to become a doctor at home, and a shift into nursing.) Such Kiribati policies are part of the broader aim of national education policy that, in the 2009 words of President Anote Tong, would make i-Kiribati more 'acceptable' migrants to enable them to 'migrate with dignity' rather than as 'environmental refugees'. The Speaker of the Tuvalu parliament has even stated that 'we should focus on rugby, not only for rewards and fame, but [since playing rugby] can also pave the way for permanent residency in New Zealand' (quoted in Shen and Binns 2012: 77). This shift to the deliberate export of skilled workers has potential ramifications for development.

Migration of Pacific islanders especially to work in military and security assignments in the Middle East, Afghanistan and most recently Syria represents a particular recent case of contract labour migration. Fiji has sent hundreds of soldiers for peacekeeping duties in Iraq since late 2004, and smaller numbers have gone from Tonga, Samoa and Vanuatu

(Connell and Negin 2012). Additionally over 2000 Fijians serve in the British Army, 60 of whom have been sent to Iraq as part of British Army deployments. The various American SISI, including Palau, American Samoa and the Virgin Islands, have similarly provided military. In 2005 there were 300 soldiers from American Samoa, 100 from the Northern Marianas and 60 from Guam (Hezel 2005). In 2011 around 200 Palauan men and women, out of a population of fewer than 20 000, were serving in the US military. In addition to military and peacekeepers, Pacific Islanders have moved to the Middle East to serve in security services, and in back-up positions in catering, construction and other areas, despite exploitation by recruiters and others (Alexeyeff 2008, Stillman 2011). One estimate had over 1000 Fijians working in Iraq and Kuwait as soldiers, security guards, drivers and labourers in the mid-2000s (Maclellan 2007). Individuals have been actively recruited by such organizations as the British-based Global Risk Strategies, which opened a branch in Fiji to recruit ex-military personnel for contracts around the world, drawn by salaries of $1700 a month – vastly more than was available at home – but occasionally at extreme social cost.

Temporary agricultural employment overseas has similarly been sought and sponsored by SIS governments. In the 1970s New Zealand developed a South Pacific Work Permit Scheme that provided temporary employment opportunities for islanders, mainly from Fiji and Samoa, since New Zealand was unable to obtain a workforce for unskilled work (Levick and Bedford 1988). That lapsed but in 2004, after the Fiji Minister of Labour, Kenneth Zinck, had finalized contracts for migrant workers to go to Kuwait, he commented that: 'We need to look for overseas companies to employ locals to combat poverty and unemployment. [We are] already holding talks about sending workers to either Australia or New Zealand to work in apple farms' (*Fiji Sun*, 20 December 2004). Three years later agreement was reached for a guestworker scheme for five Pacific SIS, initiating a new migration era.

The development of seasonal labour migration schemes in the Pacific closely followed the experience of Canada's SAWP, although that too had precedents in the migration of temporary labour from the eastern Caribbean to the Florida canefields beginning in the 1940s (Wood and McCoy 1985). The Pacific scheme even had an uneasy resonance with nineteenth century labour migration, especially from Vanuatu (then the New Hebrides) to work on sugar plantations in Australia: a process and era known as 'blackbirding' (Connell 2010b), that became even more remarkable when islanders returned in 2012 to the Queensland cane fields. Early SAWP migrants came from Jamaica (and also Mexico), but later from smaller Caribbean SIS. During the 1990s and 2000s numbers increased to about 18–20 000 per year, about half of whom were from island states, mainly the largest ones (World Bank 2006). Caribbean island SIS had relatively few places per capita. In 2007 New Zealand implemented a Recognized Seasonal Employer (RSE) scheme allowing a quota of 5000 guestworkers (which later grew to 8000) for the agricultural industry, for up to eight months in a year. The scheme provided workers with opportunities for repeated 'circular migration' through being able to return in subsequent years. Five countries – Kiribati, Samoa, Tonga, Tuvalu and Vanuatu – had priority, with Solomon Islands added in 2010. Continued pressure from island states resulted in Australia introducing a similar scheme in 2008, the Pacific Seasonal Worker Pilot Scheme (PSWPS), offering up to 2500 places for workers from Kiribati, PNG, Tonga and Vanuatu, with the workers guaranteed at least six months' employment. In 2011 Nauru, Samoa, Solomon Islands and Tuvalu also became potential participating countries.

Between 2007 and 2010 about 15 900 seasonal workers, mainly men, travelled from the Pacific to New Zealand under the RSE scheme. Most were from Vanuatu (over 7000), Samoa and Tonga. By mid-2011 numbers participating in the Australian scheme had just passed 500 with Tongans making up a substantial majority of these. Participating countries were overwhelmed by demand with thousands signing up for only hundreds of jobs. Households of guestworkers were usually poorer than non-participating households in terms of incomes and access to services (such as piped water and electricity). Most earned sums several times those available at home (Connell and Hammond 2009). Remittances reached high levels, usually much more than half the migrants' earnings, and were used for paying bills (electricity and telephone), home renovation (including access to water, solar panels and permanent floors) and their children's education, with lesser amounts sent for churches, medical bills and vehicle or land purchase.

Both schemes proved pro-poor, because of local selection procedures, and regionally equitable (Gibson et al. 2008, Blanco 2009, Gibson and McKenzie 2011a). Savings supported welfare gains through improved housing and education, even if domestic agricultural productivity may have declined slightly during the contract labour period. Benefits were less in smaller states such as Kiribati and Tuvalu where numbers were fewer, transport costs greater, workers found difficulty in adjusting to very different lands, climates and employment experiences, hence earnings were smaller (Bedford et al. 2010). The Caribbean and both Pacific schemes achieved 'quadruple wins' – for workers, companies and countries of origin and destination – but not on any significant scale, compared with SIS populations. Demand for places far exceeded supply.

The limitations of short-term labour migration are evident in the minimal structural changes, and the absolute dependence on the fluctuating needs of Canadian, Australian and New Zealand farmers. The prospects for larger numbers of agricultural places are small, though care-giving and tourism provide possibilities. Contemporary guestwork has made a valuable but limited contribution to island development.

The emergence of regional trading blocs and agreements in different parts of the world, loosely modelled on the experiences of NAFTA (the North American Free Trade Agreement) and the EU, is likely to result in greater regional migration. The international migration of skilled workers is linked to the General Agreement on Trade in Services (GATS), established in 1995 through the WTO to liberalize international service trade, including the movement of so-called 'natural persons'. A key component of GATS involves the harmonization and mutual recognition of professional qualifications across borders, which will increase migration opportunities. The establishment of the CSME at the end of 2008 has led to greater mobility within the Caribbean region, notably from poorer to richer states, indicating 'the danger that each country competes with the other with the risk of selling what they have at bargain basement prices' (Caribbean Commission on Health and Development 2006: 77). GATS objectives have yet to be implemented to involve most other SISI, although in the Pacific they have enabled the migration of nurses from Solomon Islands to Vanuatu, and outcomes may be similar to those in the Caribbean: more international migration with the main benefits accruing to more developed countries. A critic of such policies has remarked:

> A development policy which plans on exporting key segments of a country's skilled people, such as nurses, teachers and agricultural workers actually risks treating people as commodities for export. It does not take account of the investment in health and education of these citizens from birth to adulthood including costs of loss of leadership, entrepreneurship, technical and creative skills in the homes, communities and at the national regional levels. (Wiltshire 2010: 70)

Such wide-ranging 'externalities' reflect the complexities of migration, where migration both erodes and nourishes the fabric of society.

THE OUTWARD URGE

Migration has become more complex and more entrenched in island cultures and has shifted the demographic balance further overseas. Globalization has extended the number of destinations, brought longer migration chains and been negotiated by governments rather than simply being the outcome of household decisions. Contracted migration of guestworkers and of some skilled workers, such as soldiers and nurses, increasingly trained for migration, has demarcated the greater selectivity and increasing significance of migration. Migration has become a means of redistribution of human resources to maximize the benefits for extended families. Remittances are crucial, especially in smaller states and remote islands, and are a source of sustainability. Beyond that, they are part social insurance (for possible return), part welfare provision and part investment. More formalized transfers of remittances, enabling their more effective use as temporary capital, have been suggested, both globally and for Pacific island states (Brown et al. 1995, Jayaraman et al. 2010) but have been resisted both by migrants seeking to maximize the household utility of remittances (and distrustful of governments), and conscious that remittances are private money, and by governments disinclined to disturb remittance flows (a 'golden goose' with a huge nest egg). However the combination of mobile phones, ATM cards and social networking has recently quickly increased the value and utility of remittances.

Countries without particular legal migration opportunities have increasingly sought them. Kiribati and Tuvalu have drawn attention to the particular difficulties of sustaining populations in small, densely populated states, with few resources, while threatened by environmental change. Tuvalu especially has frequently sought migration and resettlement opportunities overseas (Connell 2003a: 94–5). Such requests were rejected in Australia, but New Zealand provided a migration quota of 75 people per year for both countries. Demand is generally increasing, hence Bedford has argued that 'the most contentious issue confronting Australia and New Zealand in the Pacific during the next half century will be how to cope with pressure for an emigration outlet from Melanesia' (2003: 37). The rise of illegal migration further emphasizes demand. Many illegal Fijian and various Polynesian overstayers are in Australia, New Zealand and the United States (Scott 2003), Cape Verdeans are in the United States and France (Carling 2002), Comorians in Mayotte and France, and Caribbean islanders in more developed Caribbean states, Canada and the United States. Numerous islanders have died in desperate migration attempts.

Despite its growing significance, resistance to migration is widespread. Rhetorically and practically, many people wish to remain where they are. Despite relocation to Rabi in phases after 1945, when phosphate mining ended in 1979, small numbers of Banabans returned to their 'home island', despite its isolation, poor communications and dismal development prospects. Bikinians regularly seek to return to a 'home' few have now experienced (see Chapter 5). 'Home' exerts powerful influences even in less desperate circumstances. As a Samoan taxi driver told me in 2011: 'I love my country. I have my taxi. My wife works for a bank. We have a house. That's enough.' Many have stayed or (as in this case) returned from overseas, and seek to play some part in national development. People stay and return for many reasons – a job is available, a business can be started, the climate, Christianity, family obligations, the absence of racism, peace, the security of land, status and, for some, an easier life.

Yet migration to larger places is pervasive and many fewer return than anticipate doing so. For over half a century, migration has provided real gains, above all from the reliable flow of remittances. State regulation – insofar as it exists – is more concerned with encouraging than preventing migration. The Barbados Minister of Health observed in 2005: 'Our people are our greatest resource, and we must allow them to sell their skills abroad, thereby assisting in earning valuable foreign exchange' (quoted in Lewis 2001: 91). Similarly a former Tongan Prime Minister has stated: 'Skilled workers are our best export. I'd rather people go overseas with some skills. At least they don't just sweep the floors' (Feleti Sevele, pers. comm., 2011). In Samoa remittance flows that have enabled better schools and churches have resulted from village leaders 'acting globally … only possible because of the size and wealth of the village's nodes, or sub-villages, created by migration' (Macpherson and Macpherson 2009: 91). The greater self-sufficiency that would follow a decline of migration and remittances would be difficult and painful; intimations of that followed the GFC. There may be inequalities, frustrations and unwelcome social change, but migration has provided opportunities and change would have proceeded in the same direction, but more slowly. Migration, at least in the Caribbean, may have 'sapped the nurturing of a collective sense of identity' (Thorndike 1985: 4), being bound up with transnational ties, but identity may also be nourished by migration. Remittances nurture households rather than nations and return migration is for individual and household goals. Collectively these transnational emotional and monetary ties nonetheless sustain the SISI. In a world where public sector solutions are against the tide of contemporary history, any residual will to discourage migration has disappeared. Indeed quite the converse; as the opening quotation from the Prime Minister of St Vincent suggests, migrants may be heroic figures.

International migration has deferred and mitigated, but not resolved, development issues. Migration opportunities necessarily depend on the policies of possible destination states, and are largely structured according to skills. The possibility of blocked migration in the future, a situation long present in public debate (Macpherson 1992b, Shankman 1993), emphasizes the potential problem of high rates of natural increase should emigration be curtailed. Migration is both a measure of uneven development, with perceptions that SISI are lagging, and a consequent search for and diversification of opportunities for personal, household and national growth. It is only exceptionally depopulation but rather the expansion of island realms, a broadening of opportunities and the creation of diasporas (Hau'ofa 1994). As SISI and metropolitan destinations change, so the nature of migration changes. The following chapter examines environmental change in SISI, and the symbiotic links between environmental and economic change, and ultimately how they influence island futures.

7. Environmental change

> Islands are candles, to be snuffed out one by one only for those island-living species which are territorially bound and unable either to adapt to, or resist or retire in the face of innovation. The concept of insularity would have had as little meaning to the Pacific islanders as to the Vikings: the sea was as much part of their living space as was the land. It is not insular environment that makes an island population vulnerable, or size, but the adaptive and mobility characteristics of the population concerned, and the real degree of isolation. Creatures that can move, retire into or recruit from other environments have far greater survival capacity than those who have no choice but to remain passively where they are.
>
> Brookfield, 1980: 25

Most SISI are tropical and lie in tectonically unstable areas prone to significant hazards such as cyclones (hurricanes/typhoons), volcanoes and earthquakes. Such natural disasters engender considerable economic losses. Since the 1970s not only has there has been an increase in the number of natural hazards but developing countries have borne the brunt of them, extending the gap between 'rich' and 'poor' within and between countries. Between 1970 and 2002 more than three-quarters of all natural disasters took place in the developing world, sometimes repeatedly striking the same countries. Dominica, for example, has experienced nine different hurricanes since 1984 (Strobl 2012). The catastrophic human and economic impacts of such hazards have been locally much more significant than 'evolutionary' changes exemplified in climate change, accelerated erosion or increasing salinity. This chapter examines environmental change and what this means for sustainable livelihoods. It covers both dramatic events such as cyclones (and less frequent volcanic eruptions) that disrupt, change and even end people's lives, and less tangible changes, such as climate change (usually conceived as global warming), and considers whether evolutionary changes will become more significant and devastating. Vulnerability of SISI to long-term environmental change, such as soil degradation, is barely discussed here, since data are scarce. Many of the environmental issues experienced in SISI are similar to those in low-lying tropical coastal environments, such as the

Ganges and Mekong deltas, despite the vast differences in scale and numbers affected.

Low islands are particularly at risk from such climate-related disasters as cyclones and droughts. The significance of a crude Insularity Index, where coastlines are related to land area, indicates that atolls (and atoll states) are most vulnerable to climate change (and to cyclones) having long sinuous coasts and small land areas (Nunn and Kumar 2006). Many atolls and all atoll states have high population densities and significant concentrations of vulnerable people. High islands are at greater risk from earthquakes and volcanic activity, though these are usually less problematic. Hazards can overwhelm substantial proportions of entire islands and SISI, places of retreat and safety are few, while governments (with limited resources and management capacity) are not easily able to manage major disasters. In the past 15 years at least 34 of the 46 tropical SIDS have experienced cyclones (Méheux et al. 2007) while many are located on active plate boundaries, and vulnerable to earthquakes and vulcanism. As many as 13 SIS have been included amongst the world's 25 most disaster prone countries.

In a number of SISI significant human-induced landscape changes and modifications have been long established, notably in the dramatic transformations beginning around 300 years ago that changed many Caribbean and Indian Ocean islands from subsistence-oriented ecological systems to what amounted to sugar plantations. In several such places, well documented for Rodrigues (Mauritius), and more qualitatively for Carriacou (Grenada), changes brought risks of social and/or ecological collapse at least as early as the late eighteenth century (Richardson 1975, Bunce et al. 2010). Collapse devastated the rather smaller Easter Island (Rapanui). Some coral reef degradation in the Caribbean had occurred even in the pre-Columbus era, just as had land degradation in southern Melanesia, where grassland replaced forests (Spriggs 1986, Jackson 1997) and various islands were only intermittently populated. Climate changes involving both droughts and cyclones, a dual possibility paralleled in contemporary climate change, significantly affected the nineteenth century resilience of Rodrigues island ecology and economy (Bunce et al. 2009). Data are scarce but such historic events offer some precedents for the impacts of contemporary and future change, and for resilience and adaptation.

Underpinning almost all discussions of environmental change in island realms are debates over climate change, hence it is appropriate to begin there, although contemporary climate change from global warming is yet to have a significant impact on SISI, despite claims to the contrary. Especially but not only in atoll environments, local people and others have pointed to various changes, such as erosion, flooding and contamination of agricultural systems by seawater, that have been attributed to global warming and SLR (e.g. Connell 2003a, Rudiak-Gould 2009, 2013, Kuruppu and Liverman 2011). None are necessarily the outcome of SLR but may be responses to local stresses of various kinds, from geophysical to human, including tectonic shifts, seawall construction, sand mining and deforestation (see below). Massive public discussion of climate change at international and local levels has contributed to SLR being the primary explanation for harmful, unusual or unprecedented environmental changes. Media reports frequently detect major problems following SLR, as in the Maldives where 'already 14 of 200 inhabited islands are gone' (Doherty 2012; cf. Connell 2003a, Farbotko 2005) and in Al Gore's documentary *An Inconvenient Truth* (2006) where Tuvaluans were said to have had 'to evacuate to New Zealand'. Sober but peripheral academic reports are not averse to similar statements, such as SLR having resulted in 'the evacuation of some islands in Tuvalu' (Armstrong and Read 2004: 196). None have any semblance of truth though they are scarcely atypical. Detaching science from politics, polemics, aspirations and needs has often proved difficult.

Nevertheless climate change and variability have influenced environments, livelihoods, perceptions and discourses, and will have a greater influence in the future. The chapter subsequently examines other environmental changes that have some relationship to climate, from floods to droughts, and what these mean for erosion, coral reef evolution and biodiversity. Not all environmental influences have necessarily changed in periodicity, but events like earthquakes, volcanic eruptions and tsunamis can be unusually disruptive. The chapter thus ends with these exceptional environmental influences, which can be both critical and dramatic.

CLIMATE CHANGE

Globally three large-scale climate related changes are simultaneously occurring: a rise in air and sea-surface temperatures (projected to increase by 2°C by 2050), changes in rainfall (with expectations of less frequent but more intense rainfall events, with potential impacts on flooding, and greater storminess), and consequent higher sea levels, but with significant regional variations (Lal et al. 2002, Mimura et al. 2007). Each of these varies regionally in accordance with particular, more local changes, for example during the ENSO. ENSO fluctuations have been the most important influence on climate variability in recent decades. Thus particularly high sea levels have been recorded during La Niña events, for example from July 2006 to June 2008 in most of the Pacific (Chowdhury et al. 2010). Such short-term climatic shifts must be differentiated from long-term changes. Climatic changes, at different scales, are unusually complex, while relatively few measures exist in many regions dominated by SISI to document the extent and impact of change, both directly and in the outcome of such changes for water availability, agricultural production, food security and health. However, rainfall patterns may be becoming more unpredictable, and a greater incidence of more intense cyclones, floods and droughts are accompanying outcomes. Climate variability will be problematic even in high rainfall areas, since plants and animals are adapted to particular rainfall regimes.

Climate change – global warming – is expected to result in diverse social, economic and especially environmental changes, including threats to natural habitats of fauna and flora (terrestrial and marine), loss of land, coastal erosion, increased frequency of storms and increased salinity of land. Warmer seas may inhibit patterns of ocean mixing and the supply of deepwater nutrients near the surface. Some marine species, such as turtles and those associated with coral reef ecosystems, may be at particular risk and changes in their habitats, migration and therefore distribution and numbers are likely to be good indicators of climate change (Hawkes et al. 2009). Terrestrial fauna, such as butterflies and frogs, are similar sentinel indicator species, but no studies currently exist on their changing distributions within SISI. There is some weak evidence that, at least in parts of Solomon Islands, the fruiting patterns of breadfruit have changed (Lal et al. 2009c), but much 'evidence' of climate-induced changes in the distribution and behaviour of flora and fauna is apocryphal rather than based on systematic analysis. It is difficult to attribute specific ecological and environmental changes to climatic change, however much of an 'aggravating factor' it appears. Disease and loss of habitat may be more important. Nonetheless environmental changes are currently becoming evident in some marine and coastal contexts. They are more likely to be perceived and recognized by those people who have been exposed to external discourses of climate change impacts, as in atoll states including Tuvalu, Kiribati and the Marshall Islands, and especially in the urban centres of those states (Rudiak-Gould 2010, 2011, Kuruppu and Liverman 2011, Farbotko and Lazrus 2012).

Actual measurements of climate change in most SISI are limited (and consequently 'evidence' of change is controversial and contested). Assumptions and projections are derived from regional and global estimates and measures, though these are often scarce in oceanic realms so that general circulation models lack local specificity. In the Pacific, at least, there has been a net increase in temperature since around 1800 when the Little Ice Age ended (Nunn 2007). The limited data from the region broadly agree with IPCC data on a global warming of 0.56–0.77°C between 1901 and 2001. Over the past 50 years in the Pacific warming trends have, however, varied between 0.08°C and 0.20°C per decade (Australian Bureau of Meteorology and CSIRO 2011). Similar trends are likely to have occurred in the Caribbean Sea and Indian and Atlantic Oceans, consistent with global warming.

Predictions of future climate change inevitably vary but suggest acceleration in the rate of global warming. It has been suggested that temperature changes in the South Pacific will increase by between 0.45°C and 0.82°C by the 2020s and by 0.8°C to 1.79°C by the 2050s (Barnett 2011). Temperature increases in island regions will be less than global increases since oceans will warm more slowly. The impact that warming will have on rainfall patterns is even less certain, though it is expected to be both more seasonal and more intense (Australian Bureau of Meteorology and CSIRO 2011, Barnett 2011). ENSO events may also become more frequent and intense, whereas cyclones may be less frequent but more intense. Neither temperature nor rainfall are expected to change substantially in the immediate future, though small and subtle changes may have significant results.

Oceanographic changes, notably warming seas, pose particular risks to coral ecosystems, which are highly sensitive to temperature changes. Atolls are consequently at particular risk, since reefs protect atoll islands from storm surges (and tsunamis). Warming seas are now occurring in most parts of the Pacific, for instance around Palau, providing growing evidence for climate change influencing subsurface ocean circulation (Williams and Grottoli 2010). Contemporary coral bleaching has been reported in a number of instances, where it has been primarily attributed to a rise in water temperature and extreme temperature anomalies, exacerbated by other kinds of environmental stress, such as increased turbidity, as in the Chagos Islands (BIOT) (Sheppard 1999), Kanton Atoll (Phoenix Islands, Kiribati) and Palmyra (Williams et al. 2010). Temperature fluctuations following ENSO events have been much more important than human influences in contributing to bleaching (Schleyer and Benayahu 2010, Williams et al. 2010), hence coral reefs, even in areas remote from direct human influence, are not isolated from the effects of global warming on sensitive environments (Alling et al. 2007). The ability of reefs to adapt to and cope with rising sea-surface temperatures, though widely debated, remains uncertain (see below), and the geographical extent of bleaching appears to be increasing. Further increases in water temperatures will result in more extensive bleaching, the consequent reduction in the resilience of coral atolls to other environmental changes, and the decline of the fisheries (and tourism) potential of reef environments. The effects of projected climate change (and further ENSO events) on reef environments may combine with and exacerbate human-induced changes from fishing, tourism (especially boat mooring, fishing and diving) and pollution to weaken reef resilience.

SEA-LEVEL RISE

The most anticipated, feared and controversial impact of climate change and global warming is SLR. It has also become the single most politicized environmental issue, with multiple claims from, and about, a range of locations, mainly atolls, where it is reported to be having a significant impact. In recent years, SLR is said to have posed particular problems for such atolls as the Carteret Islands and Takuu (Bougainville, PNG), where environmental pressures have long been an issue, but above all in Tuvalu, and also Kiribati and the Maldives. It has been widely reported in global media that the Carteret and Torres Islands have experienced recent SLR and devastating erosion, to the extent that resettlement is essential, and that the entire nation of Tuvalu is similarly threatened.

Carteret Islanders have been described as 'the first direct climate change refugees with islands inundated and damaged, gardens and water supplies destroyed by salt water intrusion and evacuation announced in 2005' (Roper 2009: 5). In the Carterets and neighbouring Takuu (Mortlock Islands) damage has been caused by particularly large 'king tides' (storm surges), associated with El Niño conditions and cyclones, probably exacerbated by movements of nearby plate boundaries, as in the Torres Islands (Vanuatu) (see below). The Carteret Islands have experienced food shortages of varying severity, at least since the 1960s, associated with steadily growing human and pig populations, and the increased use of coconuts for copra to generate income. Resettlement was considered at least as early as the 1960s (Mueller 1972, Connell 1990, 2012b). Attempts to find appropriate land on nearby Bougainville were thwarted by local landowners, and fewer than ten households have moved.

The harmful impacts of SLR have been most frequently reported from Funafuti, the much larger, highly urbanized main atoll of Tuvalu, where flooding of the interior of the main island (Fongafale) is widely cited as evidence of the contemporary impact of SLR (Connell 1999, 2003a).

Historical records show that the interior of this island was subject to flooding in the 1890s, though contemporary flooding has been exacerbated by the construction and sealing of an airstrip occupying as much as a quarter of the main island, sand mining, vegetation removal, depletion of the water lens (and thus compaction) accompanying rapid population growth, causeway construction and other human-induced changes (Yamano et al. 2007, Woodroffe 2008). Similarly in Tarawa (Kiribati), erosion and flooding have followed modification of shorelines by reclamation, housing, causeways, sand and gravel mining (Donner 2012). In Majuro (Marshall Islands) a 35-year record shows that erosion has been greatest in the most urbanized and most modified parts of the atoll (Ford 2012). Contemporary flooding may also be more frequent in association with storm surges. Increased storm frequency may have contributed to recent changes in the Carterets and elsewhere, and certainly to perceptions of change. On a number of similar atolls such as Woleai and Moch (FSM) and Rongelap (Marshall Islands), islanders are conscious that erosion was part of a 'normal process' – more of the same – and had long taken steps to modify and defend eroded areas and secure land elsewhere, at least as early as the 1970s (Bridges and McClatchey 2009, Lipset 2011, Pam and Henry 2012, Rubow 2013).

Recent IPCC estimates of SLR have suggested a global average of 3.1 mm/year between 1993–2009, following an increase from 1.7 mm/year between 1950–2009, with the eventual outcome of SLR of between 0.18 and 0.80 m by the end of the century (Meyssignac and Cazenave 2012, Zecca and Chiari 2012). The rate of SLR is likely to increase in any future scenario, but is necessarily uncertain partly because it is impossible to know how society will evolve in this century. Past variations in SLR are considerable, due to a range of geophysical, gravitational and oceanographic processes, while diverse measuring techniques and short-term records further complicate assessment. Considerable regional and local variations occur but relatively few measures (either in time and space) exist for most island realms. Most reliable data from the Indian and Pacific Oceans are over such short time periods that their relevance for long-term change is slight (Dunne et al. 2012). Contemporary records for the Chagos Archipelago suggest stability over the past 20 years, whereas in the central Pacific, SLRs of around the IPCC estimates are more evident (Becker et al. 2012, Dunne et al. 2012). SLR is especially pronounced in certain locations, such as the Western Pacific (Nicholls and Cazenave 2010). One study has suggested a rising trend for some Pacific island states of 3.1 mm/year, with Nauru, FSM, Fiji, Tonga and Kiribati exhibiting even more rapid rises, and thus higher than earlier IPCC predictions, but data exist for only a short time period and are also complicated by ENSO fluctuations, since during La Niña conditions SLR increases (Aung et al. 2011, 2009, Chowdhury et al. 2010, Hussein et al. 2010, Becker et al. 2012).

Rather more specifically, measures of SLR on Abaiang Atoll, Kiribati, using geomorphological techniques, suggest an average rate of SLR ranging from 0.1 to 0.2 mm/year between 1998 and 2008 (Flora et al. 2009) and on Tarawa of 1.4 mm/year between 1968 and 1998 (Biribo and Woodroffe 2010) although other measurements suggest over 4.3 mm/year between 2002 and 2009 (Rankey 2011). Funafuti has recorded a particularly high rate of 5.1 mm/year, about three times that of the global mean (Chowdhury et al. 2010, Becker et al. 2012, Meyssignac and Cazenave 2012), and significantly faster than earlier rates which, if accurate and if continued, would put much increased pressure on Tuvalu. Projections for other areas, such as the Maldives, are equally high (Mimura et al. 2007). While the Maldives have been said to have experienced recent sea-level falls because of increased evaporation and intensification of monsoonal activity in the central Indian Ocean, that has been challenged and rebutted (Morner et al. 2004, Church et al. 2006). In the Caribbean SLR was estimated to have risen on average by 2.8–5.0 mm/year during the 1990s, with a projected rise of between 10–50 cm by 2025, but distinct regional variations exist there too (Parry et al. 2007). Clearly there are both discrepancies, according to sources and analytical techniques, regional variations and considerable uncertainty, despite the overall upward trend.

In the western Pacific, SLR is complicated and compounded by tectonic shifts. The Torres Islands (Vanuatu), at the convergence of the Australian and Pacific plates, experience particularly strong vertical movements with long-term uplift rates of about 1 mm/year, while also being earthquake and cyclone prone. A single earthquake in Ranongga (Solomon Islands) lifted parts of the island by over 3 m (McAdoo et al. 2007). Between 1992 and 2010, SLR around the Torres Islands was nearly 8 mm/year, significantly above the global average, but an anomaly that was part of a decade-long and increasing trend in the western Pacific. SLR is complicated by movements during and between earthquakes. Between 1997 and 2009 the islands subsided by about 117 mm, one of the highest recorded subsidence rates in the world. Unprecedented flooding in the islands was probably primarily the result of subsidence after a 1997 earthquake while a 2009 earthquake reversed that subsidence. Thus, while SLR was relatively rapid in the western Pacific region, the principal causes of observed flooding were subsidence, related to plate movements, and earthquakes (Ballu et al. 2011, Siméoni and Ballu 2012). Similar explanations probably account for parallel changes in

Solomon Islands, notably in the Reef Islands, such as at Nupani atoll where subsidence resulted in half the agricultural land being lost to erosion in the 1940s and 1950s and islanders gradually resettling on to the nearby volcanic island of Tinakula (Davenport 1972, Birk 2012), and in Vanikoro (Becker et al. 2012, Traufetter 2012), but also in the Bougainville atolls, which are earthquake prone and close to the plate boundary, and even conceivably in Funafuti.

Climate change and SLR are expected to bring increased rates of erosion and shoreline regression especially on sand-barrier and lagoon coasts and on coral atolls and cays. Erosion has occurred throughout SISI but so far there is little evidence of it being consistently in excess of deposition and accretion in most broad coastal systems. However, where human impacts have been considerable erosion is more likely, especially where coastal vegetation is removed, hence it occurs particularly in more densely populated coastal and urban areas where it is also more visible, and thus more frequently reported (e.g. Solomon 1994, Pierce 1999). Deposition may occur elsewhere on the coast, as it more obviously does on atolls. Erosion of islands, either during powerful storms or following SLR, can lead to a reduction in the size (and water quality) of the freshwater aquifer of the island (Roy and Connell 1991), as on Majuro (Terry and Thaman 2008), or its complete disappearance from smaller islands. Even where islands demonstrate physical resilience they may be rendered uninhabitable by the diminution of available water resources (Terry and Chui 2012). Storms have both constructional and erosional impacts on atolls, depending on the sediment composition of the islands and the orientation of the storm (e.g. Bayliss-Smith 1988, Bries et al. 2004). Erosion from storm events or long-term changes is particularly evident where urbanization has occurred. Even for high islands, as in Samoa, most of the country's important physical and social infrastructure is on the coast, and as much as 80 per cent of the 400 km coastline is regarded as 'sensitive' or 'highly sensitive' to erosion, flooding or landslip (Daly et al. 2010), in a context where cyclones are not unusual. Such 'sensitivity' is even more pronounced on low islands.

The impact of SLR is certain to be greatest for low-lying atoll islands, which have been widely perceived to erode in response, despite accretion usually occurring simultaneously (Woodroffe 2008). Quantitative studies of physical changes in 27 atoll islands in three central Pacific states (FSM, Kiribati and Tuvalu) over a 19–61-year period (when SLR was measured at 2 mm/year) show that 86 per cent of atoll islands either remained the same size (43 per cent) or increased in area (43 per cent) over that time period. Only 14 per cent of the islands exhibited a net reduction in area. Many islands both increased in area, and underwent diverse physical adjustments in response to changing influences, of which SLR was just one (Webb and Kench 2010). Similarly Nupani and most other Reef Islands (Solomon Islands) were smaller in 1960 than in 2010 (Birk 2012). On Raine Island, a vegetated coral cay in the northern Great Barrier Reef, 34 per cent of the coastline experienced net retreat and 44 per cent net progradation between 1967 and 2007. Despite concerns that Raine Island was rapidly eroding, net island growth (6 per cent by area and 4 per cent by volume) had occurred. Perceptions of erosion probably reflect large morphological changes arising from seasonal, inter-annual and inter-decadal patterns of sediment redistribution rather than any net sediment loss (Dawson and Smithers 2010). On Tarawa (Kiribati) between 1968 and 1998, when the island experienced what appears to have been a net SLR of 1.4 mm/year, shoreline responses varied. Most ocean-facing beaches showed little detectable change, with some showing net erosion and a few accretion; by contrast, a significant proportion of lagoon beaches showed accretion. The factors that most influenced change were location and human influence, while substantial fluctuations also occurred in response to El Niño conditions (Biribo and Woodroffe 2011). The same conjuncture of both accretion and erosion was evident on Maiana and Aranuka Atolls (Kiribati) between 2005 and 2009; the greatest erosion was again on ocean-facing coasts (where no-one lived) or on those parts of lagoonal coast around villages that had been most influenced by human activity, notably by the construction of groynes and seawalls, and also causeways and boat channels (Rankey 2011, Woodroffe and Biribo 2011). In each of these cases, and in Majuro (Xue 2001, Ford 2012), erosion was greatest where human activity was also greatest (construction of seawalls, vegetation removal, sand mining, etc.). This also explains the widespread assumptions of generalized erosion following SLR that come predominantly from capital cities and urbanized atolls. Over an equally short time period a similar situation appears true in the Maldives (Kench and Brander 2006). Moreover various islands in the Maldives have both proved morphologically resilient in contemporary times, but also provide a geological analogue whereby they withstood an SLR of between 0.5–1 m above present sea levels between 4000–2000 BP and can thus be expected to survive predicted future climate change and SLR (Kench et al. 2005, Kench 2012). In various regions coral reef islands have exhibited much more resilience and stability under ongoing environmental changes than usually assumed. Erosion of particular island shorelines takes place in the context of physical adjustments to entire island shorelines, as erosion is balanced by deposition and progradation elsewhere. However, it remains a moot point whether the biological content of the eroded material (notably its useful soil content) is replaced by deposition of material of comparable value.

Mangrove coasts are likely to be significantly affected by SLR at the predicted rates. In Bermuda, where SLR averaged 2.8 mm/year from 1932 to 1992, mangrove areas were substantially depleted and the rate of loss is likely to increase if the rate of SLR increases (Ellison 1993; cf. Alongi 2008). Studies of earlier periods of mangrove depletion in a range of SISI, including Grand Cayman, Kosrae (FSM) and Tongatapu (Tonga), suggest that mangrove forests may only be able to cope with an SLR of rather less than 0.1 mm per year (Ellison and Stoddart 1991). As mangroves disappear so coastal protection and valuable habitats also disappear.

Global warming poses a future threat to islands, and especially to more than a hundred populated coral atolls, should SLR occur at the predicted rates (and storminess increase), especially in the western Pacific where rates are already high. However, every detailed study of climate change in ocean realms emphasizes uncertainty, even over contemporary processes, and different analytical techniques produce quite different outcomes. For Chagos at least, where SLR was relatively slight, one conclusion was that 'these low-lying coral islands should continue to be able to support human habitation as they have done for much of the last 200 years' (Dunne et al. 2012: 25). Similar conclusions over long-term viability and resilience may be more generally valid. However, current SLR is a fraction of the sea-level changes that occur as short-term impacts (king tides and storm surges) from ENSO-related fluctuations. In Tuvalu, fears of change are such that more than a decade ago emigration had already occurred in anticipation of new difficulties to follow (Connell 2003a) and some present residents have prepared for an 'out-of-country' option by having particular family members migrate and create a base abroad (Paton and Fairbairn-Dunlop 2010), a typical but unusual example of the 'transnational corporation of kin' (Chapter 6). SISI are in no position financially, geographically or politically to defend themselves against such potential threats, but if – perhaps when – the worst does occur then islanders might eventually become environmental refugees.

CORAL REEFS

While atolls are seemingly coping with contemporary climate changes, the present acceleration of SLR might transform that situation, suggesting possible parallels, at least in the Pacific, with earlier centuries when atolls and reefs failed to keep up with SLR (Nunn 2007). Islands are less vulnerable than fringing coral reefs to changes following SLR, including enhanced sedimentation and turbidity (Field et al. 2011). The role of reefs in mitigating erosion is particularly important. However, even between coral reef islands, there is considerable diversity, according to their composition, location and response to ecological disturbances (Perry et al. 2011), so that predictions of future change are unusually difficult.

Almost all SISI have associated coral reefs, hence the future of reef systems is critical for island livelihoods. Coral reefs are particularly vulnerable environments, being low-lying and largely constructed from unconsolidated sediments, and have been described as 'one of the Earth's first ecosystems to directly suffer large-scale mortality from climate change' evident in the bleaching and mortality of various coral species (Ateweberhan and McClanahan 2010: 964). Organisms that form reefs are mainly found in tropical waters, but also exist in sub-tropical areas, such as Bermuda and the Great Barrier Reef. Coral species are particularly diverse in the Indo-Pacific region, rather than in the Atlantic and Caribbean. Coral reefs are valuable ecosystems for most SISI but face an intensifying range of threats, notably overfishing, coastal development and various kinds of runoff. But, despite 60 per cent of the world's reefs being under immediate and direct threat, with the percentage of threatened reefs increasing by 30 per cent between 1997 and 2007, 'information regarding which threats affect which reefs is limited, hampering conservation efforts'. The greatest pressure on reefs since 1998 has come from an 80 per cent increase in the threat from overfishing, most significantly in the Pacific and Indian Ocean regions (Burke et al. 2011: 1, 4) and from pollution.

During the 2000s the rate of increase in the atmospheric concentration of carbon dioxide (CO_2) has exceeded the rate for the previous half century. Increased atmospheric CO_2 contributes to ocean acidification. In turn the combination of increased ocean temperatures and acidification may reduce the calcification capacities of coral reefs, forams and other sand-creating organisms, which would have adverse effects on reef building and beach recovery (Schaeffer and Hare 2010). Over the past two centuries ocean acidity has increased by over 28 per cent, and that rate is probably increasing because of greater volumes of atmospheric CO_2. Ocean warming, especially increased sea surface temperatures, notably from ENSO events, have been secondary threats to coral reefs, but most reef species are vulnerable to temperature changes. The structural integrity of reef ecosystems, and a loss of habitat, so that bleaching may occur at lower temperatures, are probable consequences, while marine ecosystems are likely to be affected 'from the bottom up'

(Miles 2009). Coral reefs that are exposed for lengthy periods, or experience episodes of unusually high temperature, also become bleached. Cyclones can devastate corals and discourage regeneration, but in the Caribbean at least they have proved capable of fast recovery so that the main stressors on regional reefs are anthropogenic (Mora 2008). Isolated reefs in semi-tropical areas, with limited genetic variation such as those of Bermuda, are particularly at risk from environmental disturbances (Brown 2008). As ecosystems, coral reefs are under stress through direct and indirect anthropogenic impacts.

Local factors, such as pollution and sedimentation, and global changes, such as ocean acidification and temperature increases, have combined to put more pressure on fragile ecosystems. Coral reefs are undergoing unprecedented declines due to overfishing, disease and eutrophication from coastal developments, especially in the Caribbean. Loss of key species, outbreaks of disease, bleaching, storm damage, temperature stress, habitat destruction and ultra-violet light, have all increased, with harmful effects, except perhaps in some of the most remote reef systems of the cooler eastern Pacific. Sedimentation resulting from poor land-use practices is a widespread threat, especially in the Pacific, because of land-use pressures, steep topography, highly erodible volcanic soils and high rainfall, and in the Caribbean, where sewage outfalls and fertilizers are more problematic. Consequently, coral sites close to river mouths, where tidal fluctuations are weak, and close to urban areas with high wastewater nutrient loadings, have been more severely degraded (Lapointe et al. 2010, Golbuu et al. 2011), but significant and increasing chemical contamination exists even in association with such small-scale activities as boat building and painting and road runoff (Jones 2011b). The coastal squeeze has intensified pressures. Caribbean reefs have decreased by 80 per cent in 30 years as a result of pollution, sedimentation, disease and overfishing (Gardner et al. 2003, Mora 2008). Even relatively pristine reef systems, as in the Turks and Caicos Islands, are threatened by development plans involving increased fishing and tourism (Dikou et al. 2009). Other Caribbean sites that are protected as relatively pristine, such as Bonaire, have experienced a system-wide decrease in corals and an increase in macroalgae (seaweed), as a result of coral bleaching, disease and storm damage (Bries et al. 2004, Stokes et al. 2010). In Bonaire particular stresses on reef systems have followed simultaneous population growth and increased tourist numbers, with associated construction activity, causing local eutrophication, sedimentation and direct physical damage to corals. Similar situations recur in many Caribbean contexts and, to a lesser extent, in the Indian and Pacific Oceans.

On the west coast of Madagascar population growth over a 40-year period was accompanied by heavy overfishing, severe reef degradation, the near total loss of some key species and consequential depletion of many reef fish. Although a mean sea surface temperature rise of about 1°C occurred over this period, and probably contributed to that decline, the Grand Récif was in much worse condition than more remote reefs. Population growth and lack of resource management or pollution controls had resulted in massively increased discharges of sewage, sediments and other pollutants, smothering and poisoning corals (Harris et al. 2010). In Majuro, lagoon pollution peaked alongside the urban area where causeway construction reduced water circulation patterns, both discouraging coral formation (Paulay and Benayahu 1999). Concerns over overfishing on coral reef ecosystems have increased in recent decades, and Pacific island states are at risk of the complete depletion of some vulnerable subsistence and commercial marine species (Hunt 2003). In Vanuatu reef mollusc populations are close to collapse in many locations (Dumas et al. 2010), and more colourful reef fish have been devastated by the aquarium trade (infamously stimulated, especially in Vanuatu, by the 2003 film *Finding Nemo*). On a smaller scale, comparable changes are evident for the township of Gizo (Solomon Islands) where migration posed a threat to adjacent coral reef fisheries because of intensified fishing (as migrants were not bound by customary obligations) which in turn further eroded customary management systems. The concentration of parrotfish, of particular significance since they are critical to the resilience of coral reefs, was greatest in customary management areas closed to fishing, than in areas open to fishing and closer to Gizo. Weaker management resulted in the rapid decline of vulnerable fisheries (Aswani and Sabetian 2009; Chapter 3).

Various other disturbances influence the structure and dynamics of shallow marine environments and coral reef ecosystems, including severe tropical storms, freshwater plumes and infestations of invertebrate corallivores. Cyclones (and the overfishing of herbivores) are particularly detrimental, since coral reef systems are more likely to undergo shifts to macroalgae (seaweed) dominance in the event of coral mass mortality (Lokrantz et al. 2010). Cyclones, even with a restricted direct impact, are significant enough to worsen coral recruitment patterns in the Southern Caribbean (Mallela and Crabbe 2009) and probably elsewhere. Increases in nutrients and decreases in herbivory (the consumption of flora without killing them) are primary drivers of changes in community structure and ecosystem function on reefs, and altering these influences can lead to increased macroalgal growth which restricts the survival, growth and recruitment of corals as well as increasing the prevalence of coral diseases. The incidence and severity of episodic disturbances have increased in recent years, as climate related disturbances compound numerous pre-existing natural and anthropogenic disturbances (Hughes et al. 2003). Increased frequency, severity and diversity of disturbances result in declines in abundance of reef-building corals, which may change the taxonomic composition of coral communities. Coral depletion and changes in coral assemblages affect reef-associated fishes and invertebrate communities (Pratchett et al. 2011), which may undermine biodiversity, ecosystem function, fisheries production, and other critical goods and services provided by coral reef ecosystems. Local alterations to consumer pressure and primary productivity may act synergistically to reduce reef resilience and increase the probability that global scale forces such as climate change and ocean acidification will drive reefs towards algal domination.

The rising temperature of global oceans threatens coral reef ecosystems, with consequent coral bleaching the principal threat, and threatens nearby beaches through a reduced supply of coral sediment, potentially accelerating erosion. Bleaching follows stress from multiple factors, and if prolonged or severe may result in partial or complete mortality. After the 1998 ENSO, which caused the strongest oceanic warming in modern history, coral bleaching and mortality became global phenomena though most severe in the western Indian Ocean region (Ateweberhan and McClanahan 2010, Cinner et al. 2012). Likewise the greater integration of small-scale societies into complex global systems has required them to manage resilience at multiple spatial and temporal scales (Lauer et al. 2013). In 2005 unusually high ocean temperatures in the tropical Atlantic and Caribbean resulted in the most severe bleaching recorded in the Caribbean basin, especially in the eastern Antilles, with over 80 per cent of corals bleached and 40 per cent dying at many sites, such as around the USVI. In 2010 similar bleaching events again occurred in the Caribbean with indications of this being a cyclical process (Miller et al. 2009, Rogers 2009, Eakin et al. 2010). By contrast, the outcome of similarly much increased sea temperatures (associated with a moderate El Niño event) and mass bleaching in the Gilbert Island chain (Kiribati) was less devastating, though the floristic composition of the reefs changed. Coral cover revived at several sites, especially where there was less human disturbance and wave activity (Donner et al. 2010). A series of disturbances to coral reefs on the north coast of Moorea (French Polynesia) over the last 30 years, has also produced frequent severe episodes of coral bleaching every 2–5 years since 1991, corresponding to positive temperature anomalies, but the impact on long-term resilience or productivity was not then clear (Pratchett et al. 2011).

Exposure to excessively high water temperatures for six months may kill a coral community and cause the excessive loss of coral species. However, the recovery of coral reefs after ENSO bleaching events, such as that of 1998, further suggests that coral systems have considerable resilience when faced with climatic change (Schleyer and Benayahu 2010); albeit where temperature change is later reversed, reefs are remote from other disturbances and no further ENSO events occur. This may be because degradation (for whatever reason) increases the abundance of disturbance-tolerant species and thus the ability of coral reefs to resist climatic changes (Côté and Darling 2010).

Recovery rates vary according to geographical location, with reefs in the Indian Ocean and western Pacific being quicker to recover than more isolated, less complex reefs in the eastern Pacific, but recovery rates are similar for broadly 'physical' and 'biological' disturbances (Graham et al. 2011). Nonetheless thermal stress and even quite limited global warming pose threats to all coral assemblages, with coral bleaching anticipated to be a greater future threat. Episodic disturbances, such as those associated with ongoing climate change, are expected to become more frequent and more severe, hence changes in the community structure of coral assemblages will become more apparent. While climate change poses an important and significant threat to the future structure and dynamics of coral reef communities, outbreaks of local invasive species, such as *Acanthaster planci*, remain causes of coral loss, as on Moorea (Pratchett et al. 2011), but relatively little is known about the role of marine invasives, including algae and corallivores, on coral reefs. Both local and global influences continue to put stress on coral reef systems.

While global warming may extend the geographical area in which coral reef development is possible (though there are few potential areas adjoining current coral regions), it is more likely to bleach and kill off corals in shallow waters where temperature increases are most likely. Reefs will be less productive (and human pressures on them may be intensified) thus the range and quality of foods that they produce will fall with negative impacts on the food security and nutrition of coastal dwellers who have routinely depended upon them (Turner et al. 2007). Mass mortality of coral reefs may lead to the loss of sand production for beaches and subsequent coastal erosion, as appears to have occurred in part of the Chagos Archipelago (Brown 2008). Such circumstances, especially on atolls, potentially threaten local livelihoods and the tourism industry.

Fishers have always had to adapt to the vagaries of weather and climate, but current rates of change are unprecedented in recent history and are accompanied by complex socio-economic changes. Overexploitation and human degradation of marine habitats have threatened fisheries and reduced their capacity to adapt to biophysical change, which will increase the degree of uncertainty in terms of temporal and spatial variations in fish populations, habitat viability and stability (Robinson et al. 2012). The mobility of oceanic fish makes management unusually difficult. Many, but not all, of the impacts of climate change are likely to be negative, emanating from sea temperature changes, lower ocean pH levels, changes in rainfall that will affect estuarine fisheries (and the spawning grounds of juvenile fish) and in ocean circulation, which will alter habitats and change the distribution and mix of species (Brander 2007, Badjeck et al. 2010, Grafton 2010). Climate change will influence activities already characterized by mismanagement and conflict, to reveal the fragility or resilience of particular management systems. Climate variability and change can also alter the local institutions that form the basis of resource management, specifically property rights, potentially leading to conflict over rights and resource access (Badjeck et al. 2010). While threats may outweigh the benefits, even short-term predictions of the relationship between diverse environmental changes and their impact on human livelihoods are largely impossible.

CYCLONES

Cyclones are frequent in most regions dominated by SISI, other than those extremely close to the equator, although cyclonic activity has become more common there (Rasmussen et al. 2009). Time-series data are too limited to give more than localized indications of the extent of any increased periodicity and severity, while regional variations are considerable and also fluctuating (Mimura et al. 2007). In the Pacific region no significant trends in either the number of cyclones or their intensity were evident from 1981 to 2007 (Australian Bureau of Meteorology and CSIRO 2011), although there is fragmentary evidence that cyclones may be becoming more intense (Webster et al. 2005, Elsner et al. 2008, Rasmussen et al. 2009, Walsh et al. 2012), probably because of the increased frequency of ENSO events. Cyclone intensity is likely to increase in the future, by 2–11 per cent by 2100, but the frequency of cyclones may fall by 6–34 per cent over the same period: fewer but more damaging cyclones (Knutson et al. 2010). The impact of changing patterns and intensity is thus uncertain, at least and probably particularly in Pacific islands; 'owing to their small land masses and widely dispersed nature within a vast expanse of ocean, whether or not individual islands sustain significant damage by the passage of tropical cyclones would seem to be governed largely by chance' (Terry and Etienne 2010: 193). Indeed it is likely that

> the South Pacific cyclone basin and the islands therein will … remain most vulnerable to episodes of anomalous cyclonic behaviour … [and] continue to experience strong but irregular interannual fluctuations in cyclone and track characteristics [which] will remain a much more dominant feature of temporal patterns than possible evolving changes in long-term average cyclone activity resulting from climate change. (Terry and Gienko 2010: 232, 223).

That may well be true elsewhere. While the extent of future cyclone activity in SISI is uncertain, greater frequency and greater intensity seem probable.

Cyclone frequency is highly variable. Tuvalu was hit by an average of three cyclones per decade between the 1940s and 1970s, but eight occurred in the 1980s (Nunn 1990) and several have occurred since then, but with varied intensity in different parts of the country. In the Cook Islands cyclone events have become more frequent and more intense since the 1970s (de Scally 2008), with five cyclones in five weeks in 2005. By contrast, the number of cyclones in the Cayman Islands peaked in the 1930s but the islands still receive about four cyclones a decade (Novelo-Casanova and Suarez 2010: 450). French Polynesia has had as many as five cyclones in a year (Dupon 1984). In 2010 an unusually large number of severe tropical cyclones struck the eastern Pacific; a concentration that was atypical and may reflect El Niño conditions that pushed cyclogenesis activity east of its usual locations (Terry and Etienne 2010). Cyclone paths, periodicities and strengths are thus fluid, though they fluctuate around established zones, and there is considerable debate on the possible effects of global warming, increased sea surface temperatures and elevated CO_2 concentrations on their incidence and strength (Terry 2007). Should cyclones become more frequent in areas previously largely free of them, the impacts are likely to be worse because of limited familiarity and preparedness.

Numerous accounts trace the devastating physical and human consequences of cyclones, through immediate physical damage, erosion and deposition, subsequent flooding (and occasionally disease), and the broad social and economic consequences for island communities and countries (Bayliss-Smith 1988, Harmelin-Vivien 1994, Spennemann 1996). The potential for greater damage and slower recovery has increased as monoculture has become more common, cassava has replaced more wind-resistant and 'famine' crops, coastal populations have grown and vegetation removed, infrastructure (such as power stations, hospitals and wharfs) constructed, population growth extended from leeward sites to windward sites (notably on atolls) and informal housing constructed (Dupon 1984, Paulson 1993, Spennemann 1996). Some cyclones cause damage across the whole country, notably in the Caribbean, as in the case of Hurricane Ivan in Grenada in 2004, one of four significant cyclones in that year, where 39 people were killed, some 85 per cent of the island was devastated and the two bases of the economy almost totally destroyed – with 80 per cent of nutmeg trees gone and 90 per cent of tourist facilities destroyed or damaged. Damage was estimated at $815 million (Nurse and Moore 2005). Looting followed and projections of an annual economic growth rate of 5 per cent were slashed to 1 per cent. The cyclone continued over Jamaica, where 17 people were killed, and Grand Cayman (Cayman Islands), where despite strict building codes (not always present in SISI, and vividly illustrated in the 2010 Haiti earthquake), a quarter or more of the buildings on the island were reported to have become uninhabitable, with 85 per cent damaged to some extent. Much of Grand Cayman remained without power, water or effective sewerage several months later, despite its being one of the most affluent UKOTS. Even after five months, barely half the pre-Ivan hotel rooms were usable. Three-quarters of the total economic impact was in damage to houses and other physical infrastructure (Brown 2008). The human impact of Ivan was, however, slight compared with 1903 and 1932 cyclones in the Cayman Islands (Novelo-Casanova and Suarez 2010). Even in relatively large and fragmented states damage can be enormous, depending on the power and path of the cyclone; in Tonga Cyclone Isaac in 1982 destroyed almost a quarter of all housing and made half the population homeless (Lewis 1995). In 1989 Hurricane Hugo destroyed 90 per cent of Montserrat's housing (Harden and Pulsipher 1992). Localized damage is enough to cause substantial problems. After two cyclones in 1985 and 1986 many families on the island of Tongoa (Vanuatu), which primarily depended on copra for their economic livelihood, had no income for two years since their coconut palms produced nothing (Warrick, cited in Lal et al. 2009c). Banana farmers in the Caribbean have frequently been without incomes for nine months after cyclones (Henry-Lee et al. 2010). Niue fishermen were unable to fish for some weeks after Cyclone Heta since boat-launching ramps had been destroyed, while the overall damage was estimated at 63 per cent of GDP (Barnett 2011, Walsh et al. 2012). Such accounts of cyclone damage indicate considerable human suffering and monetary loss.

The impact of cyclones on ecosystems may also be considerable. Boulders may be relocated and reefs damaged, usually more severe for coral atolls than high islands (Bayliss-Smith 1988, Scheffers and Scheffers 2006). Tropical storms and nearby cyclones affect forest structure, through defoliation and branch falls, and in cyclone-prone regions forests exist in a situation of ‘chronic disturbance’. Increases in cyclone frequency would increase the intensity of disturbances and shift forest composition towards ‘disturbance-specialist’ species (Webb et al. 2011). However, many insular floras, as in the Bahamas and Guam, have become highly resistant to cyclones and can regenerate quickly (Morrison and Spiller 2008, Hirsh and Marier 2002). Fauna dependent on such forests decline when their habitats are lost. In American Samoa the fruit bat population fell by 80–90 per cent after cyclones, and subsequent over-hunting (Craig et al. 1994). Cyclones have regularly affected coral reef ecosystems, which have become adapted to them. In their pristine state, coral reefs have the resilience to overcome and recover from even substantial disturbances. However, more human-induced disturbances on reef ecosystems have affected both reef habitats and their associated coral assemblages, and this general degradation may jeopardize the resilience of reefs to cyclones (McClanahan et al. 2002), especially with any greater cyclone frequency. Direct mechanical disturbances can affect live coral cover, while indirect post-cyclone disturbances follow increased terrestrial runoff and fresh water impacts (Connell 1997). Cyclone Erica in 2003, an abnormally strong cyclone for New Caledonia, had significant effects on coral reef habitats on the north-western coast, disturbing reef fish assemblages, with degrading effects on local reef ecosystems extending for at least four years. Nonetheless, under moderate anthropogenic pressure, coral reef habitats were resilient to even abnormally strong natural disturbances; they avoided major benthic shifts (e.g. algal proliferation) and exhibited a relatively quick recovery (Guillemot et al. 2010). Indeed the majority of studies of cyclone impacts on atolls have emphasized resilience and the ability to ‘bounce back’ reasonably quickly.

Some isolated islands such as Tikopia retain traditional storage methods and have been able to survive significant cyclones, without external intervention. The powerful Cyclone Zoe in 2002 devastated vegetation and human settlements, but no deaths occurred, as islanders sheltered in caves on higher ground (Rasmussen et al. 2009). A broadly similar situation occurred in Cikobia (Fiji) during and after Cyclone Tomas in 2010. By contrast, less remote islands, such as Woleai (FSM) and Bellona (Solomon Islands), have often given up time-consuming practices, such as preserving food, in favour of waiting for government support, and so become more at risk to the immediate aftermath of cyclones. So too have islands like Futuna (Wallis and Futuna) and Tubuai

(French Polynesia), both of which experienced devastating cyclones early in 2010, which destroyed the majority of island houses and crops, resulting in long waits for relief supplies (as central islands had also been affected). Occasionally cyclones have been a catalyst for migration (Chapter 5). As early as the 1950s cyclones in Paama (Vanuatu) resulted in families permanently leaving the island for urban centres. The entire population of Alamagan (Northern Marianas), albeit just 14 people, were resettled in 2009 after Cyclone Choi-wan went directly overhead and destroyed housing, vegetation and agriculture. Construction of more permanent housing and longer-term forecasting has enabled most SISI to withstand cyclones better, but has emphasized the reduced resilience of coastal populations living in vulnerable housing.

FLOODS

Floods often occur in association with cyclones. As cyclones become potentially stronger, and populations increasingly concentrated in low-lying coastal and riverine areas, their impact is greater. Most floods are associated with excess rainfall, in association with cyclones, but flooding can be exacerbated by coastal water intrusion through storm surges and high tides. Typically, in areas such as the Rewa delta (Fiji), floods have become more frequent; between 1970 and 1998, 15 tropical cyclones caused floods which intensified in both frequency and magnitude, at the same time as the affected population grew, upper watershed deforestation and monocultural agriculture have increased and river sedimentation has exacerbated flood problems (Kostaschuk et al. 2001, Lata and Nunn 2011). Rates of sediment deposition in tropical Pacific islands are some of the highest in the world, because of the high frequency of cyclones and associated extreme rainfalls (Terry 2007), contributing to the severe degradation of coastal lagoon ecosystems (Dumas et al. 2010). In many parts of Fiji, where most towns are in low-lying areas, and Samoa, where the capital, Apia, is built on the floodplain of four rivers, the lack of urban planning has resulted in many businesses and residential properties being located in flood-prone areas (see below), hence vulnerability to flooding is increasing. In 1931 the combination of cyclone and flooding resulted in the death of more than 225 people in Fiji, mainly in recently settled floodplain areas in north-west Viti Levu. One consequence was resettlement away from the floodplain, but growing social and economic pressures subsequently resulted in renewed floodplain settlement, exposing new generations to unfamiliar risks (Yeo and Blong 2010), that duly occurred in 2009 and 2012. In the 2012 floods the worst affected were displaced sugar cane farmers who had moved to the low-lying outskirts of the town of Nadi in search of employment. Here, as elsewhere, well documented for Puerto Rico (Lopez-Marrero 2010), the relatively poor were more likely to have to live in the most flood-prone areas and be unable to afford more flood-resistant homes.

Rainfall patterns have changed in SISI. In the outlying Mauritian islands of Rodrigues and Agalega there has been an overall decline in annual rainfall over the past two decades but combined with an increased number of heavy rainfall events, which have resulted in flash floods becoming more common (Gemenne and Magnan 2010: 22–23). Where deforestation has occurred, as it often has near coastal and urban areas, flash flooding is more intense and the impacts greater. A 2001 flood event, in itself not unusual, inundated large parts of the Apia urban area, within a ‘negligible time’ from the start of the intense rainfall, resulting in more than $5 million damage (Penaia et al. 2007). Hard paving results in more rapid runoff and more intense flooding. Fiji is particularly prone to floods, which have accounted for almost a third of disasters since 1970, and has a high-density lowland population. The small town of Navua is flooded about once every seven years and Nadi, one of the larger towns, and a major transport centre, is also regularly flooded. Damage from 2012 floods in Nadi was estimated at around $40 million – about 1.5 per cent of GDP (*Fiji Times*, 12 April 2012). The incidence of dengue, leptospirosis and typhoid increased.

Flooding has social and economic costs that are often prolonged through damage to infrastructure. In May 2011 the centre of the island of Tanna (Vanuatu) was inundated by two weeks of torrential rain; practically all taro plants, cassava, sweet potato and kava, the biggest income source for the population of the Middle Bush area, were totally destroyed by flooding that washed away plants, trees and roads; the last of these affecting the ability to respond. Similar tropical downpours occur on other islands. The 2009 Fijian floods, which followed a series of depressions at a time of high tides, affected over 150 000 people and cost about $13 million to the country’s sugar belt and adjoining urban areas (Ambroz 2009b, Holland 2009, Lal et al. 2009a). Almost 50 per cent of the flood-affected farm households were expected to subsequently fall below the poverty line. The floods further exacerbated the sugar industry’s unprofitability, as sugar cane suffered a decline in sugar content or was drowned. Roads and seawalls were damaged, with the sea claiming parts of what were reclaimed wetland areas, while landslides washed away farmland in hilly areas. Salt inundated farms were not expected to produce crops for at least two years (Lal, Rira et al. 2009a, Lal 2010). The need for immediate relief efforts distorted international aid flows.

Both the 2009 floods in Fiji and Cyclone Ami in 2003, which badly affected sugar growing areas on Vanua Levu, with the loss of $35 million of sugar (Lal 2003), resulted in some migration into urban areas, especially Suva, since cane growing was no longer productive, government assistance was minimal, sugar cane prices were falling and long-term land tenure and eviction issues were increasing issues (Lal 2003, Lal et al. 2009a). The most affected were poorer farmers with smaller landholdings and houses on lower ground (Lal et al. 2009a: 24). Socio-economic factors thus worsened the impacts of flooding for some population groups, whether relatively remote or poor.

Floods typify how the ultimate outcomes of natural disasters are the result of interaction across a complex web of factors including the timing, duration and intensity of the disaster, and the sensitivity to disaster of affected households, communities, commerce and governments. Natural disasters may become human disasters. After the 2012 floods in Fiji, when about 8000 people were displaced, various Fijians who had lost their homes appealed unsuccessfully to New Zealand for migration opportunities, denied after the 2006 coup (Tan 2012).

DROUGHT

Droughts may appear incongruous in tropical islands, yet occasional, sometimes prolonged droughts are common to most SISI and can significantly reduce agricultural production, worsen the quality of life (as poorer quality water has to be sourced in more distant places) and be expensive in terms of lost production or the cost of providing replacement water and food. The ironically named Cape Verde has long been drought prone, with limited access to potable water and malnutrition being aggravated by drought in the 1980s (Connell 1988). Atolls in the southern Gilbert Islands chain, such as Tamana, are particularly drought prone, and water shortages have even prompted migration (Lawrence 1992). Several Caribbean islands experience lengthy dry spells, and various Pacific islands (such as the Phoenix and southern Line Islands) have proved too dry for long-term settlement. Water shortages are the single most important reason for a large number of Pacific islands, almost all atolls, being unpopulated, and may have also accounted for the past depopulation of others. Rainfall variability is substantial on many other islands and rain cannot be relied upon (Duncan 2011). Droughts reduce recharge of freshwater lenses on atolls and limestone islands, and during extended droughts lenses become more saline or even disappear, due to rapid outflow to the sea, as in Kiribati and the Bahamas (White et al. 1999, Greaver and Sternberg 2010), hence the standard replacement supply of water, when rainwater is absent, is less potable and useful, and simply less available. In the eastern Caribbean intermittent but pervasive hazards of drought (and windblow) have been more devastating to agricultural production, at least for such crops as bananas, than seemingly more catastrophic cyclones (Shakow and O'Keeffe 1981), precisely because some plants are adapted to a particular narrow range of conditions. Even on islands where rainfall is normally heavy, significant breaks in rainfall are enough to damage food plants.

The El Niño event of 1997–98 brought droughts to many Pacific states, with consequent reduced agricultural production and food shortages, and were significant enough for the Marshall Islands to be declared a 'disaster area' with desalination plants installed in the two urban centres, and smaller water makers sent to outer islands (Wong et al. 2005). Late in 2011, a similar situation recurred across a belt of the central Pacific, with Tuvalu, Tokelau and the Cook Islands significantly affected by drought after several rainless months in the normal rainy season. In Tuvalu and Tokelau, health was affected by people drinking polluted lens water, crops failed, water and desalination plants were shipped in from as far as New Zealand, and rations imposed in each of the three states. The extent to which such circumstances may recur, and with what intensity, cannot be predicted. Future rainfall patterns are uncertain; variation in the range of projections of future rainfall is large and even the direction of change is not clear (Mimura et al. 2007: 694). Meteorological data for Fiji indicate that there has been a net shift to drier conditions since the 1950s, and that drought periods have become somewhat longer (Deo 2011). Thus, at least for Fiji, climate change or ENSO fluctuations have increased the extent and severity of extreme events, with both droughts and floods becoming more significant.

Small and remote islands are more at risk of the consequences of drought where alternative water sources are few and relief supplies may be difficult to access. Decreased rainfall both limits water availability and facilitates ocean water intrusion. Alternative water sources, other than rainfall, are invaluable in coping with drought. Permanent roofs and water tanks have reduced some problems, and provided a less polluted water supply on atolls such as Woleai (FSM). In Bermuda harvesting water run-off from roof-top catchments is mandatory. On Majuro (Marshall Islands) the airport runway has long been a significant component of water supply. Governments have increasing been perceived as the alternative source (e.g. McInnes 1986). The Vanuatu island of Aniwa was able to avoid many negative consequences from drought in 2010 because alternative wells (however inadequate) were available and government food supplies brought to the island covered the subsequent months of crop failure (Wilson 2010). The more remote and smaller Lau islands (Fiji) appear to have suffered more from drought in the same year, since fewer alternative water sources were available and external support was weak. On small islands such as Bequia, and others in the Grenadines, where the main source of supply is roof water, fluctuations in rainfall patterns mean that every year local supplies run out and must be supplemented by water shipped in by the government. Because of low household incomes, residents cannot afford to adopt alternative adaptation strategies (Cashman et al. 2010). Whole countries, such as Nauru, and most recently Tuvalu and Tokelau, have occasionally experienced water shortages, with water being brought from overseas. Most Marshall Islands atolls now have desalination plants. Nauru has three, which use a substantial proportion of the island's imported energy. Capacity to cope with drought, other than being a function of its intensity, is increasingly the outcome of government support. That support may be weak in outer islands and may be exacerbated by future changes in rainfall patterns. There is an unusual immediacy of water supply and consumption on small islands, and water stress on most SISI is likely to become greater in the future irrespective of future climate change.

WATER

Most SISI depend directly on rainwater for fresh water supplies. Few have significant rivers or lakes, and many – especially urbanized atolls and limestone islands – are vulnerable to drought. Growing populations, greater affluence and new demands have placed increased pressure on water, while inadequate and decaying infrastructure, pollution and limited ability to pay have contributed to problems, especially in urban areas. However, monitoring of freshwater resources has typically been sporadic and inconsistent. Future climate change suggests that access to fresh water will become more difficult – with greater seasonality of rainfall, increased storminess not compensated by any ability to store water, high runoff in storms and a reduction in the size of water lenses on atolls (e.g. Lal et al. 2002). Human pressures on water sources are likely to be even more crucial. Urban water supply and management are problems for most SISI.

Many rivers, lagoons and especially urban watercourses are polluted. In Fiji both groundwater and river pollution are caused by long-term fertilizer and pesticide use on sugar cane farms and from sugar mill discharges (UNEP 1999, Kumar 2010). In Tonga new commercial crops, such as squash, have resulted in substantial recent increases in the use of fertilizers and pesticides, especially on the main island of Tongatapu, suggesting a 'catastrophic eutrophication of the lagoon' where there is groundwater seepage, further emphasized by leaking septic tanks. At several sites, pesticides have also contaminated drinking water sources. In a sense, along with squash, 'Tongatapu is exporting both its soil fertility and water quality to Japan' (van der Velde et al. 2007: 464). In Admiralty Bay, Bequia (St Vincent), pollution is so severe that swimming is dangerous, with the main contributing factors being toxic chemicals used in farming and sewage. Such problems have tended to increase with accelerated urbanization and vegetation removal and only exceptionally been effectively remedied. Sewage and other discharges – the 'urban runoff' of oil – have contributed to eutrophication, pollution and loss of habitat in various places, as in Vila, Tarawa and Majuro (where the lagoon is unsuitable for swimming or fishing). In Majuro lagoon sites are contaminated by deposition of garbage directly into the lagoon and from the runoff from pig compounds constructed adjacent to it (Johnson 2011). As around Noumea (New Caledonia), eutrophication 'hotspots' are centred in the more urbanized embayments (Fichez et al. 2010). Mining runoff is a specific problem for New Caledonian reefs and lagoons (David et al. 2010). Yet pollution and eutrophication occur even in quite remote villages and islands, such as Gau (Fiji), where more 'modern' sanitation practices have been adopted and subsidized by governments to the detriment of water sources (Veiteyaki 2006).

On atolls especially growing populations and growing demand for new functions (e.g. vehicle washing, washing machines) have resulted in sometimes excessive water use, contributing to compaction of lenses (as in Funafuti, Tuvalu) and saltwater intrusion, through the upwelling of salt water, even on large limestone islands such as Guam (Diaz Arenas and Falkland 1991, UNEP 1999). Demand for water has rapidly increased in most SISI other than in exceptional places like Bermuda where it has stabilized (Rowe 2008). On Tongatapu potable water consumption increased almost tenfold over the 1970s and 1980s because of the higher standard of living, which meant the advent of flush toilets, showers and car washing (UNEP 1999: 13). Beyond domestic use, beer making, soft drink manufacture, fish processing and sugar refining are large users of water, especially in urban areas. Water charges in SISI are rarely progressive to reflect rising demands.

Demand has increased substantially where tourism has grown; a problem on islands with low rainfall, such as Anguilla. Theft of water has also been a problem there (as in unmetered usage elsewhere). Even Windward Island states like Grenada and Dominica experience water shortages (Cashman et al. 2010) exacerbated by limited rainfall (over a short season) and drought. In a number of SISI illegal and unregulated groundwater extraction has become a problem. Pitcairn experienced water shortages in 2010, with slowly rising tourist numbers, and drilling for water was considered. Several SISI have thus had to invest in expensive desalination plants (where running costs are compounded by high fuel costs). Antigua invested in desalination as early as 1970. Desalination plants are most common in Caribbean tourist destinations, such as Anguilla, Bermuda and Providenciales (Turks and Caicos), where elite tourism and second homes have necessitated new supplies.

Water shortages on atolls are exacerbated where lenses are polluted, or water is brackish and can only be used for washing. This is problematic for urbanized atolls such as Tarawa, Majuro and Funafuti, where populations are growing and waste management systems inadequate (Chapter 5). Contamination of lenses on atolls may be caused by taro pits, pit latrines, septic tanks or leachate from solid waste. Human and animal wastes seep through porous coral soil into aquifers (Storey and Hunter 2010). In the past two decades health problems, including typhoid and cholera, have been one consequence in urban centres in Kiribati, the Marshall Islands and FSM. Any combination of reduced aquifer size on atolls (perhaps through erosion), increased salinity (through greater storm and tidal surge frequency), growing population and waste pollution, will put greater pressure on water resources in the future.

The water supply situation in Tarawa is particularly problematic and relatively well documented. Tarawa primarily relies on access to shallow groundwater, with some rainwater tanks and desalination. A range of factors, notably population growth, have led to over-extraction, conflicts, severe pollution and high levels of endemic waterborne disease (White et al. 1999, 2008), that 'epitomizes the worst-case scenario in terms of urban systems complexity. It is also a distressing situation in terms of related health concerns, water scarcity and environmental problems' (Moglia et al. 2010: 1529). High levels of faecal coliform bacteria have been found in groundwater, in association with high incidences of pneumonia and diarrhoea, the leading causes of death amongst children (White et al. 1999), contributing to one of the highest infant mortality rates in the world. The water resources in Tarawa are predominantly freshwater lenses, with the water table typically 0.8 to 1.6 m below the ground surface, supplemented by rainwater. Pollution generated by the 45 000 urban residents has contaminated the freshwater lenses. Due to its small scale, limited human resources, system complexity, isolation and scarce financial resources, and the inability to finance water services by charging customers, water management capacity is inadequate, water can be polluted while sustainability of resources is uncertain and landowners are resistant to restructuring (Moglia et al. 2008). The declaration of water reserves on privately owned land has led to conflicts, illegal settlements and vandalism of public assets, while water consumption per capita is increasing towards levels in more developed nations, further threatening sustainability. Over-extraction of freshwater from groundwater sources, especially during droughts, social disruption and conflict due to displacement of traditional landowners from water reserves, emphasize that the 'water supply in Tarawa is at its limit and there are no other options other than to add new water supply sources in order to sustain a growing population' (Moglia et al. 2010: 1537). A similar situation prevails in Majuro, where some 40 per cent of water samples from the city's fresh water system have been found to be contaminated with faeces or other pollutants (Johnson 2011). All urban atolls with rising populations, scarce rainfall and few economic options experience severe problems.

Economic (and moral) regimes are rarely in place to encourage effective water conservation. Capacity to pay for access to water and sanitation services is limited, especially in expanding urban settlements, where there has been minimal government investment. Water management is made difficult by fragmentation in the water sector, inadequate water resources legislation, failures of policy and planning (and implementation), insufficient human resources (with technical and management skills) and deficient community education, awareness and participation. Reluctance and inability to pay, challenges to collecting revenue, and land tenure constraints have reduced the success of supply schemes. Dense populations with inadequate access to clean water and sanitation services pose public health problems, which may increase where funding is limited and human resources stretched.

Environmental changes affect water supply. Storms may create waves that wash over atolls, and contaminate freshwater lenses. A cyclone that swept over Pukapuka (Cook Islands) in 2005 reduced a potable lens water supply to brackish water unsuitable for drinking for ten months, while fresh food supplies were wiped out by the incursions of sea water into the taro gardens, and rainwater tanks and wells were so polluted that fresh water had to be imported (Terry and Falkland 2010). A similar cyclone event in 2003 on the much larger high island of Vanua Levu (Fiji) resulted in three-quarters of the island's water sources, of various kinds, still being contaminated a month later, because of soil, bacteria and other debris washing into rivers and entering groundwater supplies (Mosley et al. 2004). Greater severity of storms would increase the incidence of such problems. Projected impacts of climate change are likely to exacerbate problems of scarcity and equitable access (Mukheibir 2010). Again, this is particularly problematic for atolls, especially if rainfall decline and SLR (with reduction in size of freshwater lenses) were to occur simultaneously. In Kiribati this has the potential to halve the size of lenses over a relatively short time period (Kuruppu et al. 2011), exacerbating contemporary problems.

DISEASE

Access to water, environmental degradation, poor environmental management, especially in urban areas, and climate change all affect livelihoods and health. Poor environmental conditions can exacerbate illnesses, and increase the need for cooling and therefore fuel costs. Greater use of air conditioning increases greenhouse gases and, if accompanied by reduced ventilation rates, indoor pollutant concentrations. Climate change would affect urban environments, both outside, where higher temperatures, especially in urban 'heat islands', increase stress and morbidity, but also indoors, with higher temperatures and relative humidities. Urbanization at higher densities with greater fuel and vehicle usage intensifies the heat island effect and increases air pollution. Diseases such as asthma, presently significant in many SISI urban areas, may become more prevalent, yet relatively little is known about how changing climate influences urban environments in SISI.

Hazards also influence health status. During 2002–03, when Vanuatu experienced five cyclones, the incidence of malaria doubled that of the preceding year (Lal et al. 2009c). Flooding affects health, with a greater incidence of skin and vectorborne diseases. Loss of access to safe water supplies accentuates the incidence of diseases such as diarrhoea so that, in the Pacific, diarrhoea rates are positively correlated with temperature, and negatively with water availability. Such circumstances are particularly prevalent in the atoll states while El Niño has been linked to outbreaks of malaria, cholera, dengue and other infectious diseases (Singh et al. 2001). Hazards also have impacts on the increased mortality of fauna, whether native or domesticated, due to the hazard itself, food shortages, predation by animals or humans and failure to breed (Wong et al. 2005). Greater incidence of hazard events, and temperature increase, are likely to worsen health status, and increase the social and economic costs of health care.

Climate change potentially harms human health, directly and indirectly, in several ways. Heat waves can cause direct effects such as dehydration, heat exhaustion and heat stroke. Indirectly, climate change introduces new challenges for the control of infectious diseases. Climate-related ecosystem changes increase the range, seasonality and infectivity of some vectorborne diseases. Many significant diseases, including cholera and diarrhoeal diseases, as well as malaria, filariasis and dengue, are highly sensitive to both temperature and rainfall. The incidence of dengue in the Pacific and the Caribbean is correlated with both rising temperatures and ENSO events (Hales et al. 1996, Amarakoon et al. 2008) but the latter may be no more significant in the future (Lal et al. 2002). By the end of the century in the Caribbean, and by the late 2000s in the Pacific, dengue had become more widespread and was regarded as having reached 'epidemic' proportions in places. The 1998 dengue fever epidemic in Fiji cost $3–6 million (Mimura et al. 2007), and future outbreaks in small states could be more costly. Contemporary climate records from Zanzibar have revealed a strong correlation between higher temperatures, rainfall and subsequent cholera outbreaks a few months later; a 1°C rise in average monthly temperatures meant that cholera cases would double in the following four months (Reyburn et al. 2011). Rising temperatures, longer wet seasons and increased urbanization have all contributed to the global rise of dengue and other vectorborne diseases.

Anecdotal evidence suggests that malaria is moving southwards in Pacific SISI as temperature increases. This may discourage tourism in places like Vanuatu. Malaria has been recorded in Fiji and Tonga, outside the customary malaria zone, but that is likely to be a function of human mobility (Opeskin 2009). Waterborne diseases such as cholera could spread more quickly. The late 1990s El Niño events left stagnant pools in Pacific island riverbeds, and in the Caribbean similar ENSO events required more water storage facilities, some of which became breeding sites for mosquitoes. However various studies of the relationship between temperature increases and changing disease patterns have mainly revealed the complexity of change, the challenges in modelling and the difficulty of prediction (McMichael et al. 2006: 864, Costello et al. 2009).

Human influences on the epidemiology of disease are evident. Diseases such as malaria have increased in prevalence with breakdowns in health service delivery (Opeskin 2009). Outer islands with poorer services may therefore be at greater risk and the social burdens of increased disease be borne by the relatively poor. Morbidity and mortality may increase without adaptation. Epidemics have occurred in the past, including measles and flu in the early-twentieth-century Pacific, and cholera in nineteenth-century Jamaica, Barbados and St Kitts, and twentieth-century Micronesia. Overcrowded urban areas and poor public health provision in some SISI suggest that epidemics may recur. Destruction of reefs by natural factors, such as storms, bleaching or anthropogenic influences such as blasting, dredging work, construction of ports and piers and eutrophication, have led to denuded and damaged surfaces being colonized by opportunistic macro-algae that are the hosts for the dinoflagellates responsible for a growing incidence of ciguatera (Bagnis 1987, Hales et al. 1999). Ciguatera is mainly confined to a circumglobal belt between latitudes 35°N and 35°S, although there is a large variability in its occurrence. Ocean warming will extend that belt. The combination of inadequate health services, growing populations, poorly managed urbanization and climate change pose future problems.

BIODIVERSITY

Biodiversity as an environmental asset provides valuable resources for local people and, less directly, high levels of ecological resilience associated with mature and complex habitats, as well as services of global significance such as carbon storage (Danielsen et al. 2010). The extent of biodiversity is unrecorded and unknown in many SISI and in a number of places, including isolated atolls and the larger islands of western Solomon Islands, discoveries of new mammal species continue to be made occasionally (Chapter 1). While discoveries are still being made, many Solomon Islands' species are now endangered and becoming extinct (Pikacha 2008), a process that occurred much earlier in less remote and more densely populated SISI. The Caribbean monk seal is extinct and manatees are nearly gone (Jackson 1997). Although turtles are protected in most SISI, food shortages for whatever reason, but usually associated with cyclones, drought, the absence of shipping and economic downturns, may result in people turning to turtles out of perceived necessity (and because the flesh is prized), as they have done in Woleai, Kapingamarangi (Lieber 1994), Mokil (FSM) (Oles 2007b) and Tonga (*Radio Australia*, 28 July 2011). Future crises may intensify this process.

Most of the species that once existed on islands are now extinct as a result of habitat change, natural enemies, competition, disease and 'cascades of extinction', where one species depends on another for survival. Most losses occurred long ago as 'first contact' extinctions soon after initial human settlement (Gaston 2009, Costion et al. 2012). Islands are sites of 'concentrated extinction'; of some 724 known extinctions in the last 400 years, about half were of island species, especially birds (Wong et al. 2005). Nothing typifies the fate of island biodiversity more than that epitome of island extinctions: the disappearance of the dodo. Some 129 species of birds are believed to have become extinct in the last five centuries, of which 119 were island endemics; of the 93 of these for which a reason for extinction can be established, 78 were eliminated wholly or partly by introduced species. Seabirds are currently under enormous pressure with 28 per cent of world seabird species globally threatened, as bycatch in fishing or through invasive species extirpating breeding colonies (Croxall et al. 2012). Human impact is critical.

While birds are particularly at risk, other species are also endangered but less well accounted for. Freshwater fish species have been similarly devastated, and all known extinctions of land snails since 1500 AD have been on islands (Simberloff 2005, 2009). Endemic small island fauna experience relatively high risk. Flora too are equally threatened, especially on small islands where human presence has been significant, such as Pitcairn (Kingston and Waldren 2003, Kingston et al. 2003). However at Alexander Selkirk Island (Juan Fernandez, Chile) and, most notably, Rapanui, there is considerable debate on the extent and cause of extinctions, the significance of climate change (notably the 'AD 1330 event' of rapid cooling and sea-level fall) and the role of multiple introduced species in changing the floristic composition (Haberle 2003, Castro et al. 2007, Nunn 2007). By contrast, on several small southeastern Pacific islands and on Ascension, the flora have been substantially augmented following relatively recent human contact without significant losses of existing flora and with gains to overall biodiversity (Wilkinson 2004, Castro et al. 2007, Anon 2010). Such islands usually had very limited pre-contact biodiversity and small post-contact populations. More generally, invasion and extirpation have not necessarily meant overall loss since many plant species have been introduced especially during early phases of European colonialism; on some islands, notably in the Caribbean, introduced species of plants are as common as, or more common than, endemic species (Roderick and Vernon 2009, Connell 2013a).

Coincident with some endemic species facing extinction, and a partial explanation for this, is the arrival of invasive alien species (IAS) of various kinds, to the extent that 'at no time in history has the rate of biological invasion or the diversity and volume of these invaders been so high and the consequences so great' (Reaser et al. 2007: 98). Particularly voracious African snails, cane toads and fire ants are all recent arrivals in Pacific islands. IAS have become the most significant driver of species decline and species extinction (rather than hunting and habitat destruction, for example by forest clearance or coastal construction). Within the Caribbean UKOTS, the impact of invasive animals, and to a lesser extent plant species, is regarded as a greater threat to ecosystems than either damage to coral reefs or SLR (Forster et al. 2011). Even seemingly isolated South Georgia, where glaciers are retreating at an increasing rate, exposing new sites to species invasions (especially rats), which in combination with increasing temperatures that may independently render new sites favourable for alien species, has become particularly vulnerable. So too have equally cold Macquarie Island and more temperate islands such as Lord Howe (Christie 2010, Connell 2013a). Increased human mobility is accompanied by increased mobility of IAS.

Islands, especially smaller remote islands, typically have a restricted diversity of fauna and flora, and of ecological habitats, than larger landmasses. Islands also have high endemism, with many species unique to particular islands and island groups. The lone native mammal of the Falkland Islands, a small fox, became extinct many years ago (McDowall 2005). Many island species are particularly at risk (notably flightless birds, such as the New Caledonian cagou), because of small numbers, land clearance, deforestation, hunting, IAS (especially rats, snakes, dogs and cats, and some bushes and grasses) and new diseases. Many species of birds that have disappeared from the Hawaiian islands in the last two centuries were probably affected by the spread of avian pox and malaria by introduced mosquitoes. Since the development of avian malaria is temperature dependent this may be exacerbated by temperature change (Quammen 1996, Gaston 2009).

Animals, introduced accidentally or deliberately, have posed both environmental and socio-economic problems: an extraordinary range of IAS occurs in most SISI (Veitch et al. 2011). Goats have been particularly devastating on many islands, including Aldabra, the Galapagos, Alexander Selkirk Island, St Helena, Pitcairn and the Marquesas. At least two Marquesan islands have been devastated by goats and sheep: grazing animals have a particularly serious impact on environmental change (Rolett 2009). Goats (alongside fire, invasive plant species and agricultural clearance) have contributed to the disappearance of the endangered Fijian crested iguana from the main islands of Fiji, while simultaneously much larger American iguanas have invaded other Fijian islands. Introduced reindeer are a problem in South Georgia. Cats have devastated seabird colonies on several islands, including Ascension. Rats are a notorious problem on many islands. The small remote coral atoll of Henderson (Pitcairn Islands) is estimated to have 30 000 rats which have decimated four species of birds, including the Henderson petrel at its only known breeding site, with as many as 95 per cent of petrel chicks being eaten alive (so that the petrel was close to extinction); even endangered turtles nesting on the beaches were at risk (Connell 2013a).

In Fiji mongooses were introduced in the 1880s to control rats in the cane fields; while they largely failed to do that they eliminated ground frogs and several bird and lizard species. In the Comoros, civets, introduced to control earlier introduced rats, decimated the endemic snake population (Harris and Rocha 2009). Macaque monkeys, introduced to the island of Angaur (Palau) at the beginning of the twentieth century, have resulted in Angaur having less than half the number of bird species of neighbouring Peleliu Island, while the macaques reduced agricultural diversity and productivity since they raided and destroyed crops. This affected social organization since the status of women was compromised by their inability to grow enough food (Parkes 2011), emphasizing how cascade effects not merely cross animal and human ecosystems but have social and economic consequences.

Ants, quintessential 'tramp species', notably the yellow crazy ant and little fire ant, are a serious threat to biodiversity in several areas, from Bougainville through New Caledonia to Fiji, and also in the Caribbean (Ward 2008). The yellow crazy ant has reduced crab numbers and also the reproductive success of some bird species on Christmas Island (Indian Ocean) through direct predation, habitat alteration or competition for food (O'Dowd et al. 2003). On Bird Island (Seychelles), it displaced sooty terns, the main tourist attraction and economic mainstay of the island (Feare 1999). The omnivorous fire ant has been blamed for reducing species diversity and the abundance of flying and tree-dwelling insects, eliminating spider populations and destroying sites for nesting seabirds.

The accidental introduction around 1950 of the brown tree snakes to Guam, an island with abundant prey resources and no natural predators, has become a 'textbook example' of the complex devastation caused by IAS. It resulted in the eventual extinction of most forest birds and other vertebrate species (including 10 of the 12 indigenous bird species, 6 of 10 native lizard species and 2 of 3 fruit bats), thousands of electrical power outages affecting private, commercial and military activities (literal 'brown-outs'), widespread loss of domestic birds and pets, the increase of spiders in forests (because of the loss of their predators) and emotional trauma to residents. The economic cost of rectifying damage and preventing further spread was considerable (Savidge 1987, Rodda and Savidge 2007). Since islands have lower ratios of animal to plant species compared with mainlands, the brown snake also had an impact on vegetative biodiversity, since its intense predation deprived vegetation of natural pollinators such as native birds and fruit bats, with reduced natural reforestation (Mortensen et al. 2008). Economic, social and environmental costs were all substantial.

Fish may also be IAS. In several parts of the Caribbean, lionfish, accidentally introduced in the 1980s, are voracious predators that feed on reef fishes, with severe impacts on native coral reef diversity, reducing indigenous fish populations and resulting in a more algae-dominated coral community (Betancur-R et al. 2011, Lesser and Slattery 2011, Albins 2013). Introduced pathogens may also damage reef systems. Significant natural impacts on Bonaire's reefs during the 1970s and 1980s included an outbreak of white band disease, which caused the death of 90 per cent of the standing stock of *Acropora cervicornis*, and the mass mortality of *Diadema antillarum* (lime urchin) thought to be caused by a waterborne pathogen. Both events affected the entire Caribbean Basin to a greater or lesser extent (Roberts and Hawkins 1994). The declining health of the reefs reduced related biodiversity, diminished their tourist value and reduced prospects for a sustainable economy.

Introduced flora have sometimes been as devastating as fauna. An Andean tree, the velvet tree (*Miconia calvescens*) introduced into Moorea and Tahiti (French Polynesia) in 1937, referred to locally as 'green cancer', has significantly altered forest structures on both islands especially at high altitudes where eradication has been impossible, and covers about a quarter of the forested area (Reaser et al. 2007). Invasive weeds have been particular problems in several SISI, such as the Galapagos and Vanuatu, even contributing to farm abandonment (Miller et al. 2010). Invasive plants are often more effective than endemic species in colonizing sites of ecological disturbance, notably after storms (Roderick and Vernon 2009), hence may become more significant in future.

Islands closest to mainlands and dry lands, especially in the Caribbean, are particularly at risk. Birds are more likely to become extinct on islands where agricultural development has occurred than on the smallest islands (Karels et al. 2008). Both owls and rats have devastated avian species on Norfolk Island and Lord Howe Island (Australia), but some endemic species have survived on even smaller islands off both these islands. However quite isolated islands with small populations and limited biodiversity, like Ascension, have also lost significant numbers of birds, including endemic species, to cats, while some plants have lost their ecological niche to introduced thorn bushes (Royle 2001). Other isolated islands, such as Rapanui, with very limited indigenous plant and animal communities quickly became 'resource sinks' with multiple well-documented extinctions, though explanations vary considerably (Diamond 2005, Stenseth and Voje 2009, Hunt and Lipo 2009, 2011, Rull et al. 2010).

In recent decades, biodiversity loss has generally accelerated. Forestry has been an attractive proposition in relatively poor areas, with the income earned from timber being superior to that from firewood and maintained biodiversity (Murdoch 2011). Consequently the largest number of documented extinctions has occurred in the Pacific islands, which by the 1990s had more threatened species than any other world region, with roughly seven times more endangered bird species per capita than in the Caribbean (Given 1992). In Solomon Islands 102 species of birds were considered rare by the 1990s and some subsequently disappeared (Steadman 2006, Pikacha 2008). Rock iguanas have gone from much of the Caribbean. In the Turks and Caicos in the 1970s a population of 15 000 iguanas was completely destroyed within five years by a mere handful of dogs and cats introduced by hotel workers (Hudson 2007). Birds are at even greater risk. The Vincentian parrot became endangered, like some other Caribbean birds, through the illegal clearing of prime habitat areas for agriculture and charcoal production, hunting (for food or for the pet trade), predation and through storms and cyclones, in the absence of effective conservation-related legislation (Christian 1993). In Ontong Java (Solomon Islands) the number of nesting birds declined during the twentieth century following some human predation, but the main changes in avifauna followed reductions in habitat for nest sites; the copra trade resulted in more complex forests being replaced by coconut plantations and, since the 1970s, surviving stands of large trees were felled for timber and firewood, as settlement became more dispersed with the intensification of the bêche-de-mer trade (Bayliss-Smith and Christensen 2008). In Niue the decline of the Pacific pigeon followed more widespread hunting rather than extensive cyclone damage to habitat (Powlesland et al. 2008), just as lowland bird species declined in Makira and elsewhere in Solomon Islands, especially where construction of logging roads facilitated hunters' access to forests (Danielsen et al. 2010). Longline fishing has not just brought obvious reductions in the main fish species, but the loss of bycatch, such as dolphins, while the lines themselves have become places of death for seabirds such as albatrosses, and plastic ingestion from debris in the major oceanic gyres has accelerated their decline (Connell 2013a). Direct human influences can be locally as significant as the incidental arrival of IAS.

Where human change has been substantial, for example on urbanized atolls like Majuro, the terrestrial ecosystems and vegetation and flora 'are among the poorest and most degraded in the world' (Thaman 2008: 64) even though most native species are still present. With less devastating impacts, as on Cat Island (Bahamas), woodlands have shown considerable resilience to human disturbance; no species have become extinct, invasive plants have not been able to establish themselves and native trees and shrubs rapidly colonize disturbed habitats (Byrne 1980). Coastal zones have necessarily been most at risk of invasion, since boats remain a major source of arrivals, and coastal zones have substantially disturbed and degraded habitats through the 'coastal squeeze'. Faunal diversity may have declined most through the incidental arrival of IAS as much as by deliberate destruction of animals and their habitats, but both have been ubiquitous.

Indigenous biodiversity itself and knowledge of that biodiversity are declining (as traditional knowledge plays little part in formal education systems). The loss of indigenous technical knowledge, especially of littoral species, is of particular concern since attempts to link biodiversity to sustainable development become more difficult (Rigamoto et al. 2004, Thaman et al. 2010). Conservation has failed to arrest ecosystem decay and been limited by the continued introduction of new species (mainly accidentally), limited resources for funding, implementing and managing conservation programmes and inadequate knowledge of effective management practices. To some extent that also reflects diverse interpretations of biodiversity, and especially divergences between local people and external observers (Caillon and Degeorges 2007). In remote Henderson Island (Pitcairn Islands), where the endangered Henderson petrel is threatened with extinction and other birds are vulnerable, effective management programmes were abandoned for lack of government resources, a function of the invisibility of such distant places and issues in metropolises. With current trends in land degradation, overexploitation of nearshore marine resources, coastal population growth, deforestation, economic stagnation, the decline of traditional conservation practices, the relative absence of new conservation regimes, and especially increased and more rapid mobility, the loss of biodiversity is likely to continue. One consequence of the loss of indigenous species and the arrival of multiple IAS is that the taxonomic similarity between areas and islands has gradually increased – a process of 'biotic homogenisation' (Castro and Jaksic 2008). Increased temperatures will further influence biodiversity (Mimura et al. 2007), notably involving such iconic indicator species as frogs and butterflies. Animals with small geographic ranges, specific habitat requirements and difficulty dispersing are most likely to become extinct with climate change, partly because adaptation is difficult where ecological diversity is limited – and thus especially on smaller islands. One way or another the greatest influences on reducing (and increasing) biodiversity continue to be human.

CATASTROPHES: VOLCANOES, EARTHQUAKES AND TSUNAMIS

Environmental changes may sometimes be instantaneous. Volcanic activity is common in the 'ring of fire' that runs across the Pacific SISI in a chain from Solomon Islands across to Tonga (and eventually to New Zealand). A similar chain runs through the eastern Caribbean and several Indian Ocean and Atlantic islands also experience volcanic activity. Many people in SISI live close to areas of recent volcanic activity, in Tristan da Cunha, Montserrat and Dominica (which, with nine live volcanoes, is the country with the world's highest volcano density). Vanuatu has been the site of some particularly violent volcanoes; in pre-contact times some islands disappeared following volcanic and other tectonic activity (Nunn et al. 2006, Nunn 2009), and it is the Pacific island state most at risk to volcanic hazard (Simpson et al. 2011). The subterranean Kick 'em Jenny, off Grenada, has erupted at least 12 times between 1939, when it was first reported (producing 2-metre high tsunamis in Grenada), and 2001. Volcanoes are a wide-ranging threat.

Volcanic eruptions in the early twentieth century, notably in Martinique (Mont Pelée) and St Vincent (Soufrière), caused hundreds of deaths, population displacement and habitat destruction. In recent cases, such as Montserrat and Manam (PNG), continued activity after initial eruptions has effectively prevented return migration. In Montserrat, after the 1995 and 1998 eruptions which destroyed the capital city, the port, airport and effectively two-thirds of the island, and killed 19 people, some 60 per cent of the population left for other countries (mainly Antigua and the UK), and the remaining 40 per cent were relocated to the northern third of the island, the 'safe zone'. Active ash flows have continued in the southern third of the island, which is likely to remain uninhabitable for some years (Philpott 1999, Voight et al. 1999, Davison 2005, Pulsipher 2010). Similar population shifts have followed volcanic activity in Tonga, Vanuatu and the Northern Marianas (Chapter 5). Volcanoes in a number of other SISI, such as St Vincent, which last erupted in 1979, Rabaul and Karkar (PNG), Gaua, Ambae and Tanna (Vanuatu) and Karthala (Comoros), are closely monitored because of their high potential for disruptive activity.

While volcanic activity is better monitored than in the past, it is not particularly well managed in SISI, hence has the potential to cause future unexpected casualties. As recently as 1985, it had been predicted that new volcanic activity was unlikely in Montserrat, lava flows would be unimportant and would flow eastwards (Baker 1985). The complete converse occurred within a decade. Inadequate financial and physical resources (some, such as solar panels, stolen) and management, disorganization and poor risk perception have posed problems of comprehension and response to volcanic hazard in Grande Comore, while poverty and the utility of the 'sand' from lahar flows results in people staying nearby (Morin and Lavigne 2009). Typically, as has been argued for New Britain (PNG), 'the vulnerability of communities is increasing because of a growing population and increased investment, particularly in coastal areas' (Johnson 2000: 13). Volcanic activity poses certain obvious economic costs, resulting from the destruction of infrastructure, housing, etc., and damage to agriculture and other economic activities. As in Montserrat, continued volcanic activity may pose additional costs such as the need for ash masks and expensive hosing of ash from houses (Pulsipher 2010). Costs are rarely balanced through gains to tourism (although various volcanoes attract visitors, with St Lucia claiming the only 'drive-in' volcano in the world, and Tanna using one spectacular volcano as the basis of tourist development).

Both the Caribbean and Pacific SISI lie in earthquake-prone regions. Earthquakes (and cyclones) have devastated several Caribbean towns and cities in recent centuries. As the 2010 Haitian earthquake demonstrated, damage is worst where urban areas have been constructed on poorly consolidated reclaimed land and buildings do not meet hazard-resistant standards (whether for earthquakes or cyclones), while both high-rise buildings and settlements are particularly at risk. After the 1974 earthquake in Antigua, housing losses were mainly of 'buildings of traditional construction inhabited by the lowest income earners' (Tomblin and Aspinall 1975: 1559). In 2002 the Vanuatu capital, Vila, was almost cut off from the hinterland when an earthquake destroyed most access roads and bridges. Most of the damage resulted from either ground failure or inadequate structural design, especially in urban settlements (Shorten 2002). Impacts are most severe where housing is poorly constructed and resources to remedy damage are lacking. Urbanization has increased the potential for earthquake damage.

Earthquakes are frequent in many Pacific SISI, causing problems ranging from landslides, that may block rivers and destroy gardens, to collapsed housing, deaths and personal injuries. In the western Solomon Islands a 2007 earthquake created both short- and long-term food security for people on Ranongga island following the destruction of many gardens, and limited savings had to be used for food purchases. Landslides that accompanied the earthquake were more severe than in earlier earthquakes because more land had been cleared for agriculture as the population had grown; two people died and water supplies and ecosystems were disrupted. Uplift of land exceeded 3 m in places, killing coral and affecting access for transport and fishing, with the future potential being reduced by coral damage and widespread mangrove mortality (McAdoo et al. 2008). Previous earthquakes had resulted in the resettlement of villagers from more rugged areas to coastal locations and the 2007 earthquake resulted in similar interest, but concerns over land tenure and greater population densities discouraged it (McDougall et al. 2008). Few detailed studies have been made of earthquake impacts in SISI but many earthquakes have taken similar forms, with the relatively poor, subsistence-oriented people of outer islands and urban settlers being most disadvantaged.

Landslides occur in most SISI, notably in earthquake- and cyclone-prone islands, in some high-rainfall, high islands and where the timber industry has accelerated deforestation on steep slopes, as in Solomon Islands: the outcome of both tectonic activity and human intervention. Earthquake-prone islands, such as Dominica, are particularly prone to landslides, which also pose tsunami hazards (Teeuw et al. 2009). Landslides have followed cyclones, as in St Lucia, worsening the initial impact on agriculture (Henry-Lee et al. 2010). Several high-island SISI have experienced significant landslides that have become more significant through more sustained land clearance especially in marginal upland areas following, as in Pohnpei (FSM), 'continuing unsustainable population growth and the resulting growing shortage of available land' (Harmsworth and Raynor 2004: 242). In 1997 landslides following heavy rain killed 19 people, roads and rivers were damaged and dams collapsed, at some considerable cost. In relatively nearby Chuuk (FSM), a similar situation in 2002 brought 43 deaths from landslides on a cluster of small high islands in Chuuk lagoon. Fatalities appeared to be more numerous where there had been deforestation (Sanchez et al. 2009). More generally, again as in Pohnpei, such environmental problems have followed 'increasing pressure on the local community to generate economic wealth from an environmentally fragile island' (Harmsworth and Raynor 2004: 220). That was probably also true of the Chuuk lagoon islands, where rapid population growth occurred in the 1980s, but employment opportunities stagnated. Similar circumstances exist in the eastern Caribbean, where landslides are relatively frequent and triggered by heavy rainfall, but also by the construction of unplanned housing on steep hillsides surrounding urban areas, increasing susceptibility to disasters. Leakage of mains water supplies and the discharge of waste water on slopes further increases risk (Anderson et al. 2011). During Hurricane Tomas in 2010 in St Lucia, severe flooding and mudslides resulted in at least 14 deaths, with many others buried in landslides, and the southern half of the island cut off from the north. Heavy rainfall events and landslides may become more frequent with climate change.

Tsunamis are not as rare as media coverage might suggest, and most tropical SISI have had some experience of them. The original Nevis capital, Jamestown, was destroyed by an earthquake and tsunami in 1690. Out of fear of tsunamis, Pacific islanders in the Torres Islands (Vanuatu) and elsewhere once lived on inland terraces, but evangelization led to migration to the coast and greater vulnerability (Ballu et al. 2011). The massive 2005 Indian Ocean tsunami struck the Maldives (and the Andaman and Nicobar islands) and affected the Seychelles. In 1931 18 villages in Solomon Islands were swept away by a tsunami that rose 6–10 metres above high tide, and between 1952 and 1992 the country experienced three tsunamis. Pacific SISI have experienced four significant tsunamis in the past two decades (while the north coast of PNG also experienced a devastating tsunami in 1998 when 2000 people died). Parts of the western Solomon Islands were devastated in 2007 when a tsunami struck close to Gizo, the provincial capital, killing 52 people and displacing 5000. Nine people were killed in neighbouring Simbo. Damage to transport infrastructure, water supply and sanitation alone were estimated at $40 million. Former migrants from the Gilbert Islands (Chapter 5) were disproportionate victims, partly because they lacked the indigenous knowledge and experience of long-term residents such as those of Simbo who were better able to survive (McAdoo et al. 2009, Lauer et al. 2013). A further tsunami in 2010 struck nearby Rendova Island, destroying as many as 200 homes and leaving one-third of its population of about 3600 homeless. Fatalities were avoided since the tsunami was less violent, the topography was different, it occurred in daylight and memories of the earlier tsunami resulted in people quickly leaving for higher ground (McAdoo et al. 2009). A 2009 tsunami killed over 200 people, mainly in Samoa but also in American Samoa and in the Niuas (Tonga). It displaced a further 3200 people in Samoa alone. Fatalities were avoided in other nearby islands, like Futuna, despite the lack of an effective warning system, since the tsunami arrived at low tide (Lamarche et al. 2010). Anecdotal information from Tutuila (American Samoa) and Samoa suggests that no fatalities occurred after a similar tsunami in 1917, but its magnitude is unknown. However, fewer people then lived on coasts, mangroves provided greater protection and houses were made of less dangerous local materials (Dominey-Howes and Thaman 2009).

Although there has been no significant tsunami in the Caribbean for over 50 years, the historical record since 1489 lists 88 tsunamis. Guadeloupe alone has experienced 27 tsunamis in the past 400 years (Teeuw et al. 2009). The risk of potentially destructive tsunamis is greatest from the Caribbean island arc with its sub-aerial and submarine volcanoes, oversteepened volcanic islands, steep underwater slopes and the active subduction zones of the Caribbean Plate boundary marked by numerous earthquakes. Tsunamis of similar size to recorded palaeo-tsunamis would devastate low-lying coastal infrastructure (Lander et al. 2002, Scheffers 2004), and are most likely to occur in the eastern Caribbean. The subterranean volcano Kick 'em Jenny has been described as 'a prime candidate for tsunamigenic events on a potentially hazardous scale, possible affecting the whole of the eastern Caribbean region' (Smith and Shepherd 1996: 115). Both the Caribbean and Pacific islands are also vulnerable to storm surges, which may be becoming more frequent, but their periodicity and impact is not understood, partly because of a lack of tidal gauges (Chowdhury et al. 2010a, b), though there are assumptions of future greater intensity. The geological impact of tsunamis on low islands has rarely been analysed, and necessarily varies. On reef islands, like those of the Maldives, impacts were similar to those of cyclones, with reductions in island area ranging from less than 1 per cent to 9 per cent. At least in this case reef islands were geomorphologically resilient to tsunami impact (Kench et al. 2008). The human impact of tsunamis has been much greater.

ENVIRONMENTAL FUTURES

Environmental changes and hazards have influenced the livelihoods and wellbeing of islanders. Human influences have contributed to environmental change, other than the most catastrophic events, and have often, as in the degradation of land and lagoons, been the catalyst. Human impacts have directly contributed to other disastrous hazard events: in 1983 fire wiped out two-thirds of Samoa's forests; and in 1979 an epidemic of swine fever in São Tomé killed off the national pig population resulting in costly meat imports and a decline in meat consumption (UNCTAD 1985). Catastrophes cannot easily be avoided, are unpredictable and more all-embracing and devastating on small islands, occasionally resulting in temporary depopulation, or exceptionally the disappearance of islands (Nunn 2009). Almost every kind of environmental change, whether slow changes in soil fertility or climate, or more catastrophic hazards, reduces biodiversity and productivity, and makes reliance on local resources more difficult. However, the impacts of SLR have been less pronounced than widely anticipated and assumed and, in the most threatened contexts, coral atolls and reefs have proved resilient.

The more dramatic environmental changes are uncertain in their location, periodicity and impact, although some places are prone to certain hazards, whether floods, cyclones or volcanic eruptions. Environmental change and its effects cannot be easily predicted. For climate change and SLR above all, almost every scientific conclusion and prediction is hedged with uncertainty (e.g. Walsh et al. 2012). Just as the flapping wings of apocryphal Amazonian butterflies may contribute to distant chaos, so debris from the Japanese tsunami have washed up on Marshall Islands beaches, dumped plastic bottles in the northern Pacific have entered seabird food chains in the south and the accidental introduction of one new spider species to the remote coral atoll of Palmyra has set in motion complex ecological consequences (Handler et al. 2007), so uncertainties abound, whether over new transformations or of the complex impacts of old ones. Environmental determinism has no place on islands. No necessary trajectories attend environmental (or socio-economic) change, while adaptation and mitigation vary enormously.

It is not surprising that some slow-onset changes, such as climate change and SLR, have become political issues, because human influences on them occur far from most SISI and are debatable, and because perceived physical changes that may have alternative explanations can be attributed to climate change and SLR: a 'garbage can anarchy' where once separate and complex phenomena have become systematically interrelated (e.g. Connell 2003a, Corlett 2008, Hulme 2010). The notion of 'climate change refugee' (or more generally 'environmental refugee') introduced by well-meaning NGOs has become widely used not just for other Pacific islanders but seemingly globally. Though resisted in many places, including atoll states, both for its seeming denial of sovereignty and independence, and because islanders had no wish to be powerless, passive or tragic victims, even if that elicited empathy and support elsewhere (Farbotko 2005, McNamara and Gibson 2009, Farbotko and Lazrus 2012, Pam and Henry 2012), it inexorably took on a popular life of its own. Small islands provide sites where great global narratives of climate change can be comprehended and made tangible and visible, attributed to distant sources and confirmed by local 'green governmentality discourses' based on indigenous knowledge, memory and eye-witness accounts (Farbotko 2010a). Hazards may also be perceived as the outcome of divine providence (D'Aubert 1992, Morin and Lavigne 2009,

Lipset 2011, Rudiak-Gould 2013) resulting in trust in providence (notably 'God's promise not to flood the world again') rather than engagement in mitigation and adaptation. Fashionable and simplistic as many interpretations are, they stand up poorly against evolving scientific knowledge, deny local agency and invoke powerlessness.

Yet such reactions are understandable. History demands that islanders be wary. Island geographies are changing faster than ever, both through human influences (from construction to biodiversity depletion on land and in lagoons) and from the 'perfect storm' of biophysical changes, including increased cyclone intensity, and especially the largely imperceptible ocean acidification, coral bleaching, SLR and reduced biodiversity. Larger, more diversified SISI are more resilient to hazards, while smaller, more specialized and constrained atoll economies fare worse (Heger et al. 2009). Most SISI have limited planning capacity and capital to respond effectively. Short-term warming systems may be too distant and dysfunctional, especially in multi-island states, and islands far from advice and support (Hards 2009, Lamarche et al. 2010). Yet islanders have universally developed multiple means of coping with, mitigating or averting hazard. For centuries people have planted famine crops, maintained diversity of species and garden sites, constructed seawalls or moved inland and uphill in response to flood or coastal erosion (e.g. Nunn 2007, Campbell 2010, Ballu et al. 2011). Even in quite remote atolls, such as Woleai and Moch, seawalls had been built by the 1970s (perhaps the first use of concrete on Woleai). Environmental degradation and drought have brought diverse responses, ranging from cutting back agricultural areas, intensifying processes such as mulching, and more recently seeking off-farm employment and temporarily leaving farming for alternative employment (Campbell et al. 2011). Movement out of agriculture is, however, a function of economic factors as much as environmental factors, which are catalyst rather than cause.

Environmental change has reduced sustainability and necessitated greater external orientation. Growing coastal populations and higher urban densities have raised challenges to food and water security and worsened environmental risks to lagoons, reefs, beaches and mangroves. Urbanization at increasing densities has increased the costs of hazard damage. The impacts of tsunamis, storm surges, cyclones, droughts and floods have all become somewhat greater, while the severity of some of these has increased. Not only have coastal sites become more vulnerable but, as in both the Cayman Islands and Solomon Islands, migrants to coasts are unfamiliar with past hazard events and contexts and least likely to respond proactively or even actively (Tompkins et al. 2009, McAdoo et al. 2009). Coasts have experienced a loss of both protection and production. Vulnerability has increased and resilience declined.

Hazards are particularly devastating to the poor who are most at risk and least able to withstand and recover from hazard, partly through that poverty and partly because the coastal squeeze has trapped many in particularly vulnerable locations, especially in atoll states. The erosion of moral economies, where complex social ties based on reciprocal relationships enabled survival, has made urban households more vulnerable to hazard (Bayliss-Smith et al. 1988, Paulson 1993, Cinner et al. 2012). Where subsistence economies remain vibrant and social structures relatively unchanged coping with hazard is more successful.

Despite the growing significance and visibility of environmental problems management is lacking, whether in terms of information gathering, planning (and developing legislation) or implementation. Scientific studies are fragmented (and ignored) or not understood, where key human resource skills are absent, and environmental management is rarely a priority. Lagoons are less carefully scrutinized than land and problems such as microbial pollution are not particularly visible (Morrison and Kaly 2010). Even urban land-use planning is rare. Basic management issues, as in Kiribati, have failed in part because 'efforts to promote sustainable development which run counter to freedom of movement and consumerism have proved very difficult and not politically popular' (Storey and Hunter 2010: 177). Management is an uncertain phenomenon, and a luxury in terms of the environment, for most SISI. Barely regulated environmental change is inexorably part of much broader and more complex local and global systems that give agency to and shape the lives of every islander. The environment is the victim of the wider problems faced by SISI while islanders are the occasional victims of dramatic environmental changes.

8. Islands at risk?

> Where are we coming from? What are we? Where are we going?
>
> Paul Gauguin, 1897

It is trite but true that global economies and environments are constantly and quickly changing, epitomized by the frequency with which 'neo-' and 'post-' have become prefixes, alongside the uncertainty implicit in that usage. It might seem that information superhighways, Eurozone crises and nuclear meltdowns bypass SISI and that islands are a resort from and the antithesis of change. Islands, distanced from centres of population and industrialization, certainly offer the allure of seemingly complete worlds, introspective ecosystems, secured by natural boundaries. Previous chapters have shown that this is far from the truth. Islands are no bastions of stability. Change can be rapid – from environmental disaster to the unquiet electronic revolution of mobile phones and social media now altering access to services and money and transforming social relationships in what is nothing less than an IT revolution (Cave 2012). Nonetheless implicit in islands are isolation and separation, small scale and some notion of security (and independence). These are basic, yet elusive, real and metaphorical qualities of islandness from where it is only a small step to the once much trumpeted phrase 'small is beautiful' (Schumacher 1974). But 'small is beautiful' ignored the fact that some 'big' was welcome and necessary, as small islands secured links with a larger world.

Well-known structural factors impose constraints and limitations, and development options are few. Small islands were never paradise. Pitcairn was depopulated several times in its history, Palmyra and Malden (Kiribati) and Barbados similarly. Abandonment of islands may have occurred in the distant past when local fish resources were exhausted; an ancient analogue to some contemporary threats (Rongo et al. 2009). In post-contact times Rapanui has been singled out as an island 'model' of crisis and catastrophe (Diamond 2005), and the locus of environmental morality tales, though there are many parallel examples (Kirch and Rallu 2007). Even quite elaborate agricultural systems, alongside complex social controls, were inadequate to prevent occasional famine (Currey

1980). In coral atolls and smaller islands, ecosystems were impoverished and fragile, and more easily degraded, and extinctions and extirpations occurred more frequently, even at 'first contact'. Fragile ecological systems had analogies in the social world where islands could be fractured and fissiparous; even emotions had to be kept in check, and local social institutions were structured to keep differences below the surface of social life or beyond scrutiny, and with aberrations tolerated, though not always successfully (Lutz 1988, Black 1991, Birkett 1998, McDougall 2005, Marks 2008). Discontent and a degree of 'ordinary' violence were often inescapable. Yet however vulnerability and resilience are defined, islands were and are both vulnerable and resilient to multiple human and natural influences.

In pre-colonial times islands achieved sustainable development partly by maintaining extended ties, typified by linkages across atolls and between clusters of islands, to secure social ties and claim and use land elsewhere as on such densely populated islands as Paama (Vanuatu) and Kotu (Tonga) (Perminow 1993). Linkages were often marked by elaborate exchange systems such as the *kula* ring (region) in the Trobriand Islands (PNG) and the hierarchical and asymmetrical *sawei* obligations that tied atoll outer islanders to the high island of Yap in what is now FSM (Alkire 1978, Bayliss-Smith 1982, Pam and Henry 2012). Islands were rarely isolated for long and even Tikopia was part of a Pacific 'world system' for most of its 3000-year history (Kirch 1986). On Tanna the metaphors of tree and canoe aptly describe people who were both rooted in place and tradition yet involved in mobility and exchange: 'a place provides a man with his roots; a canoe traveling on a "road" grants him the allies necessary for his survival and reproduction' (Bonnemaison 1994: 321). Yet violence also marked competition for resources. Conflict and cooperation were uneasy partners. Achieving subsistence security was difficult; survival was 'a significant achievement and affluence was both rare and precarious' (Denoon 1997: 118). The remarkable ubiquity, subtlety and complexity of techniques to achieve stability is testimony both to the challenges of survival despite (usually) much smaller populations, and to the way in which islands merely seem insular.

As early as the mid-nineteenth century, population pressures on scarce resources in the Gilbert Islands were severe enough for people to be described as 'Malthusians' where achieving an adequate livelihood was unusually difficult, and warfare not uncommon. Few populated islands are as small and remote as Tikopia, yet in the 1930s:

> Young men display an extreme eagerness to see the world and numbers of them beg a passage of every boat that calls ... They want to become

> possessed of knowledge and property from which they can reap an advantage on their return – in social prestige as tellers of tales of breathless adventure which can be made to absorb the public interest [and] in the possession of prized tools and ornaments. (Firth 1936: 18–19)

In Tokelau, 'the idea of permanent migration, involving a severance of many ties with the home island and of seeking one's fortune elsewhere, is well established in Tokelau life and thought. For the past 70 years or so it appears to have been accepted that some of nearly every group of siblings must *tahe* ("emigrate") simply because the local resources are seen as insufficient' (Hooper and Huntsman 1973: 403–404). It was ever thus. More recently migrants left small eastern Fijian islands that were 'beautiful but not places to live' (Bedford 1980: 57), while boredom drove migrants from Bikini and Sikaiana (Kiste 1974, Donner 2002). Yet home had the critical virtues of familiarity, security and relative stability. Some islands, notably atolls, were more disadvantaged than others.

Subsistence-based economies have survived longer than in many parts of the world, especially in the Pacific, though diversity of production has declined, as even remote people have absorbed a cash economy, ignoring staples and importing food from distant places. In places like Kadavu (Fiji), 'a rich tradition of agrodiversity persists in adversity … because of geographic isolation and a scarcity of profitable export commodities' (Kuhlken 2007: 72; cf. Sofer 2007), and is invaluable in coping with hazard. Yet the persistence of a distinctive way of life has been as much 'the product of an externally imposed political economy [as] of conscious resistance' (Bayliss-Smith et al. 1988: 8). In contemporary times, as SISI have struggled to develop viable structures of development, vulnerability has been less the outcome of environmental hazard and more the repercussions of global economy and society. Vulnerability has increased as environments have been degraded, local food security reduced, traditional mechanisms and practices for coping with hazards disregarded and external dependence increased. In Woleai (FSM), for example, when hazards such as cyclones devastated islands, people once turned to preserved and stored foods or sought assistance from their 'traditional partners' on other atoll islands or on more distant atolls. In similar contemporary disasters islanders now turn to Yap state or the FSM government for support. Such dependence 'is sanctioned and encouraged by government action and is culturally a legitimate strategy effectively no different from former island network relationships' (McInnes 1986: 132–133). Ironically local mechanisms for management and self-reliance have been devolved upwards, from the local 'private sector' to a distant and flagging public sector. In multiple ways – chosen and imposed – islands are increasingly part of a wider world.

BLACK SWANS, UNCERTAINTY AND INSECURITY

The complexities of environmental change alone are considerable, and the uncertainties often greater simply because islands are small. For a rather larger island than most discussed here: 'The ramifications of the vulnerabilities associated with climate change at the local community level, *even for a relatively small tropical island* like Jamaica, are fraught with interpretative challenges for the researcher and policy maker alike' (Campbell et al. 2011: 156; my italics). Subtle modelling of change for the much smaller island of Rodrigues concluded that small scale was such that 'we did not fully account for the complexity [hence] the policy usefulness of models may be limited' (Bunce et al. 2009: 223, 224; see Miller et al. 2010: 422). Though others are more sanguine over the use of models, as in Dominica (Patterson et al. 2004), modelling the relationships between climate change and biodiversity has been thwarted by the few comparable studies, the absence of data and the focus on just a few species (Sieck et al. 2011). Problems of complexity, data availability and comparability make evaluation of contemporary change difficult enough and prediction of future changes almost impossible. Throughout this book that has posed problems in going beyond local microcosms of meaning. Islands are inconvenient laboratories.

Within and between small countries and islands, economic and ecological variations are considerable, and similar islands exhibit subtle differences. The environmental contexts of two Polynesian islands, Tikopia and Mangaia (Cook Islands), shared much in common but the outcomes and consequences of human impacts, in the pre-colonial era, were quite different, according to distinct human choices, though scale played a part (Kirch 1997). Subsequently, in the more diverse colonial and post-colonial eras, contexts varied enormously (in this case with the two islands becoming outliers in two quite different entities: Solomon Islands and the Cook Islands) with repercussions for environmental, economic and social change. Likewise the trajectories of culturally similar Kiribati and the Marshall Islands have diverged dramatically under different colonial and post-colonial regimes, while the French part of the small island of Saint Martin is quite different from Dutch Sint Maarten. Policies and practices have varied at a range of scales. Quite recently, the development of MPAs has had varied outcomes in places with apparently similar ecosystems and population pressures (Aswani

2002). Different exploitation regimes have been developed for something as particular as the bêche-de-mer fishery in otherwise similar and neighbouring atoll environments on the PNG–Solomon Islands border. Within adjacent atolls, although local people perceived the marine environment similarly, their managerial responses to changes brought about by population growth and consumption differed according to local characteristics of kinship and reciprocity, themselves a function of ethnicity and religion (Christensen and Mertz 2010; Chapter 3). In Dominica, inshore fisheries appear to have been resilient to change; elsewhere in the Caribbean they have usually dwindled. Multiple studies of several islands, such as Rapanui and Rodrigues, have recorded somewhat different explanations for observed changes in ecosystems.

Development policies have sometimes been based on inadequate and superficial data and, in some contexts, may be the outcome of consultants' visions and external agencies' perspectives (often shaped in continental states) rather than of informed and moderated local knowledge and wisdom. Short-term prediction of hazard events (whether cyclones, volcanic activity or tsunamis) is now better (although response to tsunami alerts, and knowledge of them, is imperfect) but management of hazards varies, so that where policies are formulated, far distant from such places as Mayotte, and in a different political and cultural milieu (Le Masson and Kelman 2011), local implementation may be weak. At various scales, there is often limited capacity to comprehend, respond and manage, despite the need for integrated management, especially on coasts where multiple uses, environments and claimants interact (Lane 2008, Fazey et al. 2010, Kenter et al. 2011, Castrejón and Charles 2013). The inability to deal effectively with hazards, for example, results from lack of technical awareness of the mechanisms contributing to the hazard, insufficient access to finances, the absence of organizations that provide a public good, coupled with high levels of poverty and unemployment in the most affected areas (Anderson et al. 2011). Human resource and information inadequacies constrain policy formation and implementation.

Projects, practices and circumstances with limited social, economic and environmental influence in large states can be extremely significant in SISI. Unpredictable hazards may cause havoc in small places, where there is little ability and room to counter, retreat and regroup. Unusual and extreme events are 'black swans': high-impact, hard to predict and rare events, often without precedent and beyond the realm of normal expectations in history, science, finance and technology (Taleb 2007) and perhaps in the environment: a social sequel to Holling's (1986) theory of surprise based on unexpected discontinuities in ecological systems. On tiny Tanga island (PNG) unusual changes came so rapidly that they 'induced appreciation and anticipation of the unexpected' (Foster 2005: 175). Not only did 'black swans' seem to become more frequent after their formal recognition in 2007 (with the GFC, the collapse of the Greek economy, the Arab Spring, a tsunami in Japan) but also such unpredictable events even began to appear normative. They implied that the world was changing too rapidly to predict and that many changes were without historical analogues.

Unpredictability had been earlier evident in SISI, both in terms of external factors and local perturbations, and with economic, political and environmental origins. The abrupt depopulation of Banaba, Diego Garcia, and Bikini were political black swans. The use of Kwajalein for missile testing and the 'Pacific Solution' proved unexpected financial windfalls for the Marshall Islands and Nauru, respectively. Montserrat effectively lost a third of both the island and its population after the 1995 eruption. The Netherlands Antilles collapsed in 2010. A year later, French Polynesia sought to be reinscribed on the United Nations' list of colonized territories. Anguilla, New Caledonia and Bougainville (PNG) may, or may not, choose independence in the near future. American Samoa and Puerto Rico debate their status. Guam may be politically united with the Northern Marianas. The Chagossians seek to return home, but BIOT may be absorbed into Mauritius. Politics has seen strategic interests shift from the cold war to control of terrorism and drug flows. While crucial economic decisions may be more predictable, their impacts may be no less dramatic. The closure of a tuna cannery in American Samoa in 2009 and stringent economic reform in the Cook Islands in 1995 devastated employment markets and precipitated new waves of migration. In 2008 Tristan da Cunha's lone fishing factory was gutted by fire; in a moment the basis of the island's economy, which provided much of the wage employment, disappeared.

What might otherwise have been black swans have been avoided. External proposals to construct an oil refinery on Dominica in the 2000s were strongly opposed locally, and in the diaspora, in order not to compromise its assumed image of 'nature island' and tourist destination, and to avoid potential environmental degradation (Nelson 2010). The refinery never materialized but the kind of disaster some foresaw for Dominica occurred in Tristan da Cunha in 2011 with the grounding of a bulk carrier. Escaping oil badly affected indigenous birds and other wildlife, with thousands of endangered northern rockhopper penguins eventually saved in a month-long operation involving most of the island population (McKie 2011). In 1993 the main staple food crop of Samoa, taro, which also brought in AU$4 million in exports, was completely wiped out 'with frightening speed' by taro blight, as had that of

Bougainville half a century earlier (Connell 1978, Moorhead 2011). In 2010 most of the St Vincent banana crop was destroyed by Hurricane Tomas when it was still the major source of income and jobs, accounting for roughly over half the workforce and exports (Rose and Mathiason 2011). Unforeseen disasters have obvious economic consequences of national importance, especially where economies are exceptionally narrow.

While most global ties are dictated by a certain logic, of trade, aid, investment, tourism and migration, some are more precipitate and unpredictable. Government idiosyncrasies are common – not least in St Helena where prisons, desalination plants and fish canneries have all threatened, but never managed to achieve, a new order (Royle 2001). A recent decision to build an airport may radically alter economy and society. Unpredictable human-induced events can be as catastrophic as natural disasters. Guadalcanal (Solomon Islands) and Bougainville (PNG) have been devastated in the past two decades by disasters (variously construed as political, economic, social and environmental). Uncertainties are particularly critical where economies are small and the future, as in Guam or Christmas Island, is largely defined by (fluctuating) strategic considerations beyond local control or influence. In the Falkland Islands issues that followed the conflict between Argentina and the UK, including territorial sovereignty, problems of decolonization, the right to self-determination and settlement of international disputes, continue to affect the future of the territory, but changed gear with the discovery of oil (Dodds and Benwell 2010). Echoes of such issues ensure uncertainties elsewhere.

Small states are vulnerable to predatory powers. After the Grenada crisis of 1983 (when the United States exercised 'gunboat diplomacy' to invade and overthrow a socialist regime) and soon after the Falklands war, 'small is beautiful' seemed ludicrously inappropriate and the Commonwealth Secretariat shortly afterwards launched the book *Small is Dangerous* (Harden 1985: see Bartmann 2007). External influence may be welcome, notably the decade-long deployment of the Australian-centred RAMSI in Solomon Islands, that intervened militarily, after serious instability and violence. Quite differently, Australia briefly took over some of the reins of government in Nauru after economic crisis produced a failed state. Various SISI have baulked at independence and full sovereignty, even when encouraged by colonial powers, not least Montserrat and Bermuda, preferring an external guarantor of security and stability. For others, such as Mayotte and the Cook Islands, dependence enables welcome migration opportunities. Micronesian states negotiated migration in their Compacts of Free Association. External intervention and linkages emphasize the fragility of small states to political divisions, ethnic rivalries and economic crises that cannot easily be resolved internally.

On the eve of independence of many SIS, Benedict concluded that 'the best solution for small territories is to look for some form of economic integration with their neighbours' (1967: 8). Soon afterwards Dommen bluntly stated that 'the particularly poor island countries are those which have failed to establish sufficiently intimate relations with a prosperous protector' (1980: 195), while Bertram argued for Tuvalu that 'the key role for the new state will now be to recruit a patron and to seek out opportunities for Tuvaluan labour and capital to penetrate the rest of the world' (1987: 29). Indeed neglectful colonialism and contemporary amnesia have meant that in many remote places, like Kadavu, both colonial absence and invisible state policy 'inhibited the evolution of capitalist development in the village society and entrepreneurial spirit among rural Fijians' (Sofer 2009: 354). More broadly 'capital created underdevelopment not because it exploited the underdeveloped world but because it did not exploit it enough' (Kay 1975: x). In the Pacific, especially, colonialism left space both for hybridity (see below) and for attempts to embrace 'the journey through Hell'. Some SIS, such as Dominica, have occasionally sought to revitalize old colonial ties for similar reasons, but most have not wished to relinquish political independence, though usually smaller territories chose to retain political ties, despite the efforts of the United Nations Special Committee on Decolonization. In the relatively large French department of Réunion it was argued that development and decolonization were possible by 'complete integration' with France, a similar position to that widely held in the French Caribbean (Aldrich and Connell 1992). For several decades there has been some interest in greater integration, amongst French overseas entities, UKOTS and especially the Dutch Caribbean, pejoratively 'colonialism by consent' (Oostindie 2005, Hintjens and Hodge 2012) but, more positively, a means of securing stability and diversity.

Critical influences on socio-economic change (and thus on environmental change), such as shifts towards a free-trade regime, are externally driven and beyond the scope of SISI to regulate or even influence. Fluctuating asymmetrical relationships typify SISI links with the wider world. Those that are affluent, such as Bermuda, depend on the global tourism and insurance industries. The Galapagos has developed through elite tourism and fisheries, a function of changing tastes and transport possibilities and the unsustainable exploitation of fisheries elsewhere, but both are vulnerable to further external changes (Kerr 2005). Aid fatigue and changing strategic orientations can threaten development assistance.

Contemporary global media have challenged local cultural identity, evident in the rise of Rastafarianism, hip-hop and basketball in Caribbean SISI, and a fixation on English Premier League football in the remotest places, but more insidiously through a host of soaps and serials in which the alien (and its advertisements) are increasingly the norm. So pervasive has American television become in Guam that, even a quarter of a century ago, 'not only does it make us homesick for places we have never been, it gives us the uneasy feeling that what we experience daily is abnormal' (Underwood 1985: 171). The simultaneous and unpredicted arrival of mobile phones and international guestwork in rural Tanna brought unprecedented social and economic changes within a couple of years: new affluence, access to education, solar panels, clean water and a written (texted) language. Yet cultural resilience and distinctiveness have not been easily undermined.

Many past predictions of change, or its absence, have failed to account for the successful independence struggle of East Timor, the succession of military coups in Fiji (and the Comoros), ethnic dissent in Solomon Islands and other crucial events, let alone the uncertainties of environmental hazard. One-off discoveries (minerals, such as oil in the Falkland Islands and São Tomé) and decisions (an airport for St Helena) can make a significant difference to sustainability. Most predictions would have seen Pitcairn depopulated decades ago, yet even the smallest and most remote islands have survived, and Pitcairn's population has stabilized (Amoamo 2011b). Given the complexity, and occasionally the rapidity, of global political changes (in trade and migration policy alone), predictions over the future of SISI are fraught with uncertainties.

STRUGGLING ECONOMIES, CHALLENGING ENVIRONMENTS?

Economic growth has been limited, trade balances are usually negative and free trade holds out little for future growth. The cost of increasing imports is considerable. Fuel price rises, more costly market access, population increase, superior resource-extraction technology and desire for prosperity – or at least tangible material goods – all threaten the resource base (Fazey et al. 2011). Fisheries have raised national incomes and agriculture remains a significant source of employment, income and food security in many SISI, especially in the Pacific, but has been virtually extinguished in some islands. Bananas and sugar cane have collapsed. Manufacturing has largely failed, but tourism, despite being notoriously fickle, provides incomes and prospects for growth. Tourism has rescued and revived such SISI as Antigua, Cape Verde and the Galapagos, as its reach has extended, though environmental pressures may threaten the new 'monoculture'. Job creation has slowed but urbanization accelerated, resulting in unemployment, relative deprivation, poverty, and a new generation of urban environmental problems. Although SISI cannot compete in global markets with their existing economic infrastructure, resources and human capital, international economic orthodoxy has constrained them to pursue policies that seek to do this. While the era of structural adjustment may have passed, integration with global markets is still held out as the path to development. Worsening poverty and inequality (and the problems of the PIGS) suggest this is not necessarily so. Governments are prone to fission, and the private sector is poorly developed (except in tourism) and civil society weak. Almost everywhere the demographic balance is shifting to the centre and overseas. Environmental and economic changes have had the greatest impact on the poor, notably in growing coastal population concentrations. Foreign aid (and/or dependent political status) boosts many SISI, but aid can be fickle and conditional.

A characteristic of SISI has been migration, marked by a 'culture of migration', despite desires to 'hold on to home'. An economic rationale underpins migration and remittances are invaluable components of national and household incomes. Other than after extreme hazard events it is impossible to recognize 'environmental refugees', even in Tuvalu where that might have been expected (Chapter 6). Most environmental changes (other than tectonic shifts) in SISI are evolutionary and slow, yet 'climate refugees' from tiny islands, as victims of global carbon emissions, have attracted media attention, while 'garbage can anarchy' has brought together separate phenomena into a discourse where local agency is denied and climate change has already made some small islands uninhabitable. By extension all small islands are imminently threatened, fragile and vulnerable, and environmental change assumes a critical role in island futures. Climate change may have become the dominant environmental narrative of this century, but it is not the only one.

Unequivocally climate change is accelerating – it is no longer so slow as to be ignored – and the pace of change is likely to quicken. Yet, whereas distant observers are more likely to perceive environmental threats to development, islanders fear population pressures and economic stagnation. Residents in Caribbean UKOTS ranked a loosely defined 'development' – physical changes, including pollution and coastal development – as the greatest threat to marine and coastal ecosystems rather than either natural hazards or climate change. By contrast, distant observers of the same UKOTS, based in the UK, focused on climate change (Forster et al. 2011). In Solomon Islands coastal communities, economic crises (such as price rises for fuel and store goods) were regarded as more significant than earthquakes or tsunamis (that many had recently experienced) or climate-related changes (such as cyclones or high tides) (Schwarz et al. 2011). Loss of social cohesion was more generally feared, as was the 'Malthusian narrative' of population increase (Fazey et al. 2011, Kenter et al. 2011, Schwarz et al. 2011). Those most familiar with making island livelihoods recognized economic and social changes and inadequate management as more critical and confronting than environmental changes. Yet the conjunction of climate change and globalization, enhancing and transforming connections across space and time, demonstrate the double jeopardy and 'double exposure' (O'Brien and Leichenko 2000) that link these phenomena and emphasize that islanders are subject to multiple external influences. Differentiating between the effects of changing climates and evolving socio-political factors on changing economies and ecosystems is difficult, though most environments are affected more by human influences than hazards.

While coastal squeeze and habitat destruction threaten local livelihoods, global warming has not proved devastating in the Carterets, Tuvalu, the Maldives or elsewhere, but El Niño and La Niña, high population densities, impoverished ecologies and few migration opportunities, reduce the possibilities of livelihood diversification and increase vulnerability. In the Galapagos the combination of El Niño, tourism and intensified fishing critically damaged local ecosystems before accelerated climate change came into play (Edgar et al. 2010). Natural disasters are nonetheless increasing in frequency, scale and impact, sometimes seen as the 'new normal' (Vidal 2011). While that may be partly a function of superior monitoring and recording, and of intensified ENSO fluctuations, in most scenarios the frequency of extreme climate-related events is expected to increase. Coral reefs and coral atolls, the most vulnerable physical environments, have demonstrated a remarkable resilience, and the ability to rebound from natural hazards and ENSO events. Indeed disturbance of such coral ecosystems may even enable greater resilience (Côté and Darling 2010). That gives grounds for hope that islanders will cope with future hazards.

Those most likely to be affected by environmental hazards live in flood-prone, marginal, usually urban areas and are without the financial capacity, social capital or social networks to respond. Nor can they move to another area, whether nationally or internationally, since migration is selective by economic status. Hazards are therefore most likely to further contribute to uneven development, enhance poverty and reduce economic growth, while poverty makes disasters more severe (Lal et al. 2009b).

Climate change consequently 'will impact poor people first and worst. It will exacerbate food shortages, reduce household incomes, and increase rates of illness for poor people and communities who are least able to cope with the impacts' (Commonwealth of Australia 2011: 35; cf. Ibarrarán and Ruth 2009, Dasgupta and Baschieri 2010, Mideksa 2010, Markyanda 2011, Wade 2011). In circumstances of limited public finance for mitigation, adaptation and reconstruction, this poses obvious problems. SISI are poorly placed to cope.

Ecology and economics are not necessarily the primary determinants of islanders' choices of where to live. Belonging to local society may be more important than income levels or career opportunities. Those who stay, as in the Pribilof islands (Alaska), 'are likely to have a high tolerance for uncertainty [and] a high personal resilience in the face of actual or anticipated changes' (Huntington et al. 2009: 126) or simply a greater attachment to place. In most SISI the local economy is not always closely linked to the environment, hence environmental changes have fewer implications than in the past. In the Faroes, for example, state subsidies slowed a more rapid migration (Hamilton et al. 2004), while few would remain in Saint Pierre and Miquelon had France not subsidized them after their fisheries collapsed. Extensive subsidy as a 'government island' is an option unavailable to most SISI. Greater diversity of income sources, including remittances (themselves subsidies), enables greater stability, especially in response to gradual environmental changes.

Political factors influence development strategies, typically in favour of anticipated economic growth at the expense of policies involving conservation and environmental management, as immediate economic gains triumph over potential future gains. Financial constraints 'dictate a reduction in staffing and operational activities' in countries like the Seychelles (Clifton et al. 2012). Indigenous environmental knowledge is disappearing. Attempts to manage national and local environments have often been ineffective. In parts of the Caribbean, notably Bermuda, they were so belated that little then remained of a 'natural environment'. Legislation to protect and conserve particular regions has been inadequate or non-existent. In Vanuatu, although the 'natural environment is critical to livelihoods and the nation's principal export industries, particularly tourism and agriculture … political commitment to the environment is lacking, the budget allocation to it is tiny, and the requirements of the Environmental Management and Conservation Act of 2002 are evaded' (ADB 2009a: 26). Likewise, over the last 30 years at least five studies have recommended the establishment of protected areas in Makira (Solomon Islands), yet 'despite the global importance of the island to biodiversity and the mounting threats facing its forests, there is still no protected area on the island' (Danielsen et al. 2010: 110, Wilkinson and Salvat 2012). Developing a different set of priorities is difficult without economic growth.

Regulating the private sector has also usually been ineffectual, most evident in logging and more widely in tourism, in the quest for profitability. Concerns exist:

> For an acknowledged water-scarce country, Barbados seems to be somewhat reluctant to take action to ameliorate the situation. The hotel sector of the economy is a significant consumer of water resources, yet the potential of this sector to reduce its burden on water resources is going unrealized because of a combination of lack of action on the part of the industry and a lack of attention and incentives on the part of the authorities. (Charara et al. 2011: 241)

However tourism has been in advance of government in some SISI because of the obvious scenic and recreational benefits from environmental management.

Rarely evident are policies and practices to mitigate climate change. Some of the 'hard' options, such as seawalls, desalination plants, reforestation, and improved water storage, are expensive and scarcely feasible in states with very long coastlines. In Majuro the cost of constructing seawalls ranges from \$3000 to \$17 500 per metre (McKenzie et al. 2006). Such astronomical costs emphasize how important retaining mangroves is, why governments should shift in favour of 'soft' measures (such as mangrove retention) and how the value of coastal ecosystems has rarely been considered in coastal management (O'Garra 2012, Sovacool 2012b). Even such 'soft' options as land-use planning are absent. Where change is neither evident nor imminent, lack of concern and initiative means that 'the failure of SIDS to invest in policies to respond to climate change may leave them poorly prepared to cope with adverse changes and also increases the probability of severe consequences' (Belle and Bramwell 2005: 33). That is equally true for other future changes outside the environmental sphere.

Small islands are particularly disadvantaged in achieving economic development and managing environmental change. The SISI most at risk are low-lying islands with weak economies and limited skilled human resources that lack capacity to respond to economic and environmental change. Atolls are particularly vulnerable to extreme environmental events even though coral reefs are more adaptable than hitherto believed and have coped with steady SLR and more precipitous environmental disturbances. Climate change may exacerbate vulnerability and put pressure on fresh water supplies. Future 'water wars' are as likely to occur on small coral atolls as in the world's largest river basins. Future environmental changes are unknowable – sea levels may change more rapidly, ocean acidification be faster than anticipated, storms more violent and so on, while new forms of mitigation and adaptation may be discovered. (In 2011 the Kiribati President was contemplating artificial islands akin to oil drilling rigs; a year later Kiribati purchased an island in Fiji.) Measures of SLR in some areas dominated by SISI, again particularly in the Pacific, are higher than twentieth-century predictions. The highest rates of SLR come from the western Pacific especially around Tuvalu: three times the global average between 1950–2009 (Becker et al. 2012). Economic shifts are equally uncertain. While most atolls are not being eroded, without sustainable management of inshore resources and with deep-water fisheries exploited by rapacious long-distance, sometimes illegal, fishing, viability comes under increased pressure. Throughout SISI populations have become increasingly urban and coastal and the coasts, often the most productive areas, are most prone to human modification. Overfishing, lagoon pollution, deforestation and agricultural contraction (in particularly productive areas) have reduced biodiversity and environmental stability. The coasts are also the settings of most hazard events. Atolls only have coasts.

The countries most at risk, being entirely archipelagos of low-lying islands, are the four independent states: Kiribati, the Maldives, the Marshall Islands and Tuvalu. Scarce international migration opportunities have resulted in densely populated urban centres where environmental and public health problems are rife, and long-term entrenched poverty widespread (Connell 1985, 1993a). Atoll SIS are characterized by little economic growth, limited service provision (especially in outer islands) and unmanageable urbanization. Economic options are few, perhaps symbolized in Kiribati by the national regard attached to seaweed exports (Thomas and Tonganibeia 2007). Unemployment is high. So too are energy costs. Achieving the MDGs seems improbable. Urban bias is more pronounced than in much larger states. Rural–urban migration has followed rising expectations of standards of living and access to education and health facilities, which in some outer islands barely exist. Early resettlement from Kiribati especially was a measure of inherent problems. Environmental change and degradation have also accounted for the remarkable extent of media interest (and wild speculation) on the 'fate' of Tuvalu especially (e.g. Connell 2003a, Farbotko 2005, 2010a, 2010b, Benwell 2011), to the extent that a 'Tuvalu syndrome' (where a state may 'geo-engineer' for its own benefit at the expense of others) has been described (Millard-Ball 2011). Other territories entirely composed of coral atolls include Tokelau, but most Tokelauans now live in New Zealand and it has chosen to remain a dependent territory, and many atolls are part of larger states with high islands, or in states with migration rights elsewhere.

Kiribati is most likely to experience future problems, because of its limited local development opportunities, rapid population growth and urban concentration, few migration opportunities and the absence of an obvious 'patron state'. In some respects, its misfortune was to be a British colony, along with Tuvalu, the 'Cinderellas of the empire' (Macdonald 1982), with ties to a nation that largely abandoned the region in the early 2000s when it withdrew bilateral aid from the Pacific. Soon after independence a report on their future concluded: 'considerable political skill will undoubtedly be needed to dampen down local expectations and aspirations that are no longer realistic' (Geddes et al. 1982: 187). Two years later a review (the Jackson Report) of Australia's Pacific aid concluded: 'Their long-term development prospects are discouraging. In view of structural problems which are beyond their control and beyond the reach of aid, Australia should make available limited opportunities for immigration from Kiribati and Tuvalu' (Australia 1984: 8). That was ignored. Kiribati has redefined its policy towards mobility as one of 'migration with dignity' where skilling migrants will make them more likely to be welcomed and to generate more remittances than unskilled workers. No other country has become so oriented to overseas employment and migration as a means of resolving economic and environmental problems, yet formal opportunities remain few, and are embedded in global moral economies and geopolitics.

Atolls have long been exceptional in the challenges they face, from their 'Malthusian' status a century ago, to more contemporary economic challenges combined with a 'perfect storm' of diverse environmental threats (Storey and Hunter 2010). The impacts of a 'perfect storm' have already been all too often 'observed' in Tuvalu, but there is no apocalypse now. Nonetheless debates have already focused on the constitutional status of states should they be drowned, when the conjuncture between permanent territory and population disappears, and over the status of 'environmental refugees' unrecognized in legal conventions (McAdam 2010). Thus in the Marshall Islands in 2011:

> Threatened by impending SLR, internal fraud schemes that reached into the highest levels of government, and potentially damaging changes to the relationship with United States, the Marshall Islands spent the year facing its own mortality … indeed its very existence within the community of nations

> … testing its ability to survive and persist as a nation physically, politically and socially. (Kupferman 2012: 150, 156)

Climate change can be a mystifying phenomenon. Certainly a 'perfect storm' is an apt metaphor for a future where environmental challenges may eventually outweigh economic and social difficulties.

IN SEARCH OF HYBRIDITY?

Almost everything discussed thus far can be loosely subsumed as globalization: the gradual, benign or brusque, uninvited and welcome, incorporation of SISI into a much wider world, whether through plantations, mines, tourism or trade stores; the politics of colonization and colonialism; more recently imposed development ideologies; the cultural effects of religion, language, television and organized sport; environmental impacts, from deforestation and overfishing to anthropocene climate change; technical changes in transport and telecommunications, global labour markets and so on. SISI have gone through multiple changes tending towards individualism and modernization. Migration, urbanization and the deterritorialization of societies have all contributed to the fragmentation of social structures.

In all this the Caribbean, and then the Indian Ocean, long preceded the Pacific, but trajectories of development have varied. One of the few 'compensating virtues' of small size has been that external powers have infrequently sought their fewer resources and people. Some well-known transnational corporations are simply absent from places too small and poor to be large markets; even McDonald's, that paragon of globalization, has reached just three SIS (Bahamas, Fiji and Samoa). Colonialism and even capitalism were often belated and somewhat tenuous, at least in the Pacific. Yet gloom hangs over degraded ecosystems and fractured cultures, unemployment, drugs, corruption, crime and uneven terms of trade: parts of the shock of the new. Such multiple impacts of problematic globalization (where peace and penicillin were forgotten) emphasized that despite there being virtues in other lives and worlds, and that for some the future lay elsewhere, holding on to home had parallel virtues and ensured a place in a known world.

The fruits of capitalism were desirable. It is unsurprising that a multitude of studies have written along the lines of one account of villagers in Pohnpei (FSM) who desire 'continued and increased access to the goods and prestige provided by employment' (Petersen 1979: 37), while in Samoa, where migration is pervasive:

> Ironic as it may seem these migrants seek in the west the very things that modern day Gauguins want to leave behind – access to material goods, jobs in the industrial sector, better education for their young, and social mobility in a society they have believed free of the traditional barriers of rank and family status that made such mobility difficult at home. (Shore 1978: xii)

Island societies could be every bit as hierarchical, and class based, as others. As people increasingly 'live from money' resources are carefully husbanded and where possible passed on. Elites and classes have emerged and become solidified. Socio-economic stratification is no longer transitory (Chambers and Chambers 2001: 232–233). The expectations of islanders gradually elide with those of other people elsewhere, as islanders consciously reject certain facets of 'tradition' perceived to be oppressive. Again, in Samoa:

> New technology, a market economy, and mass migration make individual production, profit and accumulation of wealth possible. In these circumstances, village people no longer render their service gladly to another, for they see little benefit in such a system for themselves and their children. (O'Meara 1990: 154)

Likewise in Lamotrek (FSM), by the 1980s:

> Cooperative spirit … has been weakened by the differential wealth that has begun to isolate individuals and fragment residential groups and lineages. The traditional authority and power structure … based primarily on control of land and genealogical standing, cannot help but be altered as money and external goods become new sources of power. (Alkire 1989: 188)

Even in such tiny and remote places significant local values have been eroded with loss of community and social cohesion (Kenter et al. 2011), and a decline in respect, trust, solidarity and social control. While that is all utterly familiar – monetization, modernity and individualism – it is also, if less frequently, qualified:

> The ceremonial system endures today not because Samoans are caught in a 'cult of custom' but because Samoans have adapted their customs to serve their present needs. The newest and most powerful force *supporting and modifying* Samoan custom today is the search for money … Yet they continue to maintain the wider social affairs of their extended families as important sources of pleasure, power and prestige, and as a major source of both immediate monetary income and of long-term economic security … Villagers still control custom however and they know how to manipulate it to their advantage. (O'Meara 1990: 154; my italics)

Custom and tradition, however conceptualized and ever changing, have endured. Both accidentally and strategically local cultures have been retained, and not because of any misplaced romanticism for an imagined past.

Islanders regularly express anxieties about globalization and express preferences for a familiar and functional social order. People decry the greater scale and anonymity of society and economy, and the passing of local values in the face of alien values (notably individualism and monetization). Change is incessant and unavoidable. The incursions of outsiders shifted values, interests and orientations. Religion reshaped cosmologies and social worlds and changed the scale of the spiritual world. The 'coming of the light' meant antagonism to the preceding 'days of darkness' as missionaries sought to replace customs with 'enlightened' values; much of local culture, especially its ritual and visible manifestations, simply disappeared. Yet in the mystical and intangible realm of religion, accommodation and change were inevitably complex. Money diminished social cohesion, with reduced time and interest in communal activities, whether ceremonial or practical, and increased jealousy and uneven development (e.g. Fazey et al. 2011, Schwarz et al. 2011). More intensive work experiences, and residence overseas for work and education, increased dependence on imported foods and goods and weakened local social obligations.

Societies have ineluctably but unevenly modernized, whether in material, symbolic or ideological terms. Although the pace and structure of change have varied even the most remote societies are 'exotic no more' (Besnier 2011: 6). Nor are they wholly modern, however that deeply problematic word is defined. Many people have entered into 'a complicit relationship with modernity within which they participated in politics, government and the cash economy while reclaiming their own grounds for participating in these projects of self development' (Lattas 2006: 148), so actively shaping their relationships with modernity and globalization. Active engagement has proved beneficial, even – perhaps particularly – when it perplexes and frustrates outsiders. Retaining fluid and flexible land-tenure systems, in the face of invocations towards individual ownership and sale, enabled Ranongga islanders to cope with and respond to diverse external pressures in satisfying ways, and achieve greater security in a world of change, uncertainty and ambiguity. Local ties and customary land tenure are obstacles to some forms of economic development but provide the basis for resilience and social support (Browne 2004, Macpherson and Macpherson 2009, Richmond 2011). Customary tenure insulates against the vagaries of the commercial economy.

Customary practices may be opposed by external agencies and by islanders. Social obligations were sometimes locally ridiculed as a foolish waste of entrepreneurial opportunity (Chambers and Chambers 2001). Governments, external agencies and the media in turn have sought to stimulate and shape modernity and nationhood, as in Fiji, by espousing a set of values redolent of modernization theory and the Protestant ethic, centred on independence, discipline, hard work and social mobility (Connell 2007g). In Kiribati the practice of reciprocity (*bubuti*) enabled both a general equivalence in development status and a means of coping with hazard, yet the ADB has argued that the practice 'may not be so necessary today' (ADB 2009c: 11). Similarly the equivalent Fijian practice of *kerekere* has been argued to lead to 'perverse incentives' (O'Garra 2012). While both hinder entrepreneurialism they support equity and resilience. Retaining at least some traditional agricultural and fishing practices, and associated technical knowledge, enabled people to cope with change in well understood environments. Such 'progressing with' and 'learning from' the past (Clarke 1990), by combining local knowledge and scientific understanding, provides some insurance against natural disasters and climate change, in giving proper recognition to the environment in a global world. The past was not necessarily found wanting.

While social capital was eroded, in many islands much of what is valued in local life remains: a sense of community in association with a tract of land; shared beliefs and values; rituals and relationships; a rough equality of material conditions; recognition (if not necessarily respect); reciprocity, exchange and some degree of community control over the means of production. Such values, that enable localized autonomy and a degree of self-reliance, are not readily transferable into a more globalized world (McDougall 2005). Equally, many western values and institutions have proved fragile 'foreign flowers' (Larmour 2005) doomed to disappear or be transformed in distant places. Globalization has left no flat world but, as islands immediately suggest, a 'bumpy' fragmented world where differences and distinctiveness remain.

Islanders have sought to combine cultural continuity with social change and economic development. Excursions into capitalism, in whatever form, were cautious essays in economic hybridity that did not contest or reject the expansion of capitalism (Connell 2007d, 2007f, Gegeo 1998, Yang 2000), but sought more successful and more culturally sensitive forms of accommodation to it. Change brought desirable additions to local culture, whether in diet, music, mobility or spiritual beliefs, well evident in medical practices, where traditional medicine is combined with modern biomedical practices into an integrated holistic system, sometimes formally recognized by governments and providing health and social benefits (e.g. McGrath 1999, Macfarlane and Alpers 2010). Legal systems can be similarly hybrid. Evolving practices may be little more than exercising, as in Martinique, a 'Creole economics' that is 'the family of local economic strategies that sidestep the legal claims of the French state' that are both rational, culturally specific and enable economic autonomy, and in the particularity of Martinique, 'an effective strategy of cultural resistance' and a means 'to subvert French authority' (Browne 2004: 48, 213). Traditional goods sent as 'reverse remittances' to the diaspora have symbolic and nostalgic value in strengthening social relationships, and emphasizing and encouraging reciprocity (Donner 2002, Alexeyeff 2004b). Village stores can be sources of social support to the village and kin as much as means of profit making. Even where islanders have been enthusiastic participants in production for cash, as in the Pacific bêche-de-mer fishery, it has taken quite different forms according to local cultural distinctions. Local languages, including Creole, in the Caribbean and Cape Verde, and rotating credit associations in Barbados and the Bahamas (Stoffle et al. 2009, Challinor 2012), are resources, cultural shields, safety nets and claims to identity, not mere limitations, enabling a degree of autonomy and a sense of identity and belonging.

Resistance to the outside world sometimes appeared either obtuse and stubborn, or a form of militant cultural conservatism. Vanatinai islanders (Milne Bay, PNG) deliberately resisted coercive external pressures through steadfast opposition to participation in the cash economy for which they:

> would sometimes give pragmatic and quite logical reasons: too much work for too little money, too far to the nearest trader, no shipping … But the most common response was the flat statement, *Abotewa* (I don't want to) … refusing to work for cash is … a reaction against pressure to give up self-sufficiency and autonomy for greater dependence on the world cash economy. (Lepowsky 1991: 224-5)

Likewise amongst Mandok (Morobe, PNG) islanders: 'we might make more money if we cash cropped, but that is not who we are' (Pomponio 1992: xix). In such remote islands disengagement also followed unmet expectations and the perceived unfairness of rewards. On Ambrym villagers:

> have retained their high level of self-sufficiency and resilience, and are thankful to live in such a peaceful society … Free of any European presence, they seem very comfortable in and with their relative isolation, remarkably

> unaffected by any outside pressures on them to be anything other than what they already are. (Tonkinson 2004: 194)

The Kwaio of Malaita (Solomon Islands) were said to have 'kept their commitment to their ancestors and the past, in the face of Christianity, copra and cash, and nowadays, tourists transistors and Toyotas' (Keesing 1982: 39). Yet few others share such a perhaps fortuitous and congenial accommodation with the outside world. Indeed Ambrym and Malaita too are embedded in wider networks of production and exchange and many have migrated from there. Similar responses and orientations occur partly because economic opportunities are few, so investing in custom avoids 'intensive self-exploitation in agricultural activity' and provides respect and autonomy (Evans 2001: 17–18). However, cultural pride and conservatism emanate from more remote islands where colonial and post-colonial incursions have been weak and opportunities to benefit from the wider world have been few and fragmented. Neither there nor elsewhere do people wish to retreat to a world without store clothing, metal tools, outboard motors and penicillin. In Ambae, not far from Ambrym, 'islanders have one foot somewhat tentatively in the capitalist market-place and the other squarely in a lifestyle based on values and relationships that are very different from those underlying a capitalist way of life' (Rodman 1993: 183). Put another way: 'People's attitude to modernization is less a measure of their conservatism than of their search to innovate in the face of great uncertainty' (Meltzoff and LiPuma 1986: 61). Where they could, islanders have wisely kept their options open, often in response to frustrations of participation, to maintain an 'exit option' when prices or conditions are unsatisfactory.

Growing needs and pressures on land and sea have meant that islanders have sought diversity. 'Occupational multiplicity' is increasingly normal whether in rural or urban areas: a 'natural survival strategy' (Baldacchino and Bertram 2009) that could even be combined with 'environmental multiplicity' (Stoffle and Minnis 2007). On drought-prone Tamana (Kiribati):

> Life is a risk-spreading activity not a rigid adherence to one particular strategy. The aim is to keep as many options for the future open as possible; plant lands for subsistence foods and cash, tend *babai* [swamp taro], maintain fishing skills and equipment, capitalise on educational opportunities as a means of access to employment, adopt family limitation techniques and augment resources through resettlement. In the face of uncertainty and limited resources, this is the best hope for a sustainable future. (Lawrence 1992: 55)

Similar 'portfolios' of livelihood strategies have emerged in Bellona (Reenberg et al. 2008), the Lau islands (Turner et al. 2007), Ambae (Rodman 1987b) and elsewhere (Chapter 3) with risk-spreading constituting a 'security-centred survival algorithm' (Brookfield 1975) for negotiating away from a poverty of opportunity.

Symbiosis may thrive in a plurality of practices. In Vanua Lava (Vanuatu) some men cultivate 'traditional' taro species introduced by mythic ancestors; others make copra in coconut plantations introduced by colonialists.

> Taro-men need copra-men to maintain open commercial roads and also small shops where Western goods such as rice, tinned fish, matches and kerosene for lamps can be found. They are the links with the outside world. Taro-men are also essential to the village: … they conserve a piece of [the village's] heritage and … keep alive the memory of the ancestors … a garden of taro can be compared to an archival library of relations and events. (Caillon and Degeorges 2007: 2923, 2926)

Change has been neither rapid, unilineal or homogeneous: indeed so often a confusing, experimental exercise, with multiple entanglements. Local structures and outside structures, in whatever context, are neither fixed nor separate; interplay between the local and the global lie at the heart of the struggle for a compatible, autonomous modernity.

Notions of individuality, social obligation and virtuous relations with kin and community are contested in a wider commercial world. Tensions and vested interests shaped accommodation, which could never be wholly successful, or appropriate for all, hence the pervasiveness of migration. As Tonkinson has remarked of religion, but so widely true, the local context 'blunted its more inconvenient or incompatible edges, blended in some of the more amenable elements from "tradition" and smoothed it into a faith they are comfortable about living with' (2004: 186). Numerous Melanesian islands have an extraordinary and expanding diversity of religions, from animism through ecumenical and indigenous forms of Christianity to Islam, reflecting and exacerbating splits in social groups through new sectarian conflicts, and in opposition to the 'elite' establishment religions that failed to secure appropriate social change and action (Eriksen 2000, Flower 2008, McDougall 2009). Every facet of modernity took new and multiple forms in different contexts.

Hybridity is not purely a rural phenomenon, the resort of villagers irremediably tied to the land, but is as significant in urban areas. Even the most predatory transnational corporations are combined with local empowerment in 'the productive friction of global connections – the awkward, unequal, unstable and creative qualities of interconnection across difference' (Tsing 2005: 30–31) that respond to flows of goods, ideas, money and people. Forms of hybridity and syncretism have shaped evolving political institutions, involving combinations of western models of statehood (parliaments and free elections) and customary governance (chiefs and elders, customary law) (Brown and Gusmao 2009, Cummins 2010, Richmond 2011). Indeed, '[m]odern rule-of-law systems have their own drawbacks: they are cumbersome, slow, excessively formal, and expensive, requiring staffing by highly trained professionals. Customary justice by contrast is low-cost, flexible, and in line with cultural expectations' (Fukuyama 2008: 28–29). Political legitimacy is a local phenomenon.

Local societies, however, are imperfect, with inefficient, patriarchal and nepotistic practices that were scarcely benign. Indeed egalitarianism, unity and solidarity are sometimes merely a 'plausible fiction' (Hooper 1993). Gender relations were often imbued with inequality and violence, leaders despotic and punishments draconian. Trade and marriage ties could just as easily become warfare. Neighbours could as easily be enemies. The past was not some 'merrie Melanesia' (Filer 1990). Considerable difficulties have accompanied nation building in conditions of economic and political uncertainty, where complex cultural differences are not simply awkward vestiges of historic practices, but mark real divisions. Local people could despise others – such as the *repo taro* – simply for having different languages and cultures, and thus being different and inferior. Even in small unitary states such as Grenada, the 'rigid stratification system and the preoccupation of the elite with race, class and status seems to have kept the society divided to a point where it affected the development of national institutions' (Benoit 2011: 577). Creating nations from distinctive cultures, or where local cultures emerged from long histories of identity crisis, with highly stratified social classes and status based on race, wealth and income, challenged SIS into constant negotiation and refinement.

Optimism that people can reject what is harmful, retain what is valuable and avoid what is destructive may, however, be no more than a pious hope. Developing effective MPAs based on indigenous knowledge and practices has proved difficult enough where coastal populations are growing, knowledge is limited, management weak and markets beckoning (Chapter 3). Indeed, resurgence of localism can produce instability, as in Solomon Islands, where 'complex processes of social political and economic change are recast [by outsiders] as predominantly technical and administrative challenges' (Dinnen 2009: 77; cf. Huffer and So'o 2005, Li 2007), although violence was a function of cultural identity, but also of poverty, migration and dissent over land tenure and inequity (Allen

2012). Moral economies, providing support for the vulnerable, have been weakened by loss of leadership, fragmentation, declining social cohesion and reduced autonomy. The public ethic where individual behaviour is monitored and restrained by the group (as in MPAs) has withered as the scale of social life has expanded. Inequalities, poverty and competition for scarce assets are more evident.

The Ambrymese, content to retain established values and practices and remain at home, are increasingly rare. By contrast the young men of the much smaller island of Panapompom (Milne Bay, PNG) not only play football, but are anxious to play (and dress for it) in a way that demonstrates they have mastered the complexities of a colonial introduction, and have acquired a moral order unconcerned with 'custom or culture' (Rollason 2011). Cultural change becomes vivid in generational shifts, with the young taking up new patterns of consumption and recreation, in education, food, drink, drugs, music, sport and dress, that stress distinct identities and moral codes. In the Cook Islands contemporary songs are informed by experiences derived from tourism, migration, consumption and politics, becoming 'part of an ambivalent discourse in which signs of global capitalism are both admired and derided' (Alexeyeff 2004a: 152–153). Such cultural shifts translate into tensions between the young seeking 'modernity', older people preferring the status quo and the desire of children to be dutiful. Hybridity and syncretism are invariably uneasy, with generational and geographical tensions and complexities.

Yet a plethora of studies have eulogized local knowledge (while decrying its disappearance) and its value in such activities as building effective structures of law and governance, developing more effective fisheries management programmes, building hazard-resistant homes and more generally coping with hazards – in some combination with scientific knowledge (Huffer and So'o 2005, Mercer et al. 2007, Lauer and Aswani 2010, Richmond 2011, Brewer 2013, Hamilton et al. 2012). Knowledge of the past, and of indigenous values, epistemologies and moral economies, is not mere romanticism and resistance to globalization, but is essential to social and economic development. Yet hybridity cannot resolve everything; culture cannot always be triumphant. Local 'knowledge' can be dogma rather than 'science' and may misinterpret past and present phenomena (Connell 2003a, Brewer 2013), yet local knowledge and environmental practices can nevertheless mitigate and soften the blows of environmental change. Accelerated climate change introduces new dimensions and new uncertainties. Hybridity is only new in its transnational dimensions and complexities, and in the need to respond to the speed and uncertainty of change.

Island development and change might well be seen through a cultural prism where culture is no threat to development but a core element of it (Curry 2003). Yet 'culture' can seem all too retrospective and amorphous for proponents of institutional change and economic growth. It may be difficult for many SISI to sustain a 'cultural turn' but in their interests to do so, despite being at variance with development strategies imposed from outside. Hybridity and livelihood diversification insulate islands and islanders from environmental and economic shocks. Identity and culture offer knowledge, security, hope and continuity within change.

FROM THE ENDS OF THE WORLD

Throughout SISI the old orders of subsistence (and plantation) agriculture and fisheries have been transformed. Societies have become more complex, diverse and heterogeneous. Tourism and urbanization have arrived, occupations and livelihoods have universally diversified, farmers and fishers are aging (and their work no longer holds status or provides 'fast money'), occupational multiplicity is more common and household incomes have shifted to off-farm sources. Lives are more mobile, livelihoods less local, dependency rates have increased, remittances (and urban contacts) become more valuable, and all accompanied by social change (Rigg 2006). Yet a strong moral preference is often attached to rural life and activities, exemplified by notions of 'safety nets', not least by governments and aid donors, linked to concerns over the rise of urban settlements. Islands are so often presented to the world as idyllic spaces, part of a natural order where nature is verdant and bountiful, migration implausible, community strong and urban blight absent (Connell 2003c, Kahn 2011). While households do indeed have attachments to land, villages and islands, and maintain at least vestiges of a moral economy, they are neither wedded to the land nor immobile. Stability was never 'normal', however much islands appeared to have fixed boundaries and distinct cultures.

Diversification and mobility characterize contemporary island lives. New opportunities (squash and passport sales) and multiple occupations (and incomes) are eagerly sought. Those who are better off have gained access to off-farm employment within or beyond the village, hence the rise of urbanization (and commuting), however bleak urban employment opportunities might be. Pluriactivity has marked successful adaptation to change, even in small and remote islands, such as the Lau islands and Pitcairn. Yet poverty and the lack of skills, capital, contacts and knowledge are critical obstacles to livelihood diversification (Cinner et al.

2012). Outer islands and remote villages may be disadvantaged. Diversity is not always possible (hence the empty huts of forlorn ecotourism projects, toothless MPAs and unsustainable logging) – contemporary components of 'a long history of flexible adaptation' (Perminow 1993) and 'opportunistic pragmatism' (Baldacchino 2010) that could not always succeed. A poverty of opportunity, rather than measurable income poverty, typifies SISI.

Occupational multiplicity and 'strategic flexibility' (Baldacchino 2010) are manifest in migration. The best opportunities for diversification lie elsewhere, in larger places and larger countries. Migration typifies and enhances livelihood diversification: reducing risk and maximizing opportunities by spreading household 'portfolios' regionally and globally, increasing household incomes (and reducing poverty through remittances becoming social support), and developing human resources. In a world where lives are now lived at very different scales, recourse to distant support is a strategy just as legitimate as earlier local and regional processes of sharing and reciprocity (McInnes 1986, Pam and Henry 2012). Only the scale has changed. Throughout SISI individuals, households and states have chosen to elaborate on their ties with distant metropolitan states, as a particularly effective form of diversity.

> [I]t thus becomes clear that 'sustainability' in the late twentieth century need not be bound to one arbitrary spatial unity. People and economies, and thus ecologies, are bound together in a global system. Politically motivated demands that development be contained within the nation-state are neither practicable nor … sustainable. (Evans 1999: 160–161)

Migration – a bottom-up globalization – is not merely a 'reluctant necessity' (Wilson 2010), resulting from the growing demand for cash incomes for 'needs', such as school fees and medicines, but enables satisfying increasing 'wants'. That has been well expressed in the agency of Ponam islanders:

> Ponam is not a society passively allowing or suffering the migration of its members. Rather it is a society which actively regulates this migration, and it does so in a way which helps secure a substantial remittance income … 'children are our garden and we survive by eating the fruit'. (Carrier 1984: 42, 49)

But 'reluctance' is expressed in elaborate arrangements to hold on to 'home' by safeguarding land ownership, as in Mokil (Oles 2007a), Tikopia (Rasmussen et al. 2009) and Naikeleyaga (Fiji) (Felgentreff 1996), even when return seems improbable, as on Pitcairn. A host of studies demonstrate that alienation from home is unwelcome and widely feared, as islands represent a 'safe haven' and familiar place in times of need (e.g. Gemenne and Magnan 2010: 49). The economic role of land has weakened as its social and strategic role has grown.

No island realm is likely to become an 'empty quarter' (Ward 1989) even in the most difficult environmental and economic circumstances, despite an almost imperceptible but seemingly irreversible contraction of the global margins of settlement and the seas of islands. As in other parts of the world, most people simply want to remain where they are, even on the most remote islands, as long as they have reasonable access to incomes, amenities and acceptable lifestyles. Remittances make remaining in small places possible, without giving up certain 'traditional' values and enabling access to 'modern' needs. Yet what constitutes an acceptable lifestyle has evolved with globalization, hence considerable permanent migration has ensued.

Shorn of mendicancy and necessitating the agency of households, a MIRAB economy makes sense and is no mirage. It has ensured the viability of many tiny islands, but accompanied a gradual transformation from subsistence to subsidy, most evident in outer islands and politically dependent SISI. So crucial had remittances become, even by the 1980s, that in smaller SISI Bertram argued that a focus on production was misplaced since 'aid, philately and migrant remittances are not merely supplements to local incomes, they are the foundations of the modern economy' (1986: 809). Philately has fallen by the wayside, but aid and remittances have not. While the MIRAB economy enabled islands to remain viable it shifted the locus of power and status away from them, challenging the social order and the retention of older values: village social relationships were drained of their former significance. Migration, markets and remittances both contribute to and challenge hybridity and come at a cost – distant human resources, divided families and 'barrel children' – as diasporas shape household and national development. Maintaining the transnational corporation of kin is a massive social achievement but a continued challenge amidst the deterritorialization of nationhood, home and identity. Remittances can be a burden on migrants, however altruistic or self-interested they may be, and their intergenerational future is uncertain. A tenuous balance exists between upward social mobility and the retention of cultures. As a 'culture of migration' became more widespread, notions of 'home' have changed, and both at 'home' and elsewhere, cultures have become ever more hybrid and syncretic.

Amidst such changes, culture still 'sits in places' (Escobar 2001) and loyalties to both places, with their distinctive environments, and people are vital. An emphasis on both locality and globalization does not bind together antithetical phenomena. Whilst identities may be hybrid and hyphenated, they may also, as in the case of Samoans in New Zealand, represent successful cosmopolitans 'at ease in multiple worlds, rather than natives of place torn by new and multiple allegiances' (Yi-Fu Tuan, quoted in Connell 1995). Indeed, it has been said in Samoa that it 'is not a place but an identity' and all Samoans, wherever they may be, are heirs to both land and identity. Contraction and compression of the world through migration and IT have widened international ties. While cultures shift, citizenship changes and new loyalties emerge, economic and social ties to home have remained remarkably resilient, enduring long beyond their predicted demise. That is not surprising: remittances and retention of cultural identity are forms of social insurance that preserve the possibility of return from an uncertain world.

Patently, islanders have limited agency to choose their own destinies and development: 'Cape Verde's situation of deprivation and poverty does not allow it to be a self-sustaining and autonomous nation. However, this has not precluded it from imagining itself as a nation moving towards a utopian future' (Rego 2008: 157). Nonetheless, while development may be the outcome of a wide-ranging asymmetry 'even the weakest of actors have some power to colour, undermine, resist or ultimately overturn structures imposed from above' (Pelling and Uitto 2001: 55). Flexibility offers numerous possibilities, but no account of development in, for example, the coral atolls of Micronesia, suggests a wealth of choice and a range of positive outcomes. While 'post-development' theorists criticize the goal of development as involving an extension of western control, the accommodations made by islanders involve what Knauft (2002) has termed 'locally modern', with outcomes in keeping with very different conceptualizations of culture and scale. All SISI have policy options that would enhance resilience and enable a greater degree of local autonomy: agricultural development policies that stress diversity and food security, education that reflects local cultures, transport and energy policies that move away from expensive non-renewable resources, regional development (infrastructure and service provision) and fostering a private sector. None of these are easy. As Browne (2004) has shown for Martinique, it is unnecessary to have a past to return to or to invent a tradition, but simply to adjust appropriately to the present. Even on Réunion, settled only in the seventeenth century, 'an island whose cultures, identities and histories have long been repressed, dismissed, rejected, derided' (Vergès 2003: 173), cultural distinctiveness offers possibilities for local development. While the allure of modernity may seem unidirectional and has certainly challenged the foundations of islander ethos and cultural identity it has rarely overwhelmed locality. Local development has both a past and a future.

WARM WINDS OF CHANGE

Diversity characterizes SISI and generalization is difficult. The most challenging development problems face several SIS that are also LDCs – Comoros, São Tomé, Samoa, Solomon Islands, Tuvalu, Kiribati and Vanuatu – all relatively remote from industrialized states, composed of archipelagos, mostly with limited tourist prospects, limited aid flows and emerging migration flows and routes, but above all with little room to manoeuvre. In islands where societies, economies and environments are changing, it is ironic that metaphorical black swans and Amazonian butterflies may be critical influences on change, and usher in new uncertainties. Islanders and their governments grapple uncertainly with their destinies: reacting, resisting and reconciling – but usually achieving a form of negotiated interdependence, even where political, economic and environmental power is asymmetrical, neglectful and biased against them. That is mixed with the tangled webs of personal and national motivation, private and public ambition, variable aspirations – and above all diverse opportunities – all in the shade of a changing climate.

All economies face difficult times, while the world may be on the verge of experiencing peak oil, fish, timber and water, all consequences of growing populations, needs and wants, amidst the 'brutal tectonics of neo-liberal globalization' (Davis 2006: 174). Trade, migration and aid opportunities are not granted willingly and unconditionally. Islanders have had to make their own way in the world, selectively appropriating what they can. On the tiny Polynesian atoll of Kapingamarangi, more than 300 km from its nearest neighbour, on the southernmost fringes of Pohnpei state in the predominantly Micronesian FSM:

> [s]ustainability depends on the maintenance of steady if not growing cash flow through government salaries and stipends, copra production, public-works projects, handicraft production, and the success of planned commercial fishing … The maintenance of this cash flow depends on levels of funding for the Federated States of Micronesia, which depends on the relationship [with] the United States, between the Federated States and the consortium of countries sponsoring fishing fleets in Micronesia, between the Federated States and the Asian Development Bank and so forth. The decisions affecting these relationships are made in Washington, Suva, and Tokyo, and lastly, on Pohnpei, where the national government determines disbursement to the state.

> The proportion of funding going to the atoll depends on the negotiating skills of its representative to the Pohnpei State Legislature, where Kapingamarangi is one of eleven municipalities competing for a shrinking pool of dollars. (Lieber 1999: 39)

Not only are such processes unknown or at best poorly understood at the local level, but also Kapingamarangi residents can do little to influence them and, as one of the most remote islands in the Pacific and one of the smallest in Pohnpei, their ability to access central funding is weak. So too is the ability of FSM to access United States' funding. Sustainability thus also depends on the ability of islanders to migrate, find jobs and send remittances home. That too, whilst also a function of decisions made elsewhere, is one where islanders themselves – even on the smallest, most remote islands – necessarily play a part in a very political shaping of ecology.

Global circumstances constantly change. Twenty-five years ago it was evident that:

> Fiji's membership of the Africa-Caribbean-Pacific group assures it of the benefits which come from a conviction that it is politically necessary to keep the economies of these countries buoyant to avoid their descent into poverty, revolution and a shift in their allegiances. The survival of Fiji as a relatively prosperous, relatively stable cornerstone in a strategically delicate South Pacific is of importance to all the larger circum-Pacific countries, and to countries as far away as Europe. (Brookfield 1987: 56–57)

But that geopolitical era has gone, agricultural exports to the EU have faded and Fiji has subsequently experienced coups. Like other SIS, it has been forced to find new means of survival, beyond the agricultural exports that once seemed guaranteed. As protection and preferences have faded, asymmetry has increased further, whether in access to education, science and knowledge, or in negotiations over trade, tourism or fisheries, sometimes – as in the banana wars and internet gambling – emphasized by United States' exceptionalism (Cooper 2012). It has been suggested that tuna fisheries offer no real development prospects for Pacific SISI (or SISI elsewhere) and that a more positive future development rests on emigration and remittances, alongside fishery leases and aid (Hannesson 2008). Governments are stymied in their efforts to generate development, resulting in improbable economic choices (.tv, internet gambling), a widespread hope for tourism, and dependence on aid. Agriculture and fisheries have limitations; tourism and migration are more attractive options.

A wide-ranging study of agriculture in the Pacific islands in 1980 concluded that 'a no-growth state of affairs' was the only realistic future for island states, unless they were to permanently rely on foreign aid, and therefore that 'a policy of unrestricted temporary or permanent migration … would seem to be the only one that gives the people of the small island countries the chance to choose individual paths of development within the wide community' (Castle 1980: 136). The Jackson Report reached the same conclusion. While such conclusions then seemed gloomy and were ignored in metropolitan states, islanders and their governments have chosen migration although doors have only been reluctantly opened to them. Deterritorialization has become realistic and practical. Self-reliance is dated; for Kiribati: 'A greater degree of self-reliance is possible though probably at a price. It will not be achieved without further sacrifice in terms of foregone consumption and restricted aspirations' (Pollard 1987: 23). A quarter of a century later such invocations seem implausible. Like 'appropriate education', self-reliance is unlikely to be acceptable and even its lingering rhetorical pursuit has long seemed no more than a reflection of the aspirations that must suffice where growth is unlikely: 'a ritual for exorcising the devil of dependence' (Joseph 1978: 223). Disadvantageous geographies and unhelpful ecologies, weak economies, frictions of distance, overburdened bureaucracies, urban unemployment, fractured social networks and uneven development have challenged notions of national, let alone sustainable, self-reliant development.

Migration offers less risk than dependence on aid or commodity prices, and is at least as sustainable as any other approach to national development, however much it is subject to international regulation. Do-it-yourself development has become the 'new self-reliance'. For SISI sustainability is global. International migration and 'occupational multiplicity' have extended families, risks and opportunities ever further, as livelihood strategies have expanded. Technological change has created new connectivities: island communities are simultaneously being united in virtual space and becoming fragmented across the globe. As opportunities have been sought more widely, local economic and ecological diversity has declined. Despite social costs, both short- and long-term migration have largely been beneficial for participating individuals, households and countries. Natural increase continues and a 'population surplus' is constantly being exported, where possible, to ensure adequate economic survival. Perhaps, as in Ontong Java, future remittances will become 'the most important coping strategy in response to climate-related impacts on food availability' (Rasmussen et al. 2009: 10). But that particular future is not imminent. Nonetheless a diasporic, multi-local future is emerging, actively sought by both households and countries.

Since, by definition, SISI have small populations, migration has been and will be limited in absolute numbers, and governed by evolving policies and practices in potential destinations. Island states are in no position financially, geographically or politically to defend themselves against potential threats, whether economic, political or environmental, but if – perhaps when – the worst occurs then islanders may become a new stream of 'environmental refugees' to metropolitan states. They will be well positioned for that, having actively created transnational kinship connections: extended global versions of the regional 'coral clusters' that linked islanders in pre-colonial times.

The smallest islands, and the smallest states, may be losing population but most people seek to retain the security of local connections. Individuals, households and SIS spread their human resources, diversify, choose educational strategies that create human capital and favour migration, make choices and seize opportunities where they can – both holding on to home and seeking to embellish that home with the fruits of education, migration and diversification. It may be, as more generally in the 'Rural South', that the best means of achieving pro-poor development has less to do with supporting agricultural development 'and more to do with endowing poor people with skills so that they can *escape* from farming and, perhaps, *escape* from the countryside' (Rigg 2006: 196; my italics), and escape from small islands. It is an uncomfortable conclusion that superior development opportunities lie elsewhere, while ironically small islands are still capable of producing foods that a growing world population needs.

Migration is rarely absolute, unambivalent or final: it is not a cause and consequence of a definite break with a cultural life that is part of a disappearing history, but a partial and conditional state, characterized by ambiguity and indeterminacy. Remittances, return migration and, now, mobile telephony, Facebook and the internet hold households together. Migration separates and connects, stabilizes and disrupts: migrants are 'leaving in order to stay' and migration has become a way of 'maintaining the family by leaving the family … and a means of going away without leaving' (Nietschmann 1979), enabling cultures to change slowly and provide a satisfying and secure alternative and base for return. Islands are no anachronism to be easily abandoned. Remittances both safeguard and erode the vitality of places; they conserve and dissolve, destabilize and sustain. There is no real paradox in migration and culture pulling in different directions in a global deterritorialized world. The culture of migration enables the retention of culture. Belonging at home is personal and political, intimate and strategic. Home-building and transnationalism are complementary.

Islands are small and small places are vulnerable, but experience has made islanders resourceful and resilient. Remoteness and isolation have enabled continuity as well as change. SISI vary enormously – from small islands like Sint Maarten, seemingly overwhelmed by tourism, to much smaller islands overwhelmed by nothing at all. In 1991 I spent three months on the remote Micronesian atoll of Woleai. On my first day I walked slowly round the atoll; it took much less than an hour as I pondered how I would cope in such a small place. Towards the end of my stay one of the chiefs requested me to talk in his house at the other end of the village, fully 300 metres away. My first reaction was to resent being asked to travel so far. The island had grown large in three months as it more obviously became the home of several hundred diverse people shaping their own futures in various ways – the same people who, when I had asked them about climate change, tended to respond 'We have dealt with everything that God had thrown against us so far; we shall survive whatever happens in the future'. Islands and islanders have survived, often rather well, through developing connections. In ever changing guises, that will require maintaining the trees, building strong canoes and creatively responding to changes in the weather and in the global economy.

References

Abbott, D. (2007) Poverty and pro-poor policies for Pacific Island countries. *Asia-Pacific Population Journal*, **22**(3), 59–74.

Abbott, D. and Pollard, S. (2004) *Hardship and Poverty in the Pacific*. Manila, Philippines: Asian Development Bank.

Abdulkadri, A., Cunningham-Myrie, C. and Forrester, T. (2009) Economic burden of diabetes and hypertension in CARICOM states. *Social and Economic Studies*, **58**, 175–197.

Adam, P. (2000) Experience and hopes for telehealth in Tokelau. *Pacific Health Dialog*, **7**(2), 61–62.

Ahlburg, D. and Levin, M. (1990) *The North East Passage. A Study of Pacific Islander Migration to American Samoa and the United States*. Canberra, Australia: ANU National Centre for Development Studies.

Akee, R. (2010) Who leaves? Deciphering immigrant self-selection from a developing country. *Economic Development and Cultural Change*, **58**, 323–344.

Akesson, L. (2008) The resilience of the Cape Verdean migration tradition. In L. Batalha and J. Carling, eds, *Transnational Archipelago. Perspectives on Cape Verdean Migration and Diaspora*. Amsterdam: Amsterdam University Press, pp. 269–283.

Akesson, L., Carling, J. and Drotbohm, H. (2012) Mobility, moralities and motherhood: navigating the contingencies of Cape Verdean lives. *Journal of Ethnic and Migration Studies*, **38**, 237–260.

Albins, M. (2013) Effects of invasive Pacific red lionfish *Pterois volitans* versus a native predator on Bahamian coral-reef fish communities. *Biological Invasions*, **15**, 29–43.

Aldrich, R. and Connell, J. (1992) *France's Overseas Frontier. Départements et Territoires D'Outre–Mer*. Cambridge: Cambridge University Press.

Aldrich, R. and Connell, J. (1998) *The Last Colonies*. Cambridge: Cambridge University Press.

Alexandraki, K. (2005) Preference erosion: cause for alarm? *Finance and Development*, **42**(1), 26–29.

Alexeyeff, K. (2004a) Sea breeze: globalization and Cook Islands popular music. *Asia Pacific Journal of Anthropology*, **5**, 145–158.

Alexeyeff, K. (2004b) Love food: exchange and sustenance in the Cook Islands diaspora. *Australian Journal of Anthropology*, **15**, 68–79.

Alexeyeff, K. (2008) Neoliberalism, mobility and Cook Islands men in transit. *Australian Journal of Anthropology*, **19**, 136–149.

Alexeyeff, K. (2009) Travelling parties: Cook Islanders' transnational movement. In H. Lee and S. Francis, eds, *Migration and Transnationalism: Pacific Islands Perspectives*. Canberra, Australia: ANU E Press, pp. 91–102.

Alexeyeff, K. (2011) Bingo and budgets: gambling with global capital in the Cook Islands. In M. Patterson and M. Macintyre, eds, *Managing Modernity in the Western Pacific*. Brisbane, Australia: University of Queensland Press, pp. 201–230.

Alkire, W. (1978) *Coral Islanders*. Arlington Heights, IL: AHM Press.

Alkire, W. (1989) *Lamotrek Atoll and Inter-Island Socio-Economic Ties*. Prospect Heights, IL: Waveland Press.

Allen, M. (2012) Land, identity and conflict on Guadalcanal, Solomon Islands. *Australian Geographer*, **43**, 163–180.

Alling, A., Doherty, O., Logan, H., Feldman, L. and Dustan, P. (2007) Catastrophic coral mortality in the remote Central Pacific Ocean: Kiribati, Phoenix Islands. *Atoll Research Bulletin,* **551**, 1–19.

Allison, W. (1996) Snorkeler damage to reef corals in the Maldive Islands. *Coral Reefs*, **5**, 215–218.

Alongi, D. (2008) Mangrove forests: resilience, protection from tsunamis, and responses to global climate change. *Estuarine, Coastal and Shelf Science*, **76**, 1–13.

Amarakoon, D., Chen, A., Rawlins, S., Chadee, D., Taylor, M. and Stennett, R. (2008) Dengue epidemics in the Caribbean: temperature indices to gauge the potential for onset of dengue. *Mitigation and Adaptation Strategies for Global Change*, **13**, 341–357.

Ambroz, A. (2009a) Tuvalu Technical Report. A preliminary analysis of extracting aggregate from the Funafuti lagoon. Project Report No. 137, Suva: SOPAC.

Ambroz, A. (2009b) Ba Floods. Economic Costs. January 2009. Technical Report No. 425, Suva: SOPAC.

Amnesty International (2011) *'Where is the Dignity in That?' Women in Solomon Islands Slums Denied Sanitation and Safety*. London: Amnesty International.

Amoamo, M. (2011a) Remoteness and myth making: tourism development on Pitcairn Island. *Tourism Planning and Development*, **8**, 1–19.

Amoamo, M. (2011b) The mitigation of vulnerability. Mutiny, resilience and reconstitution: a case study of Pitcairn Island. *Shima*, **5**, 69–93.

Anderson, M., Holcombe, E., Blake, J., Ghesquire, F. et al. (2011) Reducing landslide risk in communities: evidence from the Eastern Caribbean. *Applied Geography*, **31**, 590–599.

Andréfouet, S., Charpy, L., Lo-Yat, A. and Lo, C. (2012) Recent research for pearl oyster aquaculture management in French Polynesia. *Marine Pollution Bulletin*, **65**, 407–458.

Anon. (2010) Another green world. *The Economist*, 18 December, 145–147.

Anon. (2012) The failed states index. *Foreign Policy,* **194**, 86–91.

Antonsich, M. (2010) Searching for belonging: an analytical framework. *Geography Compass*, **4**, 644–659.

Archer, E. (1985) Emerging environmental problems in a tourist zone: the case of Barbados. *Caribbean Geography*, **2**, 45–55.

Armstrong, H. and Read, R. (1998) Trade and growth in small states: the impact of global trade liberalization. *World Economy*, **21**, 563–585.

Armstrong, H. and Read, R. (2000) Comparing the economic performance of dependent territories and sovereign micro-states. *Economic Development and Cultural Change*, **48**, 285–306.

Armstrong, H. and Read, R. (2002) The phantom of liberty? Economic growth and the vulnerability of small states. *Journal of International Development*, **14**, 435–458.

Armstrong, H. and Read, R. (2003) The determinants of economic growth in small states. *The Round Table*, **368**, 99–124.

Armstrong, H. and Read, R. (2004) Small states and island states: implications of size, location and isolation for prosperity. In J. Poot, ed., *On the Edge of the Global Economy*. Cheltenham, UK and Northampton, MA: Edward Elgar, pp. 191–223.

Asafu-Adjaye, J. (2007) Liberalising trade in the agriculture sector of a small island state: the case of Fiji. *The World Economy*, **30**, 1550–1567.

Asian Development Bank (2002) *Kiribati: Monetization in an Atoll Society.* Manila, Philippines: ADB.

Asian Development Bank (2004) *Pacific Region Environmental Strategy 2005–2009.* Manila, Philippines: ADB.

Asian Development Bank (2009a) *Vanuatu Economic Report 2009. Accelerating Reform*. Manila, Philippines: ADB.

Asian Development Bank (2009b) *Strengthening Pacific Fragile States. The Marshall Islands Example*. Manila, Philippines: ADB.

Asian Development Bank (2009c) *Kiribati's Political Economy and Capacity Development*. Manila, Philippines: ADB.

Aswani, S. (1999) Common property models of sea tenure: a case study from the Roviana and Vonavona Lagoons, New Georgia, Solomon Islands. *Human Ecology*, **27**, 417–453.

Aswani, S. (2002) Assessing the effects of changing demographic and consumption patterns on sea tenure regimes in the Roviana Lagoon, Solomon Islands. *Ambio*, **31**, 272–284.

Aswani, S. and Sabatian, A. (2009) Implications of urbanization for artisanal parrotfish fisheries in the Western Solomon Islands. *Conservation Biology*, **24**, 520–530.

Aswani, S. and Vaccaro, I. (2008) Lagoon ecology and social strategies: habitat diversity and ethnobiology. *Human Ecology*, **36**, 325–341.

Ateweberhan, M. and McClanahan, T. (2010) Relationship between historical sea-surface temperature variability and climate change: induced coral mortality in the western Indian Ocean. *Marine Pollution Bulletin*, **60**, 964–970.

Athukorala, P. (2004) Trade policy making in a small island economy: the WTO review of the Maldives. *The World Economy*, **27**, 1401–1419.

Atkins, J., Mazzi, S. and Easter, C. (2000) *A Commonwealth Vulnerability Index for Developing Countries. The Position of Small States*. London: Commonwealth Secretariat.

Atkinson, J. (2010) China–Taiwan diplomatic competition and the Pacific Islands. *The Pacific Review*, **23**, 407–427.

Aung, T., Singh, A. and Prasad, U. (2009) A study of sea-level changes in the Kiribati area for the last 16 years. *Weather*, **64**, 203–206.

Aung, T., Singh, A. and Maryam, D. (2011) Sea level changes in the Pacific region and impacts of the 2009 El Niño in Fiji Waters. *South Pacific Journal of Natural and Applied Sciences*, **29**, 26–30.

Australia (1984) Report of the Committee to Review the Australian Overseas Aid Program. Canberra, Australia: AGPS.

Australian Bureau of Meteorology and CSIRO (2011) *Climate Change in the Pacific: Scientific Assessment and New Research*. Canberra, Australia: Australian Bureau of Meteorology and CSIRO.

Australian Continuous Improvement Group (2011) Norfolk Island Public Service Review, Melbourne, Australia: Australian Continuous Improvement Group.

AusAID (2009) *Tracking Development and Governance in the Pacific*. Canberra, Australia: AusAID.

Axline, A. (1986) *Agricultural Policy and Collective Self-Reliance in the Caribbean*. Boulder, CO: Westview.

Bacchus, M. (2008) The education challenges facing small nation states in the increasingly competitive global economy of the twenty-first century. *Comparative Education*, **44**, 127–145.

Badjeck, M., Allison, E., Hall, A. and Dulvy, N. (2010) Impacts of climate variability and change on fishery-based livelihoods. *Marine Policy*, **34**, 375–383.

Bagnis, R. (1987) Ciguatera fish poisoning: an objective witness of the coral reef stress. In B. Salvat, ed., *Human Impacts on Coral Reefs: Facts and Recommendations*. Papeete, French Polynesia: Antenne de Tahiti Museum, pp. 241–253.

Bainton, N. (2010) *The Lihir Destiny. Cultural Responses to Mining in Melanesia*. Canberrra, Australia: ANU E Press.

Baker, P. (1985) Volcanic hazards on St Kitts and Montserrat, West Indies. *Journal of the Geological Society*, 142, 279–295.

Baldacchino, G. (2010) *Island Enclaves. Offshoring Strategies, Creative Governance, and Subnational Island Jurisdictions*. Montreal and Kingston, Canada: McGill-Queen's University Press.

Baldacchino, G. (2012) Islands and despots. *Commonwealth and Comparative Politics*, **50**, 103–120.

Baldacchino, G. and Bertram, G. (2009) The beak of the finch: insights into the economic development of small economies. *The Round Table*, **98**, 141–160.

Ballu, V., Bouin, M.-N., Simeoni, P., Crawford, W. et al. (2011) Comparing the role of absolute sea-level rise and vertical tectonic motions in coastal flooding, Torres Islands (Vanuatu). *Proceedings of the National Academy of Sciences*, **108**, 13019–13022.

Banks, G., Murray, W., Overton, J. and Scheyvens, R. (2012) Paddling on one side of the canoe? The changing nature of New Zealand's Development Assistance Programme. *Development Policy Review*, **30**, 169–186.

Barange, M., Allen, I., Allison, E., Badjeck, M.-C. et al. (2010) Predicting the impacts and socio-economic consequences of climate change on global marine ecosystems and fisheries: the QUEST-Fish framework. In R. Ommer, R. Perry, P. Cury and K. Cochrane, eds, *World Fisheries. A Social-Ecological Analysis*. Oxford: Wiley-Blackwell, pp. 31–59.

Barbour, P. and Terry, J. (1998) The hidden economic costs of soil erosion: a case study of the ginger industry in Fiji. *Journal of South Pacific Agriculture*, **5**, 1–18.

Barcham, M. (2009) Cleaning up the Pacific: anti-corruption initiatives. *Australian Journal of International Affairs*, **63**, 249–267.

Barclay, K. (2010) Impacts of tuna industries on coastal communities in Pacific Island countries. *Marine Policy*, **34**, 406–413.

Barker, J. (2000) Hurricanes and socio-economic development on Niue Island. *Asia Pacific Viewpoint*, **41**, 191–205.

Barnett, J. (2011) Dangerous climate change in the Pacific Islands: food production and food security. *Regional Environmental Change*, **11**, S229–S237.

Barnett, J., Lambert, S. and Fry, I. (2008) The hazards of indicators: insights from the Environmental Vulnerability Index. *Annals of the Association of American Geographers*, **98**, 102–119.

Barrett, D. (2008) Renewable energy and energy efficiency in small states. In J. Strachan and C. Vigilance, eds, *Sustainable Development in Small Island Developing States*. London: Commonwealth Secretariat, pp. 44–64.

Barrientos, A. (2010) Vulnerability and social protection in small island states: the case of Grenada. *Social and Economic Studies*, **59**, 3–30.

Barrow, C. (2008) Sexual identity, HIV and adolescent girls in Barbados. *Social and Economic Studies*, **57**(2), 7–26.

Barrow-Giles, C. (2011) Democracy at work: a comparative study of the Caribbean State. *The Round Table*, **100**, 285–302.

Bartlett, C., Pakoa, K. and Manua, C. (2009) Marine reserve phenomenon in the Pacific islands. *Marine Policy*, **33**, 673–678.

Bartlett, C., Maltali, T., Petro, G. and Valentine, P. (2010) Policy implications of protected area discourse in the Pacific islands. *Marine Policy*, **34**, 99–104.

Bartmann, B. (2007) Island war and security. In G. Baldacchino, ed., *A World of Islands*. Charlottetown, Canada: University of Prince Edward Island Institute of Island Studies, pp. 295–324.

Batista, C., Lacuesta, A. and Vicente, P. (2012) Testing the 'brain gain' hypothesis: micro evidence from Cape Verde. *Journal of Development Economics*, **97**, 32–45.

Bautista, L. (2010) *Steadfast Movement Around Micronesia: Satowan Enlargements Beyond Migration*. Lanham, MD: Lexington.

Bayliss-Smith, T. (1982) *The Ecology of Agricultural Systems*. Cambridge: Cambridge University Press.

Bayliss-Smith, T. (1988) The role of hurricanes in the development of reef islands, Ontong Java Atoll, Solomon Islands. *Geographical Journal*, **154**, 377–391.

Bayliss-Smith, T., Bedford, R., Brookfield, H. and Latham, M. (1988) *Islands, Islanders and the World. The Colonial and Post-Colonial Experience of Eastern Fiji*. Cambridge: Cambridge University Press.

Bayliss-Smith, T. and Christensen, A. (2008) Birds and people on Ontong Java Atoll, 1906–2008: continuity and change. *Atoll Research Bulletin*, **562**, 1–36.

Bayliss-Smith, T., Gough, K., Christensen, A. and Kristensen, S. (2010) Managing Ontong Java: social institutions for production and governance of atoll resources in Solomon Islands. *Singapore Journal of Tropical Geography*, **31**, 55–69.

Becken, S. (2005) Harmonising climate change adaptation and mitigation: the case of tourist resorts in Fiji. *Global Environmental Change*, **15**, 381–393.

Becken, S., Hay, J. and Espiner, S. (2011), The risk of climate change for tourism in the Maldives. In J. Carlsen and R. Butler, eds, *Island Tourism: Sustainable Perspectives*. Wallingford, UK: CABI, pp. 72–84.

Becker, M., Meyssignac, B., Letetrel, C., Llovel, W., Cazenave, A. and Delcroix, T. (2012) Sea level variations at tropical Pacific islands since 1950. *Global and Planetary Change*, **80**, 85–98.

Bedford, R. (1980) Demographic processes in small islands: the case of internal migration. In H. Brookfield, ed., *Population–Environment Relations in Tropical Islands: The Case of Eastern Fiji.* Paris: UNESCO, pp. 27–59.

Bedford, R. (2003) Doomsday revisited: a perspective on Pacific populations in the early twenty-first century. *Development Bulletin*, **62**, 34–38.

Bedford, R., Bedford, C. and Ho, E. (2010) Engaging with New Zealand's recognized seasonal employer work policy: the case of Tuvalu. *Asian and Pacific Migration Journal*, **19**, 421–445.

Beine, M., Docquier, F. and Schiff, M. (2008) *Brain Drain and its Determinants: A Major Issue for Small States*. IZA Discussion Paper No. 3398, Bonn, Germany: IZA.

Bell, J., Johnson, J. and Hobday, A., eds (2009) *Vulnerability of Tropical Pacific Fisheries and Aquaculture to Climate Change*. Noumea, New Caledonia: SPC.

Belle, N. and Bramwell, B. (2005) Climate change and small island tourism: policy maker and industry perspectives in Barbados. *Journal of Travel Research*, **44**, 32–41.

Benedict, B., ed. (1967) *Problems of Smaller Territories*. London: Athlone Press.

Benedict, M. and Benedict, B. (1982) *Men, Women and Money in Seychelles*. Berkeley, CA: University of California Press.

Bennell, P. and Oxenham, J. (1983) Skills and qualifications for small island states. *Labour and Society*, **8**, 13–37.

Benoit, O. (2011) The question of national identity and the institutionalization of the visual arts in Grenada. *Nations and Nationalism*, **17**, 561–580.

Benwell, R. (2011) The canaries in the coalmine: small states as climate change champions. *The Round Table*, **100**, 199–211.

Bernal, R. (2010) The dragon in the Caribbean: China–CARICOM economic relations. *The Round Table*, **99**, 281–302.

Bernal, R. (2011) Improving the participation of small developing countries in the governance of the multilateral trading system. In C. Deere Birkbeck, eds, *Making Global Trade Governance Work for Development*. Cambridge: Cambridge University Press, pp. 231–258.

Bertaud, A. (2002) A rare case of land scarcity: the issue of urban land in the Maldives. Mimeo, Male, Maldives.

Bertram, G. (1986) Sustainable development in South Pacific microeconomies. *World Development*, **14**, 809–822.

Bertram, G. (1987) The political economy of decolonization and nationhood in small Pacific societies. In A. Hooper, S. Britton, R. Crocombe, J. Huntsman and C.Macpherson, eds, *Class and Culture in the South Pacific*. Suva, Fiji: USP, pp. 16–31.

Bertram, G. (1999) The MIRAB model twelve years on. *The Contemporary Pacific*, **11**, 105–138.

Bertram, G. (2006) Introduction: the MIRAB model in the twenty-first century. *Asia Pacific Viewpoint*, **47**, 1–13.

Bertram, G. and Watters, R.F. (1985) The MIRAB economy in South Pacific microstates. *Pacific Viewpoint*, **26**, 497–520.

Besnier, N. (2004) Consumption and cosmopolitanism: practicing modernity at the second-hand marketplace in Nuku'alofa, Tonga. *Anthropological Quarterly*, **77**, 7–45.

Besnier, N. (2011) *On the Edge of the Global. Modern Anxieties in a Pacific Island Nation*. Stanford, CA: Stanford University Press.

Besnier, N. (2012) The athlete's body and the global condition: Tongan rugby players in Japan. *American Ethnologist*, **39**, 491–510.

Betancur-R, R., Hines, A., Arturo Acero, P., Orti1, G. et al. (2011) Reconstructing the lionfish invasion: insights into Greater Caribbean biogeography. *Journal of Biogeography*, **38**, 1281–1293.

Betzold, C., Castro, P. and Weieler, F. (2011) AOSIS in the UNFCCC negotiations: from unity to fragmentation? Working Paper No. 72, Zurich: Center for Comparative and International Studies.

Birch-Thomsen, T., Reenberg, A., Mertz, O. and Fog, B. (2010) Continuity and change: spatiotemporal land use dynamics on Bellona Island, Solomon Islands. *Singapore Journal of Tropical Geography*, **31**, 27–40.

Biribo, N. and Woodroffe, C. (2011) Rates and trends of historical shoreline changes around reef islands, Tarawa Atoll, Kiribati. Unpublished paper to Institute of Australian Geographers Conference, Wollongong, Australia, 6 July.

Birk, T. (2012) Relocation of reef and atoll communities as an adaptation to climate change? Learning from experience in the Solomon Islands. In K. Hastrup and K. Olwig, eds, *Climate Change and Human Mobility: Global Challenges to the Social Sciences*. Cambridge: Cambridge University Press, pp. 81–109.

Birk, T. and Rasmussen, K. (2013) Migration from atolls in the context of climate change adaptation: insights from Solomon Islands. *Population and Environment*.

Birkett, D. (1998) *Serpent in Paradise*. London: Picador.

Bishop, M. (2010) Tourism as a small-state development strategy: pier pressure in the Eastern Caribbean. *Progress in Development Studies*, **10**, 99–114.

Black, P. (1991) Surprised by common sense: local understandings and the management of conflict on Tobi, Republic of Belau. In K. Avruch, P. Black and J. Scimecca, eds, *Conflict Resolution in Cross-cultural Perspective*. New York: Greenwood Press, pp. 145–164.

Blanco, C. (2009) *Tongan Guestworkers in Australia*. Unpublished BA Honours thesis, University of Sydney, Australia.

Block, A. and Krausner, P. (1987) Masters of paradise island: organized crime, neo-colonialism and the Bahamas. *Dialectical Anthropology*, **12**, 85–102.

Blom-Cooper, L. (1986) *Turks and Caicos Islands Commission of Inquiry 1986*. London: HMSO.

Blouin, C. and Debnath, P. (2011) CARICOM return migration and brain circulation: case study of Caribbean-born nurses. *Canadian Foreign Policy Journal*, **17**, 101–114.

Boccagni, P. (2011) The framing of return from above and below in Ecuadorian migration: a project, a myth, or a political device? *Global Networks*, **11**, 461–480.

Boina, A. (2011) The impact of the diaspora on the society of the Grand Comoro. In T. Curtis, ed., *Islands as Crossroads. Sustaining Cultural Diversity in Small Island Developing States*, Paris: UNESCO, pp. 159–176.

Bonnemaison, J. (1994) *The Tree and the Canoe. History and Ethnography of Tanna*. Honolulu, Hawaii: University of Hawaii Press.

Borovnik, M. (2006) Working overseas: seafarers' remittances and their distribution in Kiribati. *Asia Pacific Viewpoint*, **47**, 151–161.

Borredon, L. (2011) Crime and joblessness dog Guadeloupe. *Guardian Weekly*, 30 December, 8.

Bouchet, P., le Guyader, H. and Pascal, O., eds (2011) *The Natural History of Santo*. Paris: Publications Scientifiques de Muséum National d'Histoire Naturelle.

Bourdet, Y. and Falck, H. (2006) Emigrants' remittances and Dutch disease in Cape Verde. *International Economic Journal*, **20**, 267–284.

Bourne, C. (2009) Economic growth, poverty and income inequality. *Journal of Eastern Caribbean Studies*, **34**, 21–48.

Bower, R. and Overmars, M. (2001) Report of visit to Ebeye, Kwajalein, Marshall Islands. SOPAC Preliminary Report 134, Suva: Fiji.

Brandao, F. and Zoomers, A. (2010) Sodade di Nhos Terra. The development potential of return migration to Cape Verde, *International Development Planning Review*, **32**, 267–289.

Brander, K. (2007) Global fish production and climate change. *Proceedings of the National Academy of Sciences*, **104**, 19709–19714.

Brau, R., Di Liberto, A. and Pigliaru, F. (2011) Tourism and development: a recent phenomenon built on old (institutional) roots? *The World Economy*, **34**, 444–472.

Bremner, J. and Perez, J. (2002) A case study of human migration and the sea cucumber crisis in the Galapagos Islands. *Ambio*, **31**, 306–310.

de Bres, J. and Campbell, R. (1975) *Worth their Weight in Gold*. Auckland, New Zealand: Auckland Resource Centre for World Development.

Bresson, G. and Logossah, K. (2011) Crowding out effects of ship tourism on stay-over tourism in the Caribbean: non-parametric panel data analysis. *Tourism Economics*, **17**, 125–158.

Brewer, T. (2013) Dominant discourses, among fishers and middlemen, of the factors affecting coral reef fish distributions in Solomon Islands. *Marine Policy*, **37**, 245–253.

Bridges, K. and McClatchey, W. (2009) Living on the margin: ethnoecological insights from Marshall Islanders at Rongelap atoll. *Global Environmental Change*, **19**, 140–146.

Brierley, J. (1985) Idle land in Grenada: a review of its causes and the PRG's approach to the problem. *Canadian Geographer*, **29**, 298–309.

Bries, J., Debrot, A. and Meyer, D. (2004) Damage to the leeward reefs of Curaçao and Bonaire, Netherlands Antilles from a rare storm event: Hurricane Lenny, November 1999. *Coral Reefs*, **23**, 297–307.

Briguglio, L. (1995) Small island states and their economic vulnerabilities. *World Development*, **23**, 1615–1637.

Bristol, M. (2010) Brain drain and return migration in CARICOM: a review of the challenges. *Caribbean Studies*, **38**, 129–146.

Britton, S. (1991) Tourism, capital and place: towards a critical geography of tourism. *Society and Space*, **9**, 451–478.

Broad, K. and Sanchirico, J. (2008) Local perspectives on marine reserve creation in the Bahamas. *Ocean and Coastal Management*, **51**, 763–771.

Brookfield, H. (1975) Multum in Parvo: questions about diversity and diversification in small developing countries. In P. Selwyn, ed., *Development Policy in Small Countries*. London: Croom Helm, pp. 54–76.

Brookfield, H. (1978) The environment of informality: a comparison between the Caribbean and Pacific. In P. Rimmer, D. Drakakis-Smith and T. McGee, eds, *Food, Shelter and Transport in Southeast Asia and the Pacific*. Canberra, Australia: ANU, pp. 257–274.

Brookfield, H. (1979) Land reform, efficiency and rural income distribution: contributions to an argument. *Pacific Viewpoint*, **20**, 32–52.

Brookfield, H. (1980) Introduction: the conduct and the findings of the interdisciplinary Fiji project. In H. Brookfield, ed., *Population–*

Environment Relations in Tropical Islands: The Case of Eastern Fiji. Paris: UNESCO, pp. 13–28.

Brookfield, H. (1987) Export or perish: commercial agriculture in Fiji. In M. Taylor, ed., *Fiji: Future Imperfect*. Sydney, Australia: Allen and Unwin, pp. 46–57.

Brown, A. and Gusmao, A. (2009) Peacebuilding and political hybridity in East Timor. *Peace Review*, **21**, 61–69.

Brown, K., Turner, K., Hameed, H. and Bateman, L. (1997) Environmental carrying capacity and tourism development in the Maldives and Nepal. *Environmental Conservation*, **24**, 316–325.

Brown, N. (2008) *Climate Change in the UK Overseas Territories*. Peterborough, UK: Joint Nature Conservation Committee.

Brown, R. (1998) Do Pacific Island migrants' remittances decline over time? Evidence from Tongans and Western Samoans in Australia. *The Contemporary Pacific*, **10**, 107–151.

Brown, R., Jimenez, E., Leeves, G. and Connell, J. (2006) Cents and sensibility: the economic benefits of remittances. In World Bank, *At Home and Away*. Washington, DC: World Bank, pp. 47–99.

Brown, R. and Connell, J. (1993) The global flea market: migration, remittances and the informal economy in Tonga. *Development and Change*, **24**, 611–647.

Brown, R. and Connell, J. (2004) The migration of doctors and nurses from South Pacific Island nations. *Social Science and Medicine*, **58**, 2193–2210.

Brown, R., Connell, J. and Jimenez, E. (2013) Migrants, remittances, poverty and social protection in the South Pacific, Fiji and Tonga. *Population, Space and Place* (in press).

Brown, R., Leeves, G. and Prayaga, P. (2013) *Sharing Norms and Remittances beyond Migrants' own Households: Pacific Islanders in Urban and Regional Australia* (in press).

Brown, R., Foster, J. and Connell, J. (1995) Remittances, savings and policy formation in the Pacific Island States. *Asian and Pacific Migration Journal*, **4**, 169–185.

Browne, K. (2004) *Creole Economics. Caribbean Cunning under the French Flag*. Austin, TX: University of Texas Press.

Bryant, J. (1990) Rotuman migration and Fiji: a response to uneven development. In J. Connell, ed., *Migration and Development in the South Pacific*. Canberra, Australia: ANU National Centre for Development Studies, pp. 136–150.

Buckius, C., Albert, S., Tibbetts, I. and Udy, J. (2010) Effects of diel activity and harvesting pressure on the diversity and biomass of sea cucumbers in Marovo Lagoon, Solomon Islands. *Environmental Management*, **45**, 963–973.

Bunce, M., Mee, L., Rodwell, L. and Gibb, R. (2009) Collapse and recovery in a remote small island: a tale of adaptive cycles or downward spirals? *Global Environmental Change*, **19**, 213–226.

Burke, L., Reytar, K., Spalding, M. and Perry, A. (eds) (2011) *Reefs at Risk Revisited.* Washington, DC: World Resources Institute.

Burns, P. and Cleverdon, R. (1995) Destination on the edge? The case of the Cook Islands. In M. Conlin and T. Baum, eds, *Island Tourism*. Chichester, UK: Wiley, pp. 217–228.

Byrne, R. (1980) *Man and the Variable Vulnerability of Island Life. A Study of Recent Vegetation Change in the Bahamas*. Atoll Research Bulletin No. 240, Washington, DC.

Byron, M. (1999) Migration as a way of life: Nevis and the post-war labour movement to Britain. In R. King and J. Connell, eds, *Small Worlds, Global Lives. Islands and Migration*. London: Pinter, pp. 115–136.

Cabioch, G., Banks-Cutler, A., Beck, W. et al. (2003) Continuous reef growth during the last 23 cal kyr BP in a tectonically active zone (Vanuatu, South West Pacific). *Quaternary Science Review*, **22**, 1771–1786.

Caillon, S. and Degeorges, P. (2007) Biodiversity: negotiating the border between nature and culture. *Biodiversity and Conservation*, **16**, 2919–2931.

Cambers, G. (1999) A viable solution to beach sand-mining?/Montserrat. Wise Coastal Practices for Sustainable Human Development Forum. Available at: http://www.csiwisepractices.org/?read=88.

Cambers, G. and James, A. (1994) Sandy coast monitoring: the Dominica example (1987–1992). UNESCO Reports in Marine Science No. 63. Paris: UNESCO.

Cameron, C. and Gatewood, J. (2008) Beyond sun, sand and sea: the emergent tourism programme in the Turks and Caicos Islands. *Journal of Heritage Tourism*, **3**, 55–73.

Campbell, D., Barker, D. and McGregor, D. (2011) Dealing with drought: small farmers and environmental hazards in southern St. Elizabeth, Jamaica. *Applied Geography*, **31**, 146–158.

Campbell, J. (2010) Climate-induced community relocation in the Pacific: the meaning and importance of land. In J. McAdam, ed., *Climate Change and Displacement. Multidisciplinary Perspectives*. Oxford: Hart Publishing, pp. 57–80.

Campling, L. and Rosalie, M. (2006) Sustaining social development in a small island developing state? The case of Seychelles. *Sustainable Development*, **14**, 115–125.

Carden, Y. (2003) Solid waste-level rise on atoll nation states: a less publicized environmental issue in the Republic of Kiribati. *Australasian Journal of Environmental Management*, **10**, 35–45.

Caribbean Commission on Health and Development (2006) *Report of the Caribbean Commission on Health and Development*. Kingston, Jamaica: Ian Randle.

Carling, J. (2002) *Cape Verde: Towards the End of Emigration*. Washington, DC: Migration Policy Institute.

Carling, J. (2004) Emigration, return and development in Cape Verde: the impact of closing borders. *Population, Space and Place*, **10**, 113–132.

Carling, J. and Akesson, L. (2009) Mobility at the heart of the nation: patterns and meaning of Cape Verdean Migration. *International Migration*, **47**, 123–155.

Carling, J. and Batalha, L. (2008) Cape Verdean migration and diaspora. In L. Batalha and J. Carling, eds, *Transnational Archipelago. Perspectives on Cape Verdean Migration and Diaspora*. Amsterdam: Amsterdam University Press, pp. 13–31.

Carlsen, J. (1999) Tourism impacts on small islands: a longitudinal study of community attitudes to tourism on the Cocos (Keeling) Islands. *Pacific Tourism Review*, **3**, 25–35.

Carlsen, J. and Butler, R. (2011) Introducing sustainable perspectives in island tourism. In J. Carlsen and R. Butler, eds, *Island Tourism: Sustainable Perspectives*, Wallingford, UK: CABI, pp. 1–7.

Carrier, J. (1984). *Education and Society in a Manus Village*. Port Moresby, Papua New Guinea: Education Research Unit, University of Papua New Guinea.

Cashman, A., Nurse, L. and John, C. (2010) Climate change in the Caribbean: the water management implications. *Journal of Environment and Development*, **19**, 42–67.

Cassels, S. (2006) Overweight in the Pacific: links between foreign dependence, global food trade, and obesity in the Federated States of Micronesia. *Globalization and Health*, **2**(10), 1–8.

Cassels, S., Curran, S. and Kramer, R. (2005) Do migrants degrade coastal environments? Migration, natural resource extraction and poverty in North Sulawesi, Indonesia. *Human Ecology*, **33**, 329–363.

Cassidy, F. and Brown, L. (2010) Determinants of small Pacific Island tourism: a Vanuatu study. *Asia Pacific Journal of Tourism Research*, **15**, 143–153.

Castellani, F. (2007) International skilled migration: the Caribbean experience in perspective. *Social and Economic Studies*, **56**(4), 165–206.

Castle, L. (1980) The economic context. In R. Ward and A. Proctor, eds, *South Pacific Agriculture*. Canberra, Australia: ANU, pp. 107–136.

Castrejón, M. and Charles, A. (2013) Improving fisheries co-management through ecosystem-based spatial management: the Galapagos Marine Reserve. *Marine Policy*, **38**, 235–245.

Castro, S. and Jaksic, F. (2008) Role of non-established plants in determining biotic homogenization patterns in Pacific Oceanic Islands. *Biological Invasions*, **10**, 1299–1309.

Castro, S., Muñoz, M. and Jaksic, F. (2007) Transit towards floristic homogenization on oceanic islands in the south-eastern Pacific: comparing pre-European and current floras. *Journal of Biogeography*, **34**, 213–222.

Cave, D, (2012) *Digital Islands: How the Pacific's ICT Revolution is Transforming the Region*. Sydney, Australia: Lowy Institute for International Policy.

Challinor, E. (2012) Mobile communications and belongings: Cape Verdeans in Portugal. *Identities*, **19**, 303–319.

Chambers, K. and Chambers, A. (2001) *Unity of Heart. Culture and Change in a Polynesian Atoll Society*. Prospect Heights, IL: Waveland Press.

Chandra, D. (2003) International migration from Fiji: gender and human development issues. *Asian and Pacific Migration Journal*, **13**, 179–204.

Charara, N., Cashman, A., Bonnell, R. and Gehr, R. (2011) Water use efficiency in the hotel sector of Barbados. *Journal of Sustainable Tourism*, **19**, 231–245.

Chauvet, L., Collier, P. and Hoeffler, A. (2010) Paradise lost: the costs of state failure in the Pacific. *Journal of Development Studies*, **46**, 961–980.

Chazine, J.-M. (2005) Of atolls and gardens. In S. Tcherkézoff and F. Doauire-Marsaudon, eds, *The Changing South Pacific. Identities and Transformations*. Canberra, Australia: Pandanus Books, pp. 207–229.

Chevalier, E. (2010) *Le climat, facteur d'emigration dans les pays du Sud? Etude des dynamiques spatiales et territoriales des districts d'Eten (Pérou) et de l'atoll de Funafuti (Tuvalu).* Mémoire de recherches réalisé en vue de l'obtention du Master GAELE, Université Paris Sorbonne.

Chowdhury, M., Barnston, A., Guard, C., Duncan, S. et al. (2010) Sea-level variability and change in the US-affiliated Pacific Islands: understanding the high sea levels during 2006–2008. *Weather*, **65**, 263–268.

Chowdhury, M., Chu, P., Zhao, X., Schroeder, T. and Marra, J. (2010) Sea-level extremes in the US-affiliated Pacific Islands: a coastal hazard scenario to aid in decision analyses. *Journal of Coastal Conservation*, **14**, 53–62.

Christensen, A. (2011) Marine gold and atoll livelihoods: the rise and fall of the bêche-de-mer trade on Ontong Java, Solomon Islands. *Natural Resources Forum*, **35**, 9–20.

Christensen, A. and Mertz, O. (2010) Researching Pacific island livelihoods: mobility, natural resource management and nissology. *Asia Pacific Viewpoint*, **51**, 278–287.

Christian, C. (1993) The challenge of parrot conservation in St Vincent and the Grenadines. *Journal of Biogeography*, **20**, 463–469.

Christie, D. (2010) South Georgia: threats posed by climate change, and mitigations. In M. Pienkowski, O. Cheesman, C. Quick and A. Pienkowski, eds, *Making the Right Connections: a Conference on Conservation in UK Overseas Territories, Crown Dependencies and Other Small Island Communities.* Isle of Man, UK: UK Overseas Territories Conservation Forum, pp. 145–150.

Chung, J. (1987) Fiji, land of tropical cyclones and hurricanes: a case study of agricultural rehabilitation. *Disasters*, **11**, 40–48.

Church, J. and White, N. (2006) A 20th century acceleration in global sea-level rise. *Geophysical Research Letters*, **33**(1), L01602.

Chutaro, B. (2005) Social and economic baseline survey: Jenrok Village, Majuro [Republic of Marshall Islands]. Technical Report No. 15. Apia: SPREP IWP-Pacific.

Cinner, J. (2009) Migration and coastal resource use in Papua New Guinea. *Ocean and Coastal Management*, **52**, 411–416.

Cinner, J. (2010) Poverty and the use of destructive fishing gear near east African marine protected areas. *Environmental Conservation*, **36**, 321–326.

Cinner, J., Marnane, M. and McClanahan, T. (2005) Conservation and community benefits from traditional coral reef management at Ahus Island, Papua New Guinea. *Conservation Biology*, **19**, 1714–1723.

Cinner, J. and McClanahan, T. (2006) Socioeconomic factors that lead to overfishing in small-scale coral reef fisheries of Papua New Guinea. *Environmental Conservation*, **33**, 3–80.

Cinner, J., McClanahan, T., Graham, N., Daw, T. et al. (2012) Vulnerability of coastal communities to key impacts of climate change on coral reef fisheries. *Global Environmental Change*, **22**, 12–20.

Clarke, P. and Jupiter, S. (2010) Law, custom and community-based natural resource management in Kubulau District (Fiji). *Environmental Conservation*, **37**, 98–106.

Clarke, W. (1978) Learning from the past: traditional knowledge and sustainable development. *The Contemporary Pacific,* **2**, 233–253.

Clarke, W. and Thaman, R. (1997) Incremental agroforestry: enriching pacific landscapes. *The Contemporary Pacific*, **9**, 121–148.

Clegg, P. and Gold, P. (2011) The UK Overseas Territories: a decade of progress and prosperity? *Commonwealth and Comparative Politics*, **49**, 115–135.

Clifton, J., Etienne, M., Barnes, D., Barnes, R. et al. (2012) Marine conservation policy in Seychelles: current constraints and prospects for improvement. *Marine Policy*, **36**, 823–831.

Collier, M. (2002) The effects of political corruption on Caribbean development. LACC Working Paper No. 5. Miami, FL: Florida International University.

Collier, P. (2007) *The Bottom Billion*. Oxford: Oxford University Press.

Commonwealth of Australia (2011) *An Effective Aid Program for Australia*. Canberra, Australia: Commonwealth of Australia.

Connell, J. (1978) The death of taro: local response to a change of subsistence crops in the Northern Solomon Islands. *Mankind*, **11**, 445–452.

Connell, J. (1985) Islands on the poverty line. *Pacific Viewpoint,* **26**, 463–473.

Connell, J. (1986) Small states, large aid: the benefits of benevolence in the South Pacific. In P. Eldridge, D. Forbes and D. Porter, eds, *Australian Overseas Aid: Future Directions*, Sydney and London: Croom Helm, pp. 57–78.

Connell, J. (1987a) *New Caledonia or Kanaky? The Political History of a French Colony*. Canberra, Australia: ANU National Centre for Development Studies Pacific Research Monograph No. 18.

Connell, J. (1987b) 'Trouble in paradise': the perception of New Caledonia in the Australian Press. *Australian Geographical Studies*, **25**, 54–65.

Connell, J. (1988a) *Sovereignty and Survival. Island Microstates in the Third World*. University of Sydney Department of Geography Research Monograph No. 3.

Connell, J. (1988b) The end ever nigh: contemporary population changes on Pitcairn Island. *GeoJournal*, **16**, 193–200.

Connell, J. (1990) The Carteret Islands: precedents of the greenhouse effect. *Geography*, **75**, 152–154.

Connell, J. (1991a) Wallis and Futuna: stability and change at the ends of empire. In R. Aldrich, ed., *France, Oceania and Australia: Past and Present*. Sydney, Australia: Department of Economic History, University of Sydney, pp. 91–116.

Connell, J. (1991b) Compensation and conflict: the Bougainville copper mine, Papua New Guinea. In J. Connell and R. Howitt, eds, *Mining and Indigenous Peoples in Australasia*. Sydney: Sydney University Press, pp. 55–76.

Connell, J. (1992) The backdoor of bureaucracy: employment and development in Yap State and Woleai Atoll, Federated States of Micronesia. In R. Baker, ed., *Public Administration in Small and Island States*. West Hartford, CT: Kumarian Press, pp. 174–192.
Connell, J. (1993a) Climatic change: a new security challenge for the atoll states of the South Pacific. *Journal of Commonwealth and Comparative Politics*, **31**, 173–192.
Connell, J (1993b) Anguilla: the tourist trajectory in an island microstate. *Caribbean Geography*, **4**, 131–138.
Connell, J. (1994) The Cayman Islands. Economic growth and immigration in a British colony. *Caribbean Geography*, **5**, 51–66.
Connell, J. (1995) In Samoan worlds. Culture, migration, identity and Albert Wendt. In R. King, J. Connell and P. White, eds, *Writing Across Worlds*. London: Routledge, pp. 263–279.
Connell, J. (1999) Environmental change, economic development and emigration in Tuvalu. *Pacific Studies*, **22**, 1–20.
Connell, J. (2003a) Losing ground? Tuvalu, the greenhouse effect and the garbage can. *Asia Pacific Viewpoint*, **44**, 89–107.
Connell, J. (2003b) Regulation of space in the contemporary postcolonial Pacific city: Port Moresby and Suva. *Asia Pacific Viewpoint*, **44**, 243–258.
Connell, J. (2003c) Island dreaming: the contemplation of Polynesian paradise. *Journal of Historical Geography*, **29**, 554–582.
Connell, J. (2005) A nation in decline? Migration and emigration from the Cook Islands. *Asian and Pacific Migration Journal,* **14**, 327–350.
Connell, J. (2006a) Nauru: the first Pacific failed state? *The Round Table*, **95**, 47–63.
Connell, J. (2006b) 'Saving the Solomons': a new geopolitics in the 'arc of instability'. *Geographical Research*, **44**, 111–122.
Connell, J. (2006c) 'The Taste of Paradise'. Selling Fiji and Fiji Water. *Asia Pacific Viewpoint*, **47**, 342–350.
Connell, J. (2007a) At the end of the world. Holding on to health workers in Niue. *Asian and Pacific Migration Journal*, **16**, 179–197.
Connell, J. (2007b) Towards free trade in the Pacific? The genesis of the 'Kava-Biscuit War' between Fiji and Vanuatu. *Geographical Research,* **45**, 1–12.
Connell, J. (2007c) Local skills and global markets? The migration of health workers from Caribbean and Pacific Island States. *Social and Economic Studies*, **56**, 67–95.
Connell, J. (2007d) Islands, idylls and the detours of development. *Singapore Journal of Tropical Geography*, **28**, 116–135.
Connell, J. (2007e) 'The best island on the globe': constantly constructing tourism on Niue. *Australian Geographer*, **38**, 1–14.

Connell, J. (2007f) Holding on to modernity? Siwai, Bougainville, Papua New Guinea. In J. Connell and E. Waddell, eds, *Environment, Development and Change in Rural Asia-Pacific*. London: Routledge, pp. 127–146.

Connell, J. (2007g) The *Fiji Times* and the good citizen. Constructing modernity and nationhood in Fiji. *The Contemporary Pacific*, **19**, 85–109.

Connell, J. (2008) Niue: embracing a culture of migration. *Journal of Ethnic and Migration Studies*, **34**, 1021–1040.

Connell, J. (2009a) 'I never wanted to come home'. Skilled health workers in the South Pacific. In H. Lee and S. Francis, eds, *Migration and Transnationalism: Pacific Islands Perspectives*. Canberra, Australia: ANU E Press, pp. 159–174.

Connell, J. (2009b) Bittersweet home: return migration and health workers in Polynesia. In D. Conway and R. Potter, eds, *Return Migration of the Next Generations. 21st Century Transnational Mobility*. Aldershot, UK: Ashgate, pp. 139–160.

Connell, J. (2009c) *The Global Health Care Chain: From the Pacific to the World*. New York and London: Routledge.

Connell, J. (2009d) 'We are so small'. Colonialism or autonomy in Tokelau? In G. Baldacchino and D. Milne, eds, *The Case for Non-Sovereignty. Lessons from Sub-National Island Jurisdictions*. London: Routledge, pp. 161–173.

Connell, J. (2010a) Pacific Island States in the global economy: the paradox of migration and culture. *Singapore Journal of Tropical Geography*, **31**, 115–129.

Connell, J. (2010b) From blackbirds to guestworkers in the South Pacific: Plus ça change… ? *Economic and Labour Relations Review*, **20**, 111–122.

Connell, J. (2010c) *Migration and the Globalisation of Health Care*. Cheltenham, UK and Northampton, MA: Edward Elgar.

Connell, J (2011a) *Medical Tourism*. Wallingford, UK: CABI.

Connell, J. (2011b) Elephants in the Pacific? Pacific urbanisation and its discontents. *Asia Pacific Viewpoint*, **52**, 121–135.

Connell, J. (2012a) A Tongan gravestone. *Focus*, **55**(2), Summer, 75–76.

Connell, J. (2012b) Population resettlement in the Pacific: lessons from a hazardous history? *Australian Geographer*, **43**, 127–142.

Connell, J. (2013a) Ants, rats, snakes and taro blight: contemporary invasions of the Pacific Islands. In S. Larrue, ed., *Biodiversity and Societies in the Pacific Islands*. Aix en Provence, France: Presses Universitaires de Provence, pp. 51–71.

Connell, J. (2013b) Medical tourism in the Caribbean Islands: a cure for economies in crisis? *Island Studies Journal*, **8**.

Connell, J. and Brown, R. (1995) Migration and remittances in the South Pacific: toward new perspectives. *Asian and Pacific Migration Journal*, **4**, 1–33.
Connell, J. and Brown, R. (2004) The remittances of migrant Tongan and Samoan nurses in Australia. *Human Resources for Health*, **2**(2).
Connell, J. and Brown, R. (2005) *Remittances in the Pacific. An Overview*. Manila, Philippines: ADB.
Connell, J. and Buchan, J. (2011) The impossible dream? Codes of practice and the international migration of skilled health workers. *World Medical and Health Policy*, **3**(3), 1–17.
Connell, J. and Conway, D. (2000) Migration and remittances in island microstates: a comparative perspective on the South Pacific and the Caribbean. *International Journal of Urban and Regional Research*, **24**, 52–78.
Connell, J. and Hammond, J. (2009) The new blackbirds? Vanuatu guestworkers in New Zealand. *New Zealand Geographer*, **65**, 201–210.
Connell, J. and Lea, J. (1992) 'My country will not be there'. Global warming, development and the planning response in small island states. *Cities*, **9**, 295–309.
Connell, J. and Lea, J. (2002) *Urbanisation in the Island Pacific. Towards Sustainable Development.* London: Routledge.
Connell, J. and Negin, J. (2012) The malaise of modernity: urbanization, mobility and HIV in the Pacific. *Pacific Studies*, **35**, 369–397.
Connell, J. and Pritchard, W. (1990) Tax havens and global capitalism: Vanuatu and the Australian connection. *Australian Geographical Studies*, **28**, 38–50.
Connell, J. and Rugendyke, B. (2008) Tourism and local people in the Asia-Pacific region. In J. Connell and B. Rugendyke, eds, *Tourism at the Grassroots.* London: Routledge, pp. 1–40.
Connell, J. and Soutar, L. (2007) Free trade or free fall? Trade, liberalization and development in the Pacific and Caribbean. *Social and Economic Studies*, **56**, 41–66.
Connell, J. and Stilwell, B. (2006) Merchants of medical care: recruiting agencies in the global health care chain. In C. Kuptsch, ed., *Merchants of Labour*. Geneva: ILO, pp. 239–353.
Connell, J.H. (1997) Disturbance and recovery of coral assemblages. *Coral Reefs*, **16**, S101–S113.
Conway, D. and Potter, R. (2007) Caribbean transnational return migrants as agents of change. *Geography Compass*, **1**, 24–45.
Cooper, A. (2011) *Internet Gambling Offshore. Caribbean Struggles over Casino Capitalism.* New York: Palgrave Macmillan:
Corbin, S. (2011) Medical tourism in Barbados. *Medical Tourism Magazine*, **20**, 22–25.

Corlett, P. (2008) *Stormy Weather. The Challenge of Climate Change and Displacement*. Sydney, Australia: UNSW Press.

Costello, A., Abbas, A., Allen, A., Ball, S. et al. (2009) Managing the health effects of climate change. *Lancet*, **373**, 1693–1733.

Costion, C., Liston, J., Kitalong, A., Iida, A. and Lowe, A.J. (2012) Using the ancient past for establishing current threat in poorly inventoried regions. *Biological Conservation*, **147**, 153–162.

Côté, I. and Darling, E. (2010) Rethinking ecosystem resilience in the face of climate change. *PLoS Biology*, **8**(7), e1000438, 1–5.

Coxon, A. and Munce, K. (2008) The global education agenda and the delivery of aid to Pacific education. *Comparative Education*, **44**, 147–165.

Craig, P., Trail, P. and Morrell, T. (1994) The decline of fruit bats in American Samoa due to hurricanes and overhunting. *Biological Conservation*, **69**, 261–266.

Croes, H. and Hooimeijer, P. (2010) Gender and chain migration: the case of Aruba. *Population, Space and Place*, **16**, 121–134.

Croes, R. (2006) A paradigm shift to a new strategy for small island economies: embracing demand side economics for value enhancement and long-term economic stability. *Tourism Management*, **27**, 453–465.

Crowards, T. (2002) Defining the category of 'small' states. *Journal of International Development*, **14**, 143–179.

Croxall, J., Butchart, S., Lascelles, B., Stattersfield, A., Sullivan, B., Symes, A. and Taylor, P. (2012) Seabird conservation status, threats and priority actions: a global assessment. *Bird Conservation International*, **22**, 1–34.

da Cruz, V., Fengler, W. and Schwartzman, A. (2004) Remittances to Comoros. Volume, trends, impact and implications. World Bank Africa Region Working Paper Series No. 75. Washington, DC: World Bank.

Cummins, D. (2011) Democracy or democrazy? Local experiences of democratization in Timor-Leste. *Democratization*, **17**, 899–919.

Currey, B. (1980) Famine in the Pacific: losing the chance for change. *GeoJournal*, **4**, 447–466.

Curry, G. (2003) Moving beyond postdevelopment: facilitating indigenous alternatives for 'development'. *Economic Geography*, **79**, 405–423.

Daly, M., Poutasi, N., Nelson, F. and Kohlhase, J. (2010) Reducing the climate vulnerability of coastal communities in Samoa. *Journal of International Development*, **22**, 265–281.

Danielsen, F., Filardi, C., Jønsson, K., Kohaia, V. et al. (2010) Endemic avifaunal biodiversity and tropical forest loss in Makira, a mountainous Pacific island. *Singapore Journal of Tropical Geography*, **31**, 100–114.

D'Arcy, M. (2011) Away from subsistence: food, security and resource use on Paama, Vanuatu. Unpublished BSc/BA Honours thesis, University of Sydney.

D'Arcy, P. (2006) *The People of the Sea. Environment, Identity and History in Oceania*. Honolulu, Hawaii: University of Hawaii Press.

D'Arcy, P. (2009) Variable rights and diminishing control: the evolution of indigenous maritime sovereignty in Oceania. In D. Ghosh, H. Goodall and S. Donald, eds, *Water, Sovereignty and Borders in Asia and Oceania*. London: Routledge, pp. 20–37.

Dasgupta, A. and Baschieri, A. (2010) Vulnerability to climate change in rural Ghana: mainstreaming climate change in poverty-reduction strategies. *Journal of International Development*, **22**, 803–820.

D'Aubert, A. (1992) Decision-making among Tokelauan administrators concerning climate change, sea level rise and tropical cyclones. *Proceedings of the New Zealand Geograhical Society, Auckland*, 123–127.

Davenport, M. (2006) A study of alternative special and differential arrangements for small economies. In R. Grynberg, ed., *WTO at the Margins. Small States and the Multilateral Trading System*. Cambridge: Cambridge University Press, pp. 1–8.

Davenport, W. (1972) Social organization notes on the Northern Santa Cruz Islands: the outer reef islands. *Baessler Archiv*, **20**, 11–95.

David, G., Leopold, M., Dumas, P., Ferraris, J., Herrenschmidt, J. and Fontenelle, G. (2010) Integrated coastal zone management perspectives to ensure the sustainability of coral reefs in New Caledonia. *Marine Pollution Bulletin*, **61**, 323–334.

Davies, A. (1985) Small country blues. *Meanjin*, **44**, 243–252.

Davis, M. (2006) *Planet of Slums*. London: Verso.

Davison, P. (2005) After the eruption. *The Independent*, 16 July, 26–27.

Daw, T., Robinson, J. and Graham, N. (2011) Perceptions of trends in Seychelles artisanal trap fisheries: comparing catch monitoring, underwater visual census and fishers' knowledge. *Environmental Conservation*, **38**, 75–88.

Dawson, J. and Smithers, S. (2010) Shoreline and beach volume change between 1967 and 2007 at Raine Island, Great Barrier Reef, Australia. *Global and Planetary Change*, **72**, 141–154.

De Albuquerque, K. (1991) Conflicting claims on Antiguan coastal resources: the case of McKinnon's and Jolly Hill Salt Ponds. In N. Girvan and D. Simmons, eds, *Caribbean Ecology and Economics*. St Michael (Barbados): Caribbean Conservation Association, pp. 195–205.

Delay, S. (2005) Why does aid management in Samoa succeed? *Public Administration and Development*, **25**, 433–435.

Demas, W. (1965) *The Economics of Development in Small Countries*. Montreal, Canada: McGill University Press.

DeMers, A. and Kahui, V. (2012) An overview of Fiji's fisheries development. *Marine Policy*, **36**, 174–179.

Dennis, J. (2003) *Pacific Island Seafarers. A Study of the Economic and Social Implications of Seafaring on Dependants and Communities*. Suva, Fiji: SPC.

Denoon, D. (1997) Pacific Edens? Myths and realities of primitive affluence. In D. Denoon, ed., *The Cambridge History of Pacific Islanders*. Cambridge: Cambridge University Press, pp. 80–118.

Deo, R. (2011) On meteorological droughts in tropical Pacific Islands: time-series analysis of observed rainfall using Fiji as a case study. *Meteorological Applications*, **18**, 171–180.

De Santo, E., Jones, P. and Miller, A. (2011) Fortress conservation at sea: a commentary on the Chagos marine protected area. *Marine Policy*, **35**, 258–260.

Diamond, J. (2005) *Collapse. How Societies Choose to Fail or Survive*. London: Allen Lane.

Diaz Arenas, A. and Falkland, A. (1991) Characteristics of small islands. In A. Falkland, ed., *Hydrology and Water Resources of Small Islands: A Practical Guide*. Paris: UNESCO, pp. 1–9.

Dikou, A., Ackerman, C., Banks, C., Dempsey, A. et al. (2009) Ecological assessment to detect imminent change in coral reefs of Admiral Cockburn Land and Sea National Park, Turks and Caicos Islands. *Marine Ecology*, **30**, 425–436.

Di Maria, C. and Lazarova, E. (2012) Migration, human capital formation and growth: an empirical investigation. *World Development*, **40**, 938–955.

Dinnen, S. (2009) The crisis of state in Solomon Islands. *Peace Review*, **21**, 70–78.

Dixon, J. and Jamieson, C. (2005) The cross-Pacific chicken: tourism, migration and chicken consumption in the Cook Islands. In N. Folds and B. Pritchard, eds, *Cross-continental Food Chains*. London: Routledge, pp. 81–93.

Docquier, F. and Marfouk, A. (2006) International migration by education attainment, 1990–2000. In C. Ozden and M. Schiff, eds, *International Migration, Brain Drain and Remittances*. New York: Palgrave Macmillan, pp. 151–199.

Dodds, K. and Benwell, M. (2010) More unfinished business: The Falklands/Malvinas, maritime claims, and the spectre of oil in the South Atlantic. *Environment and Planning D: Society and Space*, **28**, 571–580.

Dodhia, D. (2008) *The Emerging Debt Problems of Small States*. London: Commonwealth Secretariat.

Doherty, B. (2012) That sinking feeling. *Sydney Morning Herald*, 7 January, 5.

Dominey-Howes, D. and Thaman, R. (2009) UNESCO-IOC International Survey Team Samoa: interim report of field survey. Miscellaneous Report No. 2. Sydney: University of New South Wales Tsunami Research Centre.

Dommen, E. (1980) External trade problems of small island states in the Pacific and Indian Oceans. In R. Shand, ed., *The Island States of the Pacific and Indian Oceans*, Monograph No. 23. Canberra, Australia: ANU Development Studies Centre, pp. 179–199.

Dommen, E. and Hein, P., eds (1985) *States, Microstates and Islands*. Beckenham, UK: Croom Helm.

Domroes, M. (2001) Conceptualising state-controlled resort islands for an environment-friendly development of tourism: the Maldivian experience. *Singapore Journal of Tropical Geography*, **22**, 122–137.

Donner, S. (2012) Sea level rise and the ongoing battle of Tarawa. *Eos*, **93**, 169–171.

Donner, S., Kirata, T. and Vieux, C. (2010) Recovery from the coral bleaching event in the Gilbert Islands, Kiribati. *Atoll Research Bulletin*, **587**, 1–25.

Donner, W. (1993) Custom and modernisation on Sikaiana. *Anthropological Forum*, **6**, 541–556.

Donner, W. (2002) Rice and tea, fish and taro: Sikaiana migration to Honiara. *Pacific Studies*, **25**, 23–44.

Doulman, D. and Kearney, R. (1987) Domestic tuna industries. In D. Doulman, ed., *The Development of the Tuna Industry in the Pacific Islands Region*. Honolulu, Hawaii: East–West Center, pp. 3–31.

Dumas, P., Jimenez, H., Léopold, M., Petro, G. and Jimmy, R. (2010) Effectiveness of village-based marine reserves on reef invertebrates in Emau, Vanuatu. *Environmental Conservation*, **37**, 364–372.

Dumas, P., Printemps, J., Mangeas, M. and Luneau, G. (2010) Developing erosion models for integrated coastal zone management: a case study of the New Caledonia west coast. *Marine Pollution Bulletin*, **61**, 59–529.

Duncan, D. (2011) *Freshwater under Threat. Pacific Islands*. Bangkok: UNEP.

Dunn, L. (2011) The impact of political dependence on small island jurisdictions. *World Development*, **39**, 2132–2146.

Dunne, R., Barbosa, S. and Woodworth, P. (2012) Contemporary sea level in the Chagos Archipelago, central Indian Ocean. *Global and Planetary Change*, **82**, 25–37.

Dupon, J.-F. (1984) Where the exception confirms the rule: the cyclones of 1982–83 in French Polynesia. *Disasters*, **8**, 34–47.

Duval, D. (2004) Linking return visits and return migration among Commonwealth Eastern Caribbean migrants in Toronto. *Global Networks*, **4**, 51–68.

Eakin, C., Morgan, J., Heron, S., Smith, T. et al. (2010) Caribbean corals in crisis: record thermal stress, bleaching and mortality in 2005. *PLoS One*, **5**(11) (online).

East, A. and Dawes, L. (2009) Homegardening as a panacea: a case study of South Tarawa. *Asia Pacific Viewpoint*, **50**, 338–352.

Easter, C. (1999) Small states development: a Commonwealth Vulnerability Index. *The Round Table*, **88**, 403–422.

Easterly, W. and Kraay, A. (2000) Small states, small problems? Income, growth and volatility in small states. *World Development,* **28**, 2013–2027.

Edgar, G., Banks, S., Brandt. M., Bustamante, R. et al. (2010) El Niño, grazers and fisheries interact to greatly elevate extinction risk for Galapagos marine species. *Global Change Biology*, **16**, 2876–2890.

Eijgelaar, E., Thaper, C. and Peeters, P. (2010) Antarctic cruise tourism: the paradoxes of ambassadorship, 'last chance tourism' and greenhouse gas emissions. *Journal of Sustainable Tourism*, **18**, 337–354.

Ellis, F. (1998) Household strategies and rural livelihood diversification. *Journal of Development Studies*, **35**, 1–38.

Ellison, J. (1993) Mangrove retreat with rising sea-level, Bermuda. *Estuarine, Coastal and Shelf Science*, **37**, 75–87.

Ellison, J. and Stoddart, D. (1991) Mangrove ecosystem collapse during predicted sea-level rise: Holocene analogues and implications. *Journal of Coastal Research*, **7**, 151–165.

Elsner, J., Kossin, J. and Jagger, T. (2008) The increasing intensity of the strongest tropical cyclones. *Nature*, **455**, 92–95.

Emberson-Bain, A. (1994) Backbone of growth: export manufacturing and Fiji's tuna fish wives. In A. Emberson-Bain, ed., *Sustainable Development or Malignant Growth?* Suva, Fiji: Marama Publications, pp. 149–171.

Eriksen, A. (2009) 'New life': Pentecostalism as social critique in Vanuatu. *Ethnos*, **74**, 175–198.

Escobar, A. (2005) Economics and the space of modernity: tales of market, production and labour. *Cultural Studies*, **19**, 139–175.

Evans, M. (1999) Is Tonga's MIRAB economy sustainable? A view from the village and a view without it. *Pacific Studies*, **22**, 137–166.

Evans, M. (2001) *Persistence of the Gift. Tongan Tradition in Transnational Context*, Waterloo, Canada: Wilfred Laurier University Press.

Evans, M., Sinclair, R., Fusimalohi, C. and Liava'a, V. (2001) Globalization, diet and health: an example from Tonga. *Bulletin of the WHO*, **79**, 856–862.

Faeamani, S. (1995) The impact of remittances on rural development in Tongan villages. *Asian and Pacific Migration Journal*, **4**, 139–156.

Fairbairn, I. (1985) *Island Economies. Studies from the South Pacific*. Suva, Fiji: Institute of Pacific Studies.

Fairbairn, T. and Worrell, D. (1996) *South Pacific and Caribbean Island Economies. A Comparative Study*. Brisbane, Australia: Foundation for Development Cooperation.

Farbotko, C. (2005) Tuvalu and climate change: constructions of environmental displacement in the *Sydney Morning Herald*. *Geografiska Annaler B*, **87**, 279–293

Farbotko, C. (2010a) Wishful sinking: disappearing islands, climate refugees and cosmopolitan experimentation. *Asia Pacific Viewpoint*, **51**, 47–60.

Farbotko, C. (2010b) 'The global warming clock is ticking so see these places while you can': voyeuristic tourism and model environmental citizens on Tuvalu's disappearing islands. *Singapore Journal of Tropical Geography*, **31**, 239–253.

Farbotko, C. and Lazrus, H. (2012) The first climate refugees? Contesting global narratives of climate change in Tuvalu. *Global Environmental Change*, **22**, 382–390.

Farran, S. (2007) The 're-colonising' of Pitcairn, *Victoria University of Wellington Law Review*, **38**, 435–464.

Fay-Sauni, L., Vuki, V., Paul, S. and Rokosawa, M. (2008) Women's subsistence fishing supports rural households in Fiji: a case study of Nadoria, Viti Levu, Fiji. *SPC Women in Fisheries Information Bulletin*, **18**, 26–29.

Fazey, I., Kesby, M., Evely, A., Latham, I. et al. (2010) A three-tiered approach to participatory vulnerability assessment in the Solomon Islands. *Global Environmental Change*, **20**, 713–728.

Fazey, I., Pettorelli, N., Kenter, J., Wagatora, D. and Schuett, D. (2011) Maladaptive trajectories of change in Makira, Solomon Islands. *Global Environmental Change*, **21**, 1275–1289.

Feare, C. (1999) Ants take over from rats on Bird Island, Seychelles. *Bird Conservation International*, **9**, 95–96.

Feeny, S. and McGillivray, M. (2010) Aid and growth in small island developing states. *Journal of Development Studies*, **46**, 897–917.

Feinberg, R. (1986) The Anuta problem. Local sovereignty and national integration in the Solomon Islands. *Man*, **21**, 438–452.

Feinberg, R. (2010) Marine resource conservation and prospects for environmental sustainability in Anuta, Solomon Islands. *Singapore Journal of Tropical Geography*, **31**, 41–54.

Felgentreff, C. (1996) Migration and remittances in the eastern periphery of Fiji: the case of Naikeleyaga village. *Asian Migrant*, **9**, 21–27.

Ferguson, J. (2003) *Migration in the Caribbean: Haiti, the Dominican Republic and Beyond.* London: Minority Rights Group International.

Fichez, R., Chifflet, S., Douillet, P., Gérard, P., Gutierrez, F., Jouon, A., Ouillon, S. and Grenz, C. (2010) Biogeochemical typology and temporal variability of lagoon waters in a coral reef ecosystem subject to terrigeneous and anthropogenic inputs (New Caledonia). *Marine Pollution Bulletin*, **61**, 309–322.

Field, M., Ogston, A. and Storlazzi, C. (2011) Rising sea level may cause decline of fringing coral reefs. *Eos*, **92**, 273–274.

Fielding, D. (2010) Aid and Dutch disease in the South Pacific and in other small island states. *Journal of Development Studies*, **46**, 918–940.

Filer, C. (1990) The Bougainville rebellion, the mining industry and the process of social disintegration in Papua New Guinea. In R. May and M. Spriggs, eds, *The Bougainville Crisis*. Bathurst, Australia: Crawford House, pp. 73–112.

Finney, B. (1967) Money work, fast money and prize money: aspects of the Tahitian labor commitment. *Human Organization*, **26**, 195–199.

Firth, R. (1936) *We, the Tikopia.* London: Allen and Unwin.

Fischer, S. (2005) *Island at the End of the World. The Turbulent History of Easter Island.* London: Reaktion Books.

Fitzsimmons, C. (2009) Why dive? And why here? A study of recreational diver enjoyment at a Fijian eco-tourist resort. *Tourism in Marine Environments*, **5**, 159–173.

Flinn, J. (1994) From sea and garden to school and town: changing gender and household patterns among Pollap Atoll migrants. *Pacific Studies*, **17**, 117–131.

Flora, C., Ely, P. and Flora, A. (2009) Microatoll edge to ENSO annulus growth suggests sea level change. *Atoll Research Bulletin*, **571**, 1–10.

Flower, S. (2008) Muslims in Melanesia: putting security issues in perspective. *Australian Journal of International Affairs*, **62**, 408–429.

Foale, S. (2001) 'Where's our development?'. Landowner aspirations and environmentalist agendas in Western Solomon Islands. *Asia Pacific Journal of Anthropology*, **2**(2), 44–67.

Foley, A. (2010) Uncertainty in regional climate modelling: a review. *Progress in Physical Geography*, **34**, 647–670.

Ford, M. (2012) Shoreline changes on an urban atoll in the Central Pacific Ocean: Majuro Atoll, Marshall Islands. *Journal of Coastal Research*, **28**, 11–22.

Forster, J., Lake, I., Watkinson, A. and Gill, J. (2011) Marine biodiversity in the Caribbean UK overseas territories: perceived threats and constraints to environmental management. *Marine Policy*, **35**, 647–657.

Forster, J., Schuhmann, P., Lake, I., Wilkinson, A. and Gill, J. (2012) The influence of hurricane risk on tourist destination choice in the Caribbean. *Climatic Change*, **114**, 745–768.

Foster, R. (2005) Negotiating globalization: contemporary Pacific perspectives. *Ethnohistory*, **52**, 167–177.

Fraenkel, J. (2004) *The Manipulation of Custom. From Uprising to Intervention in the Solomon Islands*. Canberra, Australia: Pandanus.

Fraenkel, J. (2006) Political instability, 'failed states' and regional intervention in the Pacific. In J. Bryant-Tokalau and I. Frazer, eds, *Redefining the Pacific? Regionalism, Past Present and Future*. Aldershot, UK: Ashgate, pp. 127–140.

Fraenkel, J., Allen, M. and Brock, H. (2010) The resumption of palm-oil production on Guadalcanal's northern plains. *Pacific Economic Bulletin*, **25**(1), 64–75.

Frazer, I. (1987) Growth and change in village agriculture: Manakwai, North Malaita. South Pacific Smallholder Project Occasional Paper No. 11. Armidale, Australia: University of New England.

Friberg, E., Schaefer, K. and Holen, L. (2006) US economic assistance to two Micronesian nations: aid impact, dependency and migration. *Asia Pacific Viewpoint*, **47**, 123–133.

Frynas, J., Wood, G. and Soares de Oliveira, R. (2003) Business and politics in São Tomé e Principe: from cocoa monoculture to petro-state. *African Affairs*, **102**, 51–80.

Fukuyama, F. (2008) State building in Solomon Islands. *Pacific Economic Bulletin*, **23**(3), 18–34.

Furusawa, T. and Ohtsuka, R. (2006) Inter-household variations in subsistence strategies within a rural society of Roviana, Solomon Islands. *Tropics*, **15**, 29–37.

Furusawa, T. and Ohtsuka, R. (2009) The role of barrier islands in subsistence of the inhabitants of Roviana Lagoon, Solomon Islands. *Human Ecology*, **37**, 629–642.

Gagern, A. and van den Berghe, J. (2013) A critical review of fishing agreements with tropical developing countries. *Marine Policy*, **38**, 375–386.

Gailey, C. (1992a) A good man is hard to find. *Critique of Anthropology*, **12**, 47–74.

Gailey, C. (1992b) State formation, development and social change in Tonga. In A. Robillard, ed., *Social Change in the Pacific Islands*. London: Kegan Paul International, pp. 322–345.

Gaillard, C. and Manner, H. (2010) Yam cultivation on the east coast of New Caledonia: adaptation of agriculture to social and economic changes. *Australian Geographer*, **41**, 485–506.

Galbis, V. (1984) Ministate economies. *Finance and Development*, **21**(2), 36–39.

Gardner, T., Cote, I., Gill, G., Grant, A. and Watkinson, A. (2003) Long-term region-wide declines in Caribbean corals. *Science*, **301**, 958–960.

Garonna, I., Fazey, J., Brown, M. and Pettorelli, N. (2009) Rapid primary productivity changes in one of the last coastal rainforests: the case of Kahua, Solomon Islands. *Environmental Conservation*, **36**, 253–260.

Gaston, K. (2009) Extinction. In R. Gillespie and D. Clague, eds, *Encyclopedia of Islands*. Berkeley, CA: University of California Press, pp. 281–286.

Gay, D. (2005) Vanuatu's suspended accession bid: second thoughts? In P. Gallagher, P. Low and A. Stoler, eds, *Managing the Challenges of WTO Participation*. Cambridge: Cambridge University Press, pp. 590–606

Gay, J.-C. (2012) Why is tourism doing poorly in overseas France? *Annals of Tourism Research*, **39**, 1634–1652.

Geddes, W., Chambers, A., Sewell, B., Lawrence, R. and Watters, R. (1982) *Islands on the Line. Team Report*. Canberra, Australia: ANU Press.

Gegeo, D. (1998) Indigenous knowledge and empowerment: rural development examined from within. *The Contemporary Pacific*, **10**, 289–316.

Gegeo, D. (2001) Cultural rupture and indigeneity: the challenge of (re)visioning 'place' in the Pacific. *The Contemporary Pacific*, **13**, 491–507.

Gemenne, F. and Magnan, A. (2010) *The Other Migrants Preparing for Change. Environmental Changes and Migration in the Republic of Mauritius*. Geneva: International Organization for Migration.

Gewertz, D. and Errington, F. (2010) *Cheap Meat: Flap Food Nations in The Pacific Islands*. Berkeley, CA: University of California Press.

Gibson, J. (2007) Is remoteness a cause of slow growth in the Pacific? A spatial-econometric analysis. *Pacific Economic Bulletin*, **22**(1), 83–101.

Gibson, J., Boe-Gibson, G., Rohorua, H. and McKenzie, D. (2007) Efficient remittance services for development in the Pacific. *Asia-Pacific Development Journal*, **14**(2), 55–74.

Gibson, J. and McKenzie, D. (2011a) Australia's PSWPS: development impacts in the first two years. *Asia Pacific Viewpoint*, **52**, 361–370.

Gibson, J. and McKenzie, D. (2011b) The microeconomic determinants of emigration and return migration of the best and brightest: evidence from the Pacific. *Journal of Development Economics*, **95**, 18–29.

Gibson, J. and McKenzie, D. (2012) The economic consequences of 'brain drain' of the best and brightest: microeconomic evidence from five countries. *The Economic Journal*, **122**, 339–375.

Gibson, J., McKenzie, D. and Rohorua, H. (2008) How pro-poor is the selection of seasonal migrant workers from Tonga under New Zealand's Recognized Seasonal Employer Program? *Pacific Economic Bulletin*, **23**(3), 187–204.

Gillett, R. (2009) *Fisheries in the Economies of the South Pacific Island Countries and Territories*. Manila, Philippines: ADB.

Gillett, R. and Cartwright, I. (2010) *The Future of Pacific Island Fisheries*. Noumea, New Caledonia, and Honiara, Solomon Islands: SPC and Pacific Islands Forum Fisheries Agency.

Given, D. (1992) *An Overview of the Terrestrial Biodiversity of Pacific Islands*. Apia, Samoa: SPREP.

Gmelch, G. (1980) Return migration. *Annual Reviews in Anthropology*, **9**, 135–159.

Gmelch, G. (1992) *Double Passage: The Lives of Caribbean Migrants Abroad and Back Home*. Ann Arbor, MI: University of Michigan Press.

Gmelch, G. and Gmelch, S. (1997) *The Parish Behind God's Back. The Changing Culture of Rural Barbados*. Long Grove, IL: Waveland Press.

Golbuu, Y., van Woesik, R., Richmond, R., Harrison, P. and Fabricius, K. (2011) River discharge reduces reef coral diversity in Palau. *Marine Pollution Bulletin*, **62**, 824–831.

Gonschor, L. (2011) Facing land challenges in Rapa Nui (Easter Island). *Pacific Studies*, **34**, 175–194.

Goss, B. and Knudsen, D. (1994) Flexibility in offshore assembly operations: electronics assembly in the Commonwealth Caribbean. *The Developing Economies*, **32**, 210–227.

Gössling, S. and Hall, C. (2006) Uncertainties in predicting tourist flows under scenarios of climate change. *Climatic Change*, **79**, 163–173.

Gössling, S. and Schumacher, K. (2010) Implementing carbon neutral destination policies: issues from the Seychelles. *Journal of Sustainable Tourism*, **18**, 377–391.

Government of Samoa (2012) *Household Income and Expenditure Survey 2008*. Apia, Samoa: Samoa Bureau of Statistics.

Government of the Republic of Vanuatu (2002) *Vanuatu Poverty Survey: Analysis Report*. Port Vila: Vanuatu Statistics Office.

Grafton, R. (2010) Adaptation to climate change in marine capture fisheries. *Marine Policy*, **34**, 606–615.

Graham, N., Nash, K. and Kool, J. (2011) Coral reef recovery dynamics in a changing world. *Coral Reefs*, **30**, 283–294.

Graham, N., Wilson, S., Jennings, S., Polunin, N. et al. (2007) Lag effects in the impacts of mass coral bleaching on coral reef fish, fisheries and ecosystems. *Conservation Biology*, **21**, 1291–1300.

Grainger, A. (2011) Migrants, mercenaries and overstayers: talent migration in Pacific Island rugby. In J. Maguire and M. Falcous, eds, *Sport and Migration: Borders, Boundaries and Crossings*. London: Routledge, pp. 129–140.

Grant, E. (2012) Development and change on Rendova Island, Solomon Islands: the *Canarium* nut industry and the hybrid economy. Unpublished BA Honours thesis, University of the Sunshine Coast, Australia.

Greaver, T. and Sternberg, L. (2010) Decreased precipitation exacerbates the effects of sea level on coastal dune ecosystems in open ocean islands. *Global Change Biology*, **16**, 1860–1869.

Green, M. (2010) Tiny islands face change that is hard to believe in. *Sydney Morning Herald*, 27 November, 19.

Grieco, E. (2003) *The Remittance Behavior of Immigrant Households: Micronesians in Hawaii and Guam*. New York: LFB Scholarly Publishing.

de Groot, J. and Bush, S. (2010) The potential for dive tourism led entrepreneurial marine protected areas in Curaçao. *Marine Policy*, **34**, 1051–1059.

Grossman, L. (1998a) Diet and agriculture in an eastern Caribbean village. *Human Ecology*, **26**, 21–42.

Grossman, L. (1998b) *The Political Ecology of Bananas. Contract Farming, Peasants and Agrarian Change in the Eastern Caribbean*. Chapel Hill, NC: University of North Carolina Press.

Grynberg, R. (2006) Introduction. In R. Grynberg, ed., *WTO at the Margins. Small States and the Multilateral Trading System*. Cambridge: Cambridge University Press, pp. 1–8.

Grynberg, R. (2010) The trade and health debate. *Islands Business*, **36**(9), 34–35.

Grynberg, R. (2012) Healing the forum divide. *Islands Business*, **39**(8), 24–25.

Grynberg, R. and Razzaque, M. (2004) *The Trade Performance of Small States*. London: Commonwealth Secretariat.

Grynberg, R. and Silva, S. (2006) Harmonization without representation: small states, the Basel Committee, and the WTO. *World Development*, **34**, 1223–1236.

Guillaumont, P. (2009) An economic vulnerability index: its design and use for international development policy. *Oxford Development Studies*, **37**, 193–228.

Guillaumont, P. (2010) Assessing the economic vulnerability of small island developing states and the least developed countries. *Journal of Development Studies*, **46**, 828–854.

Guillemot, N., Chabanet, P. and Le Pape, O. (2010) Cyclone effects on coral reef habitats in New Caledonia (South Pacific). *Coral Reefs*, **29**, 445–453.

Haberle, S. (2003) Late quaternary vegetation dynamics and human impact on Alexander Selkirk Island, Chile. *Journal of Biogeography*, **30**, 239–255.

Hales, S., Weinstein, P. and Woodward, A. (1999) Ciguatera (fish poisoning), El Niño, and Pacific sea surface temperatures. *Ecosystem Health*, **5**, 20–25.

Hameiri, S. (2009) The region within: RAMSI, the Pacific Plan and new modes of governance in the Southwest Pacific. *Australian Journal of International Affairs*, **63**, 348–360.

Hameiri, S. (2012) Mitigating the risk to primitive accumulation: state-building and the logging boom in Solomon Islands. *Journal of Contemporary Asia*, **42**, 4056–426.

Hamilton, L., Colocousis, C. and Johansen, S. (2004) Migration from resource depletion: the case of the Faroe Islands. *Society and Natural Resources*, **17**, 443–453.

Hamilton, R., Giningele, M., Aswani, S. and Ecochard, J. (2012) Fishing in the dark: local knowledge, night spearfishing and spawning in the Western Solomon Islands. *Biological Conservation*, **145**, 246–257.

Hampton, M. and Christensen, J. (2002) Offshore pariahs? Small island economies, tax havens and the re-configuration of global finance. *World Development*, **30**, 1657–1673.

Hampton, M. and Christensen, J. (2011) Looking for Plan B: what next for island hosts of offshore finance? *The Round Table*, **100**, 169–181.

Handler, A., Gruner, D., Haines, W., Lange, M. and Kaneshiro, K. (2007) Arthropod surveys on Palmyra Atoll, Line Islands, and insights into the decline of the native tree *Pisonia grandis* (Nyctaginaceae). *Pacific Science*, **61**, 485–502.

Hanich, Q. (2012) Distributing the bigeye conservation burden in the western and central Pacific fisheries. *Marine Policy*, **36**, 327–332.

Hanich, Q. and Tsamenyi, M. (2009) Managing fisheries and corruption in the Pacific Islands region. *Marine Policy*, **33**, 386–392.

Hanich, Q., Teo, F. and Tsamenyi, M. (2010) A collective approach to Pacific islands fisheries management: moving beyond regional agreements. *Marine Policy*, **34**, 85–91.

Hannesson, R. (2008) The exclusive economic zone and economic development in the Pacific island countries. *Marine Policy*, **32**, 886–897.

Harden, C. and Pulsipher, L. (1992) Come a nasty gale: the response to Hurricane Hugo on the island of Montserrat, West Indies. *Focus*, **42**, Summer, 9–13.

Harden, S. (1985) *Small is Dangerous. Micro States in a Macro World*. London: Frances Pinter.

Hards, V. (2009) Shaken, but not stirred. The 2004 eruption of the Tristan da Cunha volcano. *Shima*, **3**, 16–32.

Harley, S., Williams, P., Nicol, S. and Hampton, J. (2011) The Western and Central Pacific tuna fishery: 2010 overview and status of stocks. Report No. 11. Noumea: SPC Tuna Fisheries Assessment.

Harmelin-Vivien, M. (1994) The effects of storms and cyclones on coral reefs: a review. *Journal of Coastal Research*, **34**, 645–656.

Harmsworth, G. and Raynor, B. (2004) Cultural considerations in landslide risk perception. In T. Glade, M. Anderson and M. Crozier, eds, *Landslide Hazard and Risk*. Chichester, UK: Wiley, pp. 219–249.

Harris, A., Manahira, G., Sheppard, A., Gough, C. and Sheppard, C. (2010) Demise of Madagascar's once great barrier reef: change in coral reef condition over 40 years. *Atoll Research Bulletin*, **574**, 1–16.

Harris, D. and Rocha, S. (2009) Comoros. In R. Gillespie and D. Clague, eds, *Encyclopedia of Islands*. Berkeley, CA: University of California Press, pp. 177–180.

Harris, V. and Goldsmith, A. (2012) Police in the development space: Australia's international police capacity builders. *Third World Quarterly*, **33**, 1019–1036.

Hastings, J., Gruby, R. and Sievanen, L. (2012) Science-based coastal management in Fiji: two case studies from the NGO sector. *Marine Policy*, **36**, 907–914.

Hau'ofa, E. (1994) Our sea of islands. *The Contemporary Pacific*, **6**, 147–161.

Hauzer, M., Dearden, P. and Murray, G. (2013) The effectiveness of community-based governance of small-scale fisheries, Ngazidja island, Comoros. *Marine Policy*, **38**, 346–354.

Hawkes, L., Broderick, A., Godfrey, M. and Godley, B. (2009) Climate change and marine turtles. *Endangered Species Research*, **7**, 137–154.

Hayes, G. (1993) 'Mirab' processes and development on small Pacific islands: a case study from the southern Massim, Papua New Guinea. *Pacific Viewpoint*, **34**, 153–178.

Hayle, C., Singh, R. and Wright, A. (2010) Planning for tourism resilience in the Caribbean. *Journal of Eastern Caribbean Studies*, **35**, 36–59.

Haynes, A. (1999) The long term effect of forest logging on the macroinvertebrates in a Fijian stream. *Hydrobiologia*, **405**, 79–87.

Hedson, J. (2000) Telecommunications in Pohnpei State, Federated States of Micronesia. *Pacific Health Dialog*, **7**(2), 46–50.

Heger, M., Julca, A. and Paddison, O. (2009) Vulnerability of small-island economies: the impact of 'natural' disasters in the Caribbean. In W. Naudé, A. Santos-Paulino and M. McGillivray, eds, *Vulnerability in Developing Countries*. Tokyo: United Nations University Press, pp. 215–248.

Hein, P. (1985) The study of microstates. In E. Dommen and P. Hein, eds, *States, Microstates and Islands*. London: Croom Helm, pp. 16–29.

Hein, P. (2004) Small island developing states: origin of the category and definition issues. In UNCTAD, ed., *Is a Special Treatment of Small Island Developing States Possible?* New York and Geneva: United Nations, pp. 1–22.

Helliwell, J., Layard, R. and Sachs, J. (2012) *World Happiness Report*. New York: Earth Institute, Columbia University.

Henry, P. (1985) *Peripheral Capitalism and Underdevelopment in Antigua*. New Brunswick, NJ: Transaction Books.

Henry-Lee, A., Watson, P., Osei, P. and Trezelle, S. (2010) Social risk management in a small state: the case of St Lucia. *Social and Economic Studies*, **59**, 61–89.

Hezel, F. (2005) The call to arms. Micronesians in the military. *Micronesian Counselor*, **57**.

Hilbish, T., Brannock, P., Jones, K. et al. (2010) Historical changes in the distributions of invasive and endemic marine invertebrates are contrary to global warming predictions: the effects of decadal climate oscillations. *Journal of Biogeography*, **37**, 423–431.

Hilborn, R., Branch, T., Ernst, B., Magnusson, A., Minte-Vera, C., Scheuerell, M. and Valero, J. (2003) State of the world's fisheries. *Annual Review of Environment and Resources*, **28**, 359–399.

Hintjens, H. and Hodge, D. (2012) The UK Caribbean Overseas Territories: governing unruliness amidst the extra-territorial EU. *Commonwealth and Comparative Politics*, **50**, 190–225.

Hirsh, H. and Marier, T. (2002) Damage and recovery of *Cycas micronesica* after Typhoon Paka. *Biotropica*, **34**, 598–602.

Hogenstijn, M. and van Middelkoop, D. (2005) Saint Helena: citizenship and spatial identities on a remote island. *Tijdschrift voor Economische en Sociale Geografie*, **96**, 96–104.

Holder, J. (2010) *Don't Burn Our Bridges. The Case for Owning Airlines*, Kingston, Jamaica: University of West Indies Press.

Holland, P. (2009) *Nadi Floods. Economic Costs. January 2009*. Technical Report No. 426. Suva, Fiji: SOPAC.

Holling, C. (1986) The resilience of terrestrial ecosystems: local surprise and global change. In W. Clark and R. Munn, eds, *Sustainable Development of the Biosphere*. Cambridge: Cambridge University Press, pp. 292–317.

Hooker, K. and Varcoe, J. (1999) Migration and the Cook Islands. In J. Overton and R. Scheyvens, eds, *Strategies for Sustainable Development. Experiences from the Pacific*. Sydney, Australia: UNSW Press, pp. 91–99.

Hooper, A. (1993) The MIRAB transition in Fakaofo. *Pacific Viewpoint*, **34**, 241–264.

Hooper, A. and Huntsman, J. (1973) A demographic history of the Tokelau islands. *Journal of the Polynesian Society*, **82**, 366–411.

Hope, K. (1986) *Urbanization in the Commonwealth Caribbean*. Boulder, CO: Westview.

Hopquin, B. and Canavate, L. (2011) Modernity on the march. *Guardian Weekly*, 22 April, 30.

Horowitz, L. (2004) Toward a viable independence? The Koniambo Project and the political economy of mining in New Caledonia. *The Contemporary Pacific*, **16**, 287–319.

Horowitz, L. (2009) Environmental violence and crises of legitimacy in New Caledonia. *Political Geography*, **28**, 248–258.

Horowitz, L. (2010) 'Twenty years is yesterday': science, multinational mining, and the political ecology of trust in New Caledonia. *Geoforum*, **41**, 617–626.

Houk, P., Rhodes, K., Cuetos-Bueno, J., Lindfield, S. et al. (2012) Commercial coral-reef fisheries across Micronesia: a need for improving management. *Coral Reefs*, **31**, 13–26.

Høyland, B., Moene, K. and Willumsen, F. (2012) The tyranny of international index rankings. *Journal of Development Economics*, **97**, 1–14.

Hudson, R. (2007) Big lizards, big problems. *Reptiles Magazine*, **15**(4), 54–61.

Huffer, E. and So'o, A. (2005) Beyond governance in Samoa: understanding Samoan political thought. *The Contemporary Pacific*, **17**, 311–333.

Hughes, T., Baird, A., Bellwood, D., Card, M. et al. (2003) Climate change, human impacts and the resilience of coral reefs. *Science*, **301**, 929–933.

Hulme, M. (2010) Cosmopolitan climates: hybridity, foresight and meaning. *Theory, Culture and Society*, **27**, 267–276.

Human Rights Watch (2012) *An Uncertain Future. Oil Contracts and Stalled Reform in São Tomé e Príncipe*. New York: Human Rights Watch.

Hunt, T. and Lipo, C. (2009) Revisiting Rapa Nui (Easter Island) 'ecocide'. *Pacific Science*, **63**, 601–616.

Hunt, T. and Lipo, C. (2011) *The Statues that Walked. Unraveling the Mystery of Easter Island.* New York: Free Press.

Hunte, K. (2011) The diaspora in Caribbean development: an untapped resource. *Small States Digest*, **1**, 10–27.

Huntington, H., Kruse, S. and Scholtz, A. (2009) Demographic and environmental conditions are uncoupled in the social-ecological system of the Pribilof Islands. *Polar Research*, **28**, 119–128.

Huntsman, J. and Kalolo, K. (2007) *The Future of Tokelau*. Auckland, New Zealand: Auckland University Press.

Husbands, W. (1986) Periphery resort tourism and tourist-resident stress: an example from Barbados. *Leisure Studies*, **5**, 175–188.

Hussein, M., Singh, A. and Aung, T. (2011) Rise and fall of sea level in Nauru area over a nodal cycle. *South Pacific Journal of Natural and Applied Sciences*, **28**, 63–68.

Hviding, E. and Bayliss-Smith, T. (2000) *Islands of Rainforest: Agro-forestry, Logging and Ecotourism in Solomon Islands*. Aldershot, UK: Ashgate.

Ibarrarán, M. and Ruth, M. (2009) Climate change and natural disasters: economic and distributional impacts. In M. Ibarrarán and M. Ruth, eds, *Distributional Impacts of Climate Change and Disasters*. Cheltenham, UK and Northampton, MA: Edward Elgar, pp. 46–66.

Inaoka, T., Matsumura, Y. and Suda, K. (2007) Tongan obesity: causes and consequences. In R. Ohtsuka and S. Ulijaszek, eds, *Health Change in the Asia-Pacific Region*. Cambridge: Cambridge University Press, pp. 127–146.

Ingham, R. and Summers, D. (2002) Falkland Islands cruise ship tourism: an overview of the 1999–2000 season and the way forward. *Aquatic Conservation: Marine and Freshwater Ecosystems*, **12**, 145–152.

Ingram, I. (2006) Adaptation in the Cook Islands. *Tiempo*, **58**, 15–17.

Ioannides, D. and Holcomb, B. (2003) Misguided policy initiatives in small-island destinations: why do up-market tourism policies fail? *Tourism Geographies*, **5**, 39–48.

Jackson, J. (1997) Reefs since Columbus. *Coral Reefs*, **16**, S23–S32.

Jacobs, B. (1975) Administrative problems of small countries. In P. Selwyn, ed., *Development Policy in Small Countries*. London: Croom Helm, pp. 134–143.

Jalan, B. ed (1982) *Problems and Policies in Small Economies*. Beckenham, UK: Croom Helm.

James, K. (1991) Migration and remittances: a Tongan village perspective. *Pacific Viewpoint*, **32**, 1–23.

James, K. (1993a) Cutting the ground from under them? Commercialization, cultivation and conservation in Tonga. *The Contemporary Pacific*, **5**, 215–242.

James, K. (1993b) Cash and kin. Aspects of migration and remittance from the perspective of a fishing village in Vava'u, Tonga. In G. McCall and J. Connell, eds, *A World Perspective on Pacific Islander Migration*. Monograph No. 6. Sydney: UNSW Centre for South Pacific Studies, pp. 359–374.

Jarvis, J. and Peel V. (2010) Flashpacking in Fiji: reframing the 'global nomad' in a developing destination. In K. Hannam and A. Diekmann, eds, *Beyond Backpacker Tourism. Mobilities and Experiences*. Bristol, UK: Channel View, pp. 21–39.

Jayaraman, T., Choong, C. and Kumar, R. (2010) Case study. Role of remittances in Tongan economy. *Migration Letters*, **7**, 224–230.

Jeffery, L. (2010) Forced displacement, onward migration and reformulation of home by Chagossians in Crawley, UK. *Journal of Ethnic and Migration Studies*, **36**, 1099–1177.

Jessen, A. and Vignoles, C. (2004) Barbados: trade and integration as a strategy for growth. INTAL-ITD Occasional Paper No. 28. Buenos Aires: Inter-American Development Bank.

Johannes, R. (1981) *Words of the Lagoon. Fishing and Marine Lore in the Palau District of Micronesia*. Berkeley, CA: University of California Press.

Johannes, R. (1998) Government-supported, village-based management of marine resources in Vanuatu. *Ocean and Coastal Management*, **40**, 165–186.

Johannes, R. (2002) The renaissance of community based marine resource management in Oceania. *Annual Review of Ecology and Systematics*, **33**, 317–331.

Johannes, R. and Hickey, F. (2002) *Evolution of Village-Based Marine Resource Management in Vanuatu between 1993 and 2003*. Suva, Fiji: UNESCO.

Johnson, G. (2011) EPA reports Marshalls water is highly contaminated. *Marianas Variety*, 2 May.

Johnson, W. (2000) Rabaul volcanoes fire 30-year passion. *AusGeo News*, **57**, April, 12–13.

Jokhan, M. (2008) Parental absence as a consequence of migration: reviewing the literature. *Social and Economic Studies*, **57**(2), 89–117.

Jolivet, A., Florence, S., Lebas, J. and Chauvin, P. (2010) Migration, health, and care in French overseas territories. *The Lancet*, 1376(9755), 1827–1828.

Jollit, I., Léopold, M., Guillemot, N., David, G., Chavanet, P. et al. (2010) Geographical aspects of informal reef fishery systems in New Caledonia. *Marine Pollution Bulletin*, **61**, 585–597.

Jones, E., Deere-Birkbeck, C. and Woods, N. (2010) *Manoeuvring at the Margins. Constraints Faced by Small States in International Trade Negotiations*. London: Commonwealth Secretariat.

Jones, P. and Lea, J. (2007) What has happened to urban reform in the island Pacific? Some lessons from Kiribati and Samoa. *Pacific Affairs*, **80**, 473–491.

Jones, R. (2011a) The local economic imprint of return migrants in Bolivia. *Population, Space and Place*, **17**, 435–453.

Jones, R. (2011b) Spatial patterns of chemical contamination in sediments of a non-industrialized but densely populated coral atoll/small island state (Bermuda). *Marine Pollution Bulletin*, **62**, 1362–1376.

Jones, S. (2009) A long-term perspective on biodiversity and marine resource exploitation in Fiji's Lau group. *Pacific Science*, **63**, 617–648.

Joseph, R. (1978) Affluence and underdevelopment: the Nigerian experience. *Journal of Modern African Studies*, **16**, 221–239.

Juhasz, A., Ho, E., Bender, E. and Fong, P. (2010) Does use of tropical beaches by tourists and island residents result in damage to fringing coral reefs? A case study in Moorea, French Polynesia. *Marine Pollution Bulletin*, **60**, 2251–2256.

Juvik, S. (1987) Joint venture rice production in the Solomon Islands: the inner workings of dependent development. *Pacific Studies*, **10**, 73–90.

Kabutaulaka, T. (2006) Global capital and local ownership in Solomon Islands' forestry industry. In S. Firth, ed., *Globalisation and Governance in the Pacific Islands*. Canberra, Australia: ANU E Press, pp. 239–257.

Kahn, M. (2011) *Tahiti Beyond the Postcard. Power, Place and Everyday Life*. Seattle, WA: University of Washington Press.

Kaplinsky, R. (1983) Prospering at the periphery: a special case – the Seychelles. In R. Cohen, ed., *African Islands and Enclaves*. Beverly Hills, CA, and London: Sage, pp. 195–215.

Karels, T., Dobson, F., Trevino, H. and Skibiel, A. (2008) The biogeography of avian extinctions on oceanic islands. *Journal of Biogeography*, **35**, 1106–1111.

Katovai, E., Burley, A. and Mayfield, M. (2012) Understory plant species and functional diversity in the degraded wet tropical forests of Kolombangara Island, Solomon Islands. *Biological Conservation*, **145**, 214–224.

Kay, G. (1975) *Development and Underdevelopment. A Marxist Analysis*. London: Macmillan.

Keegan, W., Fitzpatrick, S.D., Sealey, K., Le Febvre, M. and Sinelli, P. (2008) The role of small islands in marine subsistence stategies: case studies from the Caribbean. *Human Ecology*, **36**, 635–654.

Keesing, R. (1984) Traditionalist enclaves in Melanesia. In R. May and H. Nelson, eds, *Melanesia: Beyond Diversity*. Canberra, Australia: ANU Press, pp. 39–54.

Keller, B., Gleason, D., McLeod, E., Woodley, C. et al. (2009) Climate change, coral reef ecosystems, and management options for marine protected areas. *Environmental Management*, **44**, 1069–1088.

Kelly, D. (1986) St Lucia's female electronic factory workers: key components in an export-oriented industrialization strategy. *World Development*, **14**, 823–838.

Kelsey, J. (2004) *Big Brothers Behaving Badly. The Implications for the Pacific Islands of the Pacific Agreement on Closer Economic Relations (PACER).* Suva, Fiji: Pacific Network on Globalisation.

Kench, P. (2012) The geomorphology of Baa (South Maaalhosmadalu) Atoll and its reef islands. *Atoll Research Bulletin*, **590**, 1–28.

Kench, P. and Brander, R. (2006) Response of reef island shorelines to seasonal climate oscillations: South Maalhosmadulu Atoll, Maldives. *Journal of Geophysical Research*, **111**.

Kench, P., Nichol, S., Smithers, S., McLean, R. and Brander, R. (2008) Tsunami as agents of geomorphic change in mid-ocean reef islands. *Geomorphology*, **95**, 361–383.

Kenter, J., Hyde, T., Christie, M. and Fazey, I. (2011) The importance of deliberation in valuing ecosystem services in developing countries: evidence from the Solomon Islands. *Global Environmental Change*, **21**, 505–521.

Kerr, S. (2005) What is small island sustainable development about? *Ocean and Coastal Management*, **48**, 503–524.

Kerstetter, D. and Bricker, K. (2009) Exploring Fijians' sense of place after exposure to tourism development. *Journal of Sustainable Tourism*, **17**, 691–708.

Kingston, N. and Waldren, S. (2003) The plant communities and environmental gradients of Pitcairn Island: the significance of invasive species and the need for conservation management. *Annals of Botany*, **92**, 31–40.

Kingston, N., Waldren, S. and Bradley, U. (2003) The phytogeographical affinities of the Pitcairn Islands: a model for south-eastern Polynesia? *Journal of Biogeography*, **30**, 1311–1328.

Kirch, P. (1986) Exchange systems and inter-island contact in the transformation of an island society: the Tikopia case. In P. Kirch, ed., *Island Societies*. Cambridge: Cambridge University Press, pp. 33–41.

Kirch, P. (1997) Microcosmic histories. Island perspectives on 'global' change. *American Anthropologist*, **99**, 30–42.

Kirch, P. and Rallu, J.-L. (2007) Long-term demographic evolution in the Pacific Islands. In P. Kirch and J.-L. Rallu, eds, *The Growth and Collapse of Pacific Island Societies*. Honolulu, Hawaii: University of Hawaii Press, pp. 1–14.

Kiste, R. (1974) *The Bikinians. A Study in Forced Migration*. Menlo Park, CA: Cummings.

Klak, T., Wiley, J., Mullaney, E., Peteru, S. and Merilus, J.-Y. (2011) Inclusive neoliberalism? Perspectives from eastern Caribbean farmers. *Progress in Development Studies*, **11**, 33–61.

Knapman, B. (1976) Indigenous involvement in the cash economy of Lau, Fiji, 1840–1946. *Journal of Pacific History*, **11**, 167–188.

Knauft, B. (2002) *Exchanging the Past. A Rainforest World of Before and After*. Chicago, IL: University of Chicago Press.

Knutson, T., McBride, J., Chan, J., Emanuel, K. et al. (2010) Tropical cyclones and climate change. *Nature Geoscience*, 3, 157–163.

Koldewey, H., Curnick, D., Harding, S., Harrison, L. and Gollock, M. (2010) Potential benefits to fisheries and biodiversity of the Chagos Archipelago/British Indian Ocean Territory as a no-take marine reserve. *Marine Pollution Bulletin*, **60**, 1906–1915.

Kostaschuk, R., Terry, J. and Raj, R. (2001) Tropical cyclones and floods in Fiji. *Hydrological Sciences Journal*, **46**, 435–450.

Kothari, U. and Wilkinson, R. (2013) Global change, small island state response: restructuring and the perpetuation of uncertainty in Mauritius and Seychelles. *Journal of International Development*, **25**, 92–107.

Kronen, M. and Bender, A. (2007) Assessing marine resource exploitation in Lofanga, Tonga: one case study – two approaches. *Human Ecology*, **35**, 195–207.

Kronen, M., Meloti, A., Ponia, B., Pickering, T., Diake, S., Kama, J., Kenilolerie, P., Ngwaerobo, J. and Teitelbaum, A. (2010a) Gender and seaweed farming on Wagina Island, Choiseul Province in Solomon Islands. *SPC Women in Fisheries Information Bulletin*, **21**, 3–10.

Kronen, M., Vunisea, A., Magron, F. and McArdle, B. (2010b) Socioeconomic drivers and indicators for artisanal coastal fisheries in Pacific island countries and territories and their use for fisheries management strategies. *Marine Policy*, **34**, 1135–1143.

Kuhlken, R. (2007) Agricultural landscapes of Kadavu. Persistence and change on the Fijian periphery. In J. Connell and E. Waddell, eds, *Environment, Development and Change in Rural Asia-Pacific*. London: Routledge, pp. 56–75.

Kumar, V. (2010) Water management in Fiji. *Water Resources Development*, **26**, 81–96.

Kupferman, D. (2011) The Republic of the Marshall Islands since 1990. *Journal of Pacific History*, **46**, 75–88.

Kupferman, D. (2012) The Marshall Islands. *The Contemporary Pacific*, **24**, 150–156.

Kupu, S. (2000) Health and telecast: A milestone for Tonga in the 21st century. *Pacific Health Dialog*, **12**(2), 155–158.

Kuruppu, N. and Liverman, D. (2011) Mental preparation for climate adaptation: the role of cognition and culture in enhancing adaptive capacity of water management in Kiribati. *Global Environmental Change*, **21**, 657–669.

Kuznets, S. (1960) Economic growth of small nations. In E. Robinson, ed., *The Economic Consequences of Size of Nations*. London: Macmillan, pp. 14–32.

La Grenade, W. (2008) Balancing economic development and security: tourism and HIV/AIDS (Grenada). *Social and Economic Studies*, **57**, 27–60.

Lal, M., Harasawa, H. and Takahashi, K. (2002) Future climate change and its impacts over small island states. *Climate Research*, **19**, 179–192.

Lal, P.C. (2003) *Cyclone Ami. 13–14 January 2003*. Lautoka, Fiji: Vikas Press.

Lal, P.N. (2010) Vulnerability to natural disasters: an economic analysis of the impact of the 2009 floods on the Fijian sugar belt. *Pacific Economic Bulletin*, **25**(2), 62–77.

Lal, P.N., Rira, R. and Khatri, N. (2009a) *Floods in the Fiji Sugar Belt and Policy Implications*. Gland, Switzerland: IUCN.

Lal, P.N., Singh, R. and Holland, P. (2009b) Relationship between natural disasters and poverty: a Fiji case study. Miscellaneous Report No. 678. Suva, Fiji: SOPAC.

Lal, P.N., Kinch, J. and Wickham, F. (2009c) *Review of Economic and Livelihood Impact Assessments of, and Adaptation to, Climate Change in Melanesia*. Apia, Samoa: SPREP.

Lamarche, G., Pelletier, B. and Goff, J. (2010) Impact of the 29 September 2009 South Pacific tsunami on Wallis and Futuna. *Marine Geology*, **271**, 297–302.

Lander, J., Whiteside, L. and Lockridge, P. (2002) A brief history of tsunamis in the Caribbean. *Science of Tsunami Hazards*, **20**, 57–94.

Lane, M. (2008) Strategic coastal governance issues in Fiji: the challenges of integration. *Marine Policy*, **32**, 856–866.

Lapointe, B., Langton, R., Bedford, B., Potts, A., Day, O. and Hu, C. (2010) Land-based nutrient enrichment of the Buccoo Reef Complex and fringing coral reefs of Tobago, West Indies. *Marine Pollution Bulletin*, **60**, 334–343.

Larmour, P. (1998) Introduction. In P. Larmour, ed., *Governance and Reform in the South Pacific*. Canberra, Australia: National Centre of Development Studies, pp. 1–20.

Larmour, P. (2005) *Foreign Flowers, Institutional Transfer and Good Governance in the Pacific Islands*. Honolulu, Hawaii: University of Hawaii Press.

Larmour, P. (2009) How much corruption is there in the Pacific islands? A review of different approaches to measurement. *Pacific Economic Bulletin*, **24**(1), 144–160.

Lata, S. and Nunn, P. (2011) Misperceptions of climate-change risk as barriers to climate-change adaptation: a case study from the Rewa Delta, Fiji. *Climatic Change*, **110**, 169–186.

Lattas, A. (2006) The utopian promise of government. *Journal of the Royal Anthropological Institute*, **12**, 129–150.

Lauer, M. and Aswani, S. (2010) Indigenous knowledge and long-term ecological change: detection, interpretation, and responses to changing ecological conditions in Pacific Island communities. *Environmental Management*, **45**, 985–997.

Lauer, M., Albert, S., Aswani, S., Halpern, B., Campanella, L. and La Rose, D. (2013) Globalization, Pacific Islands, and the paradox of resilience. *Global Environmental Change*, **23**, 40–50.

Laurent, E. (2011) Priorities for small states in global trade governance. In C. Deere Birkbeck, ed., *Making Global Trade Governance Work for Development*. Cambridge: Cambridge University Press, pp. 204–230.

Lawrence, R. (1992) *Tamana Fifteen Years On: A Survey of Changes in the Household Economy since 1974*. Wellington, New Zealand: Victoria University Department of Geography.

Lazrus, H. (2009) The governance of vulnerability: climate change and agency in Tuvalu, South Pacific. In S. Crate and M. Nuttall, eds, *Anthropology and Climate Change*. Walnut Creek, CA: Left Coast Press, pp. 240–249.

Lee, D. and Smith, N. (2010) Small state discourses in the international political economy. *Third World Quarterly*, **31**, 1091–1105.

Lee, H. (2003) *Tongans Overseas. Between Two Shores*. Honolulu, Hawaii: University of Hawaii Press.

Lee, H. (2004) All Tongans are connected: Tongan transnationalism. In V. Lockwood, ed., *Globalization and Culture Change in the Pacific Islands*. Upper Saddle River, New Jersey: Pearson, pp. 133–148.

Lee, H. (2009) The ambivalence of return: second-generation Tongan returnees. In D. Conway and R. Potter, eds, *Return Migration of the Next Generations. 21st Century Transnational Mobility*. Aldershot, UK: Ashgate, pp. 41–58.

Lee, H. (2011) Rethinking transnationalism through the second generation. *Australian Journal of Anthropology*, **22**, 295–313.

Leigh, D., Frayman, H. and Ball, K. (2012) Anonymity attracts 1m companies to microstate. *Guardian Weekly*, 30 November, 5.

Le Masson, V. and Kelman, I. (2011) Disaster risk reduction on non-sovereign islands: La Réunion and Mayotte, France. *Natural Hazards*, **56**, 251–273.

Lepowsky, M. (1991) The way of the ancestors: custom, innovation and resistance. *Ethnology*, **30**, 217–235.

Lesser, M. and Slattery, M. (2011) Phase shift to algal dominated communities at mesophotic depths associated with lionfish (*Pterois volitans*) invasion on a Bahamian coral reef. *Biological Invasions*, **13**, 1855–1868.

Levick, W. and Bedford, R. (1988) Fiji labour migration to New Zealand in the 1980s. *New Zealand Geographer*, **44**, 14–21.

Lewis, J. (1985) Some perspectives on natural disaster vulnerability in Tonga. *Pacific Viewpoint*, **22**, 145–162.

Lewis, P. (2005) Unequal negotiations: small states in the new global economy. *Journal of Eastern Caribbean Studies*, **30**, 54–107.

Lewis, P. (2011) Training nurses for export: a viable development strategy? *Social and Economic Studies*, **60**, 67–104.

Li, T. (2007) *The Will to Improve: Governmentality, Development and the Practice of Politics*. Durham, NC: Duke University Press.

Lieber, M. (1994) *More Than a Living. Fishing and the Social Order on a Polynesian Atoll*. Boulder, CO: Westview.

Lieber, M. (1999) The sustainable, the expendable and the obsolete. *Pacific Studies*, **22**, 13–46.

Liki, A. (2001) Moving and rootedness: the paradox of the brain drain among Samoan professionals. *Asia-Pacific Population Journal*, **16**, 67–84.

Lima, M. and Naya, D. (2011) Large-scale climatic variability affects the dynamics of tropical skipjack tuna in the Western Pacific Ocean. *Ecography*, **34**, 597–605.

Liou, F. and Ding, C. (2002) Subgrouping small states based on socioeconomic characteristics. *World Development*, **30**, 1289–1306.

Lipset, D. (2011) The tides: masculinity and climate change in coastal Papua New Guinea. *Journal of the Royal Anthropological Institute*, **17**, 20–43.

Live and Learn Environmental Education (2006) *Promoting Sound Environmental Management in the Aftermath of the Tsunami Disaster*. Suva, Fiji, and Male, Maldives: Live and Learn Environmental Education.

Lockwood, V. (1990) Development and return migration to rural French Polynesia. *International Migration Review*, **24**, 347–371.

Lockwood, V. (1993) *Tahitian Transformation. Gender and Capitalist Development in a Rural Society*. Boulder, CO: Lynne Rienner.

Lokrantz, J., Nystrom, M., Norstrom, A., Folke, C. and Cinner, J. (2010) Impacts of artisanal fishing on key functional groups and the potential vulnerability of coral reefs. *Environmental Conservation*, **36**, 327–337.

Lomas, P. (1987) Tonga: the good word on investment. *Islands Business*, **13**(10), October, 23–30.

Long, F. (1987) 'New exports' of the Caribbean to the international economy. *Development Policy Review*, **5**, 63–72.

Lopez-Marrero, T. (2010) An integrative approach to study and promote natural hazards adaptive capacity: a case study of two flood-prone communities in Puerto Rico. *Geographical Journal*, **176**, 150–163.

Lowenthal, D. and Comitas, L. (1962) Emigration and depopulation: some neglected aspects of population geography. *Geographical Review*, **52**, 195–210.

Lui, R. (2009) Issues in water supply and sanitation of North Camp community, Ebeye, Republic of the Marshall Islands. Miscellaneous Report No. 702. Suva, Fiji: SOPAC.

Lutz, C. (1988) *Unnatural Emotions: Everyday Sentiments on a Micronesian Atoll*. Chicago, IL: University of Chicago Press.

Maass, M. (2009) The elusive definition of the small state. *International Politics*, **46**, 65–83.

McAdam, J. (2010) 'Disappearing states', statelessness and the boundaries of international law. In J. McAdam, ed., *Climate Change and Displacement. Multidisciplinary Perspectives*. Oxford: Hart Publishing, pp. 105–129.

McAdoo, B., Fritz, H., Jackson, K. et al. (2008) Solomon Islands tsunami, one year later. *Eos*, **89**(18), 169–176.

McAdoo, B., Moore, A. and Baumwoll, J. (2009) Indigenous knowledge and the near field population response during the 2007 Solomon Islands tsunami. *Natural Hazards*, **48**, 73–82.

McCall, G. (2008) Another (unintended) legacy of Captain Cook? The evolution of Rapanui (Easter Island) tourism. In J. Connell and B. Rugendyke, eds, *Tourism at the Grassroots*. London: Routledge, pp. 41–57.

McClanahan, T., Polunin, N. and Done, T. (2002) Ecological states and the resilience of coral reefs. *Conservation Ecology*, **6**(18) (online).

Macdonald, B. (1975) Secession in the defence of identity: the making of Tuvalu. *Pacific Viewpoint*, **16**, 26–44.

Macdonald, B. (1982) *Cinderellas of the Empire. Towards a History of Kiribati and Tuvalu*. Canberra, Australia: ANU Press.

Macdonald, S. and Demetrius, F. (1986) The Caribbean sugar crisis. *Journal of Interamerican Studies*, **28**, 35–58.

McDougall, D. (2005) The unintended consequences of clarification: Development, disrupting and the dynamics of community in Ranongga, Solomon Islands. *Ethnohistory*, **52**, 81–109.
McDougall, D. (2009) Becoming sinless: converting to Islam in the Christian Solomon Islands. *American Amthropologist*, **111**, 480–491.
McDougall, D. (2011) Church, company, committee, chief: emergent collectivities in rural Solomon Islands. In M. Patterson and M. Macintyre, eds, *Managing Modernity in the Western Pacific*. Brisbane, Australia: University of Queensland Press, pp. 121–146.
McDougall, D., Barry, I. and Pio, S. (2008) Disaster and recovery on Ranongga: six months after the earthquake in the Western Solomons. Mimeo, Perth, Australia.
McDowall, R. (2005) Falkland Islands biogeography: converging trajectories in the South Atlantic Ocean. *Journal of Biogeography*, **32**, 49–62.
McElroy, J. (2006) Small island tourist economies across the lifecycle. *Asia Pacific Viewpoint*, **47**, 61–77.
McElroy, J. (2011) Island demography: a review of selected Caribbean contributions. *Island Studies Journal*, **6**, 245–260.
McElroy, J. and Parry, C. (2010) The characteristics of small island tourist economies. *Tourism and Hospitality Research*, **10**, 315–328.
McElroy, J. and Pearce, K. (2006) The advantages of political affiliation: dependent and independent small-island profiles. *The Round Table*, **95**, 529–539.
Macfarlane, J. and Alpers, M. (2010) National policy for an integrated health system and local implementation: the case of Papua New Guinea and the Nasioi. *Human Organization*, **69**, 387–397.
McGillivray, M., Naudé, W. and Santos-Paulino, A. (2010) Vulnerability, trade, financial flows and state failure in small island developing states. *Journal of Development Studies*, **46**, 815–827.
McGinley, L. (2007) The Caribbean. In F. Cubbage, ed., *Forests and Forestry in the Americas: An Encyclopaedia*. Safopedia, www.encyclopediaofforestry.org/index.php/Caribbean01.
McGranahan, G., Balk, D. and Anderson, B. (2007) The rising tide: assessing the risks of climate change and human settlements in low elevation coastal zones. *Environment and Urbanization*, **19**, 17–37.
McGrath, B. (1999) Healing practice in Tonga: swimming from island to island. *Medical Anthropology Quarterly*, **13**, 483–505.
McGregor, A. (1989) The Tongan knitwear industry: a case study in export manufacturing development in the Pacific Islands. Honolulu, Hawaii: Pacific Islands Development Program.
McInnes, L. (1986) *Water on Vatulele: Environment, Community and Development*. Unpublished BSc Honours thesis, University of Sydney, Australia.

McInnes, L. and Connell, J. (1988) The world system in a Fijian store. *South Pacific Forum*, **4**, 116–121.
McIntyre, W. (2012) The Partition of the Gilbert and Ellice Islands. *Island Studies Journal*, **7**, 135–146.
McKee, D. and Tisdell, C. (1990) *Developmental Issues in Small Island Economies*. New York: Praeger.
McKenzie, E., Woodruff, A. and McClennen, C. (2006) *Economic Assessment of the True Costs of Aggregate Mining in Majuro Atoll, Republic of the Marshall Islands*. Technical Report No. 383. Suva, Fiji: SOPAC.
MacKenzie, R. (2008) Impacts of riparian forest removal on Palauan streams. *Biotropica*, **40**, 666–675.
McKie, R. (2011) Tristan da Cunha islanders rescue rockhopper penguins threatened by oil slick. *The Observer*, 17 April, 8.
McKnight, R. (1977) Commas in microcosm: the movement of Southwest Islanders to Palau, Micronesia. In M. Lieber, ed., *Exiles and Migrants in Oceania*. Honolulu, Hawaii: University of Hawaii Press, pp. 10–33.
Maclellan, N. (2007) Fiji, Iraq and Pacific Island security. *Race and Class*, **48**, 47–62.
McMichael, A., Woodruff, R. and Hales, S. (2006) Climate change and human health: present and future risks. *Lancet*, **367**, 859–869.
McNamara, K. (2009) Voices from the margins: Pacific ambassadors and the geopolitics of marginality at the United Nations. *Asia Pacific Viewpoint*, **50**, 1–12.
McNamara, K. and Gibson, C. (2009) 'We do not want to leave our land': Pacific ambassadors at the United Nations resist the category of 'climate refugees'. *Geoforum*, **40**, 475–483.
Macpherson, C. (1992a) The road to power is a chainsaw: villages and innovation in Western Samoa. *Pacific Studies*, **11**, 1–24.
Macpherson, C. (1992b) Economic and political restructuring and the sustainability of migrant remittances: the case of Western Samoa. *The Contemporary Pacific*, **4**, 109–136.
Macpherson, C. (1994) Changing patterns of commitment to island homelands: a case study of Western Samoa. *Pacific Viewpoint*, **17**, 83–116.
Macpherson, C. (2000) Oasis or mirage: the farming of black pearl in the Northern Cook Islands. *Pacific Studies*, **23**, 33–55.
Macpherson, C. (2004) Transnationalism and transformation in Samoan society. In V. Lockwood, ed., *Globalization and Culture Change in the Pacific Islands*. Upper Saddle River, NJ: Pearson, pp. 165–181.
Macpherson, C. (2008) Golden goose or Trojan horse? Cruise ship tourism in Pacific development. *Asia Pacific Viewpoint*, **49**, 185–197.

Macpherson, C. and Macpherson, L. (1987) Toward an explanation of recent trends in suicide in Western Samoa. *Man*, **22**, 305–330.
Macpherson, C. and Macpherson, L. (2009) *The Warm Winds of Change. Globalisation in Contemporary Samoa*. Auckland: Auckland University Press.
Maebuta, H. and Maebuta, J. (2009) Generating livelihoods: a study of urban squatter settlements in Solomon Islands. *Pacific Economic Bulletin*, **24**(3), 118–131.
Mahadevan, R. and Asafu-Adjaye, J. (2013) Lessons from the unfinished agenda of a small developing economy under trade and structural reforms. *Journal of International Development* (in press).
Maingot, A. (1983) Caribbean migration as a structural reality. Miami, FL: Latin American and Caribbean Center Dialogues No. 13.
Maingot, A. (1986) Coming to terms with the 'improbable revolution'. *Journal of Interamerican Studies*, **27**, 177–190.
Mallela, J. and Crabbe, M. (2009) Hurricanes and coral bleaching linked to changes in coral recruitment in Tobago. *Marine Environmental Research*, **68**, 158–162.
Malogne-Fer, G. (2004) L'aéroport à Raivavae (Australes): développement économique, migrations and identités. *Journal de la Société des Océanistes*, **119**, 193–199.
Malvatumauri National Council of Chiefs (2012) Alternative indicators of well-being for Melanesia. Vanuatu Pilot Study Report. Vanuatu: Port Vila.
Marcus, G. (1974) A hidden dimension of family development in the modern Kingdom of Tonga. *Journal of Comparative Family Studies*, **5**, 87–102.
Marcus, G. (1981) Power on the extreme periphery: the perspective of Tongan elites in the modern world system. *Pacific Viewpoint*, **22**, 48–64.
Marcus, G. (1993) Tonga's contemporary globalizing strategies: trading on sovereignty amidst international migration. In V. Lockwood, T. Harding and B. Wallace, eds, *Contemporary Pacific Societies*. Englewood Cliffs, NJ: Prentice Hall, pp. 21–33.
Marks, K. (2008) *Pitcairn. Paradise Lost*. London: Harper Collins.
Markyanda, A. (2011) Equity and distributional implications of climate change. *World Development*, **39**, 1051–1060.
Maron, N. and Connell, J. (2008) Return to Nukunuku. Employment, identity and return migration in Tonga. *Asia Pacific Viewpoint*, **49**, 168–184.
Marshall, D. (1981) Haitian migration to the Bahamas. In S. Craig, ed., *Contemporary Caribbean: A Sociological Reader*. Trinidad: College Press, pp. 101–127.

Marshall, D. (1982) Migration as an agent of change in Caribbean island ecosystems. *International Social Science Journal*, **34**, 451–467.

Marshall, M. (1979) *Weekend Warriors: Alcohol in a Micronesian Culture*. Palo Alto, CA: Mayfield.

Marshall, M. (2004a) *Namoluk Beyond the Reef. The Transformation of a Micronesian Community*. Boulder, CO: Westview.

Marshall, M. (2004b) Market highs: alcohol, drugs, and the global economy in Oceania. In V. Lockwood, ed., *Globalization and Culture Change in the Pacific Islands*. Upper Saddle River, NJ: Pearson, pp. 200–221.

Marsters, E., Lewis, N. and Friesen, W. (2006) Pacific flows: the fluidity of remittances in the Cook Islands. *Asia Pacific Viewpoint*, **47**, 31–44.

Maude, H. (1968) *Of Islands and Men*. Melbourne, Australia: Oxford University Press.

Maude, H. (1981) *Slavers in Paradise*. Canberra, Australia: ANU Press.

Mecartney, S. (2001) *Blacksands Settlement: Towards Urban Permanence in Vanuatu*. Unpublished MA thesis, University of Sydney, Australia.

Méheux, K., Dominey-Howes, D. and Lloyd, K. (2007) Natural hazard impacts in small island developing states: a review of current knowledge and future research needs. *Natural Hazards*, **40**, 429–446.

Melde, S. and Schicklinski, J. (2011) Remittances in the African, Caribbean and Pacific Countries. Background Note No. 2. Brussels: ACP Observatory on Migration.

Meltzoff, S. and LiPuma, E. (1983) A Japanese fishing joint venture: worker experience and national development in the Solomon Islands. Report No. 12. Manila, Philippines: ICLARM.

Meltzoff, S. and LiPuma, E. (1986) Hunting for tuna and cash in the Solomons: a rebirth of artisanal fishing in Malaita. *Human Organization*, **45**, 53–62.

Mercer, J., Dominey-Howes, D., Kelman, I. and Lloyd, K. (2007) The potential for combining indigenous and western knowledge in reducing vulnerability to environmental hazards in small developing states. *Environmental Hazards*, **7**, 245–256.

Mertz, O., Bruun, T.B., Fog, B., Rasmussen, K. and Agergaard, J. (2010) Sustainable land use in Tikopia: food production and consumption in an isolated agricultural system. *Singapore Journal of Tropical Geography*, **31**, 10–26.

Meyssignac, B. and Cazenave, A. (2012) Sea level: a review of present-day and recent-past changes and variability. *Journal of Geodynamics*, **58**, 96–109.

Mideksa, T. (2010) Economic and distributional impacts of climate change: the case of Ethiopia. *Global Environmental Change*, **20**, 278–286.

Miles, E. (2009) On the increasing vulnerability of the world ocean to multiple stresses. *Annual Review of Environment and Resources*, **34**, 17–41.
Millard-Ball, A. (2011) The Tuvalu Syndrome. Can geoengineering solve climate's collective action problem? *Climatic Change*, **110**, 1047–1066.
Miller, B., Breckheimer, I., McCleary, A. et al. (2010) Using stylized agent-based models for population-environment research: a case study from the Galapagos Islands. *Population and Environment*, **31**, 401–426.
Miller, J., Muller, E., Rogers, C., Waara, R. et al. (2009) Coral disease following massive bleaching in 2005 causes 60 per cent decline in coral reef cover on reefs in the US Virgin Islands. *Coral Reefs*, **28**, 925–937.
Miller, K. (2007) Climate variability and tropical tuna: management challenges for highly migratory stocks. *Marine Policy*, **31**, 56–70.
Miller, M., McClellan, D., Wiener, J. and Stoffle, B. (2007) Apparent rapid fisheries escalation at a remote Caribbean island. *Environmental Conservation*, **34**, 92–94.
Mills, B. (2004) Hucksters and homemakers. Gender responses to opportunities in the tourism market in Carriacou, Grenada. In D. Duval, ed., *Tourism in the Caribbean*. London: Routledge, pp. 257–270.
Mimura, N., Nurse, L., McLean, R., Agard, J. et al. (2007) Small islands. In M. Parry, O. Canziani, J. Palutikof, P. van der Linden and C. Hanson, eds, *Climate Change 2007: Impacts Adaptation and Vulnerability*. Cambridge: Cambridge University Press, pp. 687–716.
Mitchell, J. (2000) Violence as continuity: violence as rupture: narratives from an urban settlement in Vanuatu. In S. Dinnen and A. Ley, eds, *Reflections on Violence in Melanesia*. Sydney and Canberra, Australia: Hawkins Press and Asia-Pacific Press, pp. 189–208.
Mitchell, J. (2004) 'Killing time' in a postcolonial town: young people and settlements in Port Vila, Vanuatu. In V. Lockwood, ed., *Globalization and Culture Change in the Pacific Islands*, Upper Saddle River, NJ: Pearson, pp. 358–376.
Mitchell, J. (2011) 'Operation restore public hope': youth and the magic of modernity in Vanuatu. *Oceania*, **81**, 36–50.
Moberg, M. (2005) Fair trade and Eastern Caribbean banana farmers: rhetoric and reality in the anti-globalization movement'. *Human Organization*, **64**, 4–15.
Moberg, M. (2008) *Slipping Away. Banana Politics and Fair Trade in the Eastern Caribbean*. Oxford: Berghahn.

Moglia, M., Perez, P. and Burn, S. (2008) Water troubles in a Pacific atoll town. *Water Policy*, **10**, 613–637.

Moglia, M., Perez, P. and Burn, S. (2010) Modelling an urban water system on the edge of chaos. *Environmental Modelling and Software*, **25**, 1528–1538.

Molea, T. and Vuki, V. (2008) Subsistence fishing and fish consumption patterns of the saltwater people of Lau Lagoon, Malaita, Solomon Islands: a case study of Funaafou and Niuleni Islanders. *SPC Women in Fisheries Information Bulletin*, **18**, 30–35.

Momsen, J. (1986) Migration and rural development in the Caribbean. *Tijdschrift voor Economische en Sociale Geografie*, **77**, 50–58.

Momsen, J. (1998) Caribbean tourism and agriculture: new linkages in the global era? In T. Klak, ed., *Globalization and Neoliberalism: The Caribbean Context*. Lanham, MD: Rowman and Littlefield, pp. 115–133.

Monbiot, G. (2011) The result of spineless politics. *Guardian Weekly*, 22 July, 24.

Moore, M. (2011) Globalisation and power in weak states. *Third World Quarterly*, **32**, 1757–1776.

Moore, W. (2010) The impact of climate change on Caribbean tourism demand. *Current Issues in Tourism*, **13**, 495–505.

Moorhead, A. (2011) Lesson in diversity from Samoa's taro blight. *Partners* (ACIAR), March–May, 18–19.

Mora, C. (2008) A clear human footprint on the coral reefs of the Caribbean. *Proceedings of the Royal Society B*, **275**, 767–773.

Mora, C. (2009) Degradation of Caribbean coral reefs: focusing on proximal rather than ultimate drivers. *Proceedings of the Royal Society B*, **276**, 199–200.

Morin, J. and Lavigne, F. (2009) Institutional and social responses to hazards related to Karthala Volcano, Comoros. *Shima*, **3**, 54–71.

Morner, N.-A., Tooley, M. and Possnert, G. (2004) New perspectives for the future of the Maldives. *Global and Planetary Change*, **40**, 177–182.

Morrell, W. (2008) Special waste management in the Marshall Islands: a micro-level view of the automotive service industry's practices. In J. Terry and F. Thomas, eds, *The Marshall Islands. Environment, History and Society in the Atolls*. Suva, Fiji: USP Faculty of Oceans and Islands, pp. 252–275.

Morrison, R.J. and Kaly, U. (2010) Coastal lagoon management in three Pacific island situations: is scientific knowledge used effectively? In S. Nath, J. Roberts and Y. Madhoo, eds, *Saving Small Island Developing States*. London: Commonwealth Secretariat, pp. 172–187.

Morrison, L. and Spiller, D. (2008) Patterns and processes in insular floras affected by hurricanes. *Journal of Biogeography*, **35**, 1701–1710.

Mortensen, H., Dupont, Y. and Olesen, J. (2008) A snake in paradise: disturbance of plant reproduction following extirpation of bird flower-visitors on Guam. *Biological Conservation*, **141**, 2146–2154.

Morton, B. (2011) Under mackerel skies. *Marine Pollution Bulletin*, **62**, 2009–2010.

Morton, H. (1999) Islanders in space: Tongans online. In R. King and J. Connell, eds, *Small Worlds, Global Lives. Islands and Migration*. London: Pinter, pp. 235–253.

Mortreux, C. and Barnett, J. (2009) Climate change, migration and adaptation in Funafuti, Tuvalu. *Global Environmental Change*, **19**, 105–112.

Moseby, K., Labere, J. and Read, J. (2012) Landowner surveys inform protected area management: a case study from Tetepare Island, Solomon Islands. *Human Ecology*, **40**, 227–235.

Mosley, L., Sharp, D. and Singh, S. (2004) Effects of a tropical cyclone on the drinking-water quality of a remote Pacific island. *Disasters*, **28**, 393–405.

Mountz, A. (2011) The enforcement archipelago: detention, haunting and asylum on islands. *Political Geography*, **30**, 118–128.

Moyle, B. and Evans, M. (2008) Economic development options for island states: the case of whale-watching. *Shima*, **2**, 41–58.

Mueller, A. (1972) Notes on the Tulun or Carteret Islands. *Journal of the Papua New Guinea Society*, **6**(1), 77–83.

Mukheibir, P. (2010) Water access, water scarcity, and climate change. *Environmental Management*, **45**, 1027–1039.

Muliaina, T. (2001) Remittances, the social system and development in Samoa. In V. Naidu, E. Vasta and C. Hawksley, eds, *Current Trends in South Pacific Migration*. Asia Pacific Migration Research Network Working Paper No. 7, Wollongong, Australia, pp. 20–40.

Mullings, B. (2011) Diaspora strategies, skilled migrants and human capital enhancement in Jamaica. *Global Networks*, **11**, 24–42.

Munch, P. (1964) Culture and superculture in a displaced community: Tristan da Cunha. *Ethnology*, **3**, 369–376.

Munch, P. (1971) *Crisis in Utopia: The Ordeal of Tristan da Cunha*. New York: Crowell.

Munro, D. (1990) Transnational corporations of kin and the MIRAB system: the case of Tuvalu. *Pacific Viewpoint*, **31**, 63–66.

Munro, D. and Bedford, R. (1980) Historical backgrounds. In *A Report on the Results of the Census of Tuvalu*. Funafuti, Tuvalu: Government of Tuvalu, pp. 1–13.

Murdoch, H. (2011) Island havens. *Nelson Mail*, 2 May.

Muria, J. (1979) Guadalcanal. In I. Heath, ed., *Land Research in Solomon Islands*. Honiara, Solomon Islands: Ministry of Agriculture and Lands, pp. 41–46.

Murray, W. (2000) Neoliberal globalisation, 'exotic' agro-exports and local change in the Pacific islands: a study of the Fijian kava sector. *Singapore Journal of Tropical Geography*, **21**, 355–373.

Murray, W. and Overton, J. (2011) The inverse sovereignty effect: aid, scale and neostructuralism in Oceania. *Asia Pacific Viewpoint*, **32**, 272–284.

Mycoo, M. (2005) Minimising foreign control of land in an era of globalisation prospects for St Lucia. *Land Use Policy*, **22**, 345–357.

Naik, A., Stigter, E. and Laczko, F. (2007) *Migration, Development and National Disasters: Insights from the Indian Ocean Tsunami*. Geneva: International Organization for Migration.

Narayan, P. (2005) Did Rabuka's military coups have a permanent effect or a transitory effect on tourist expenditure in Fiji? *Tourism Management*, **26**, 509–515.

Narayan, P. and Smyth, R. (2003) The determinants of emigration from Fiji to New Zealand. *International Migration*, **41**, 33–58.

Narayan, P., Narayan, S., Prasad, A. and Prasad, B. (2010) Tourism and economic growth: a panel data analysis for Pacific Island countries. *Tourism Economics*, **16**, 169–183.

Nason, J. (1975) The effects of social change on marine technology in a Pacific atoll community. In R. Casteel and G. Quimby, eds, *Maritime Adaptations of the Pacific*. The Hague, the Netherlands: Mouton, pp. 5–38.

Naylor, R., Bonine, K., Ewel, K. and Waguk, E. (2002) Migration, markets and mangrove resource use on Kosrae, Federated States of Micronesia. *Ambio*, **31**, 340–350.

Neemia, U. (1986) *Cooperation and Conflict. Costs, Benefits and National Interests in Regional Cooperation*. Suva, Fiji: Institute of Pacific Studies.

Negin, J., Morgan, C. and Condon, R. (2012) Regional health meetings in the Pacific and their impact on health governance. *Global Health Governance*, **5**(2), 1.

Nelson, V. (2010) 'R.I.P. Nature Island': the threat of a proposed oil refinery on Dominica's identity. *Social and Cultural Geography*, **11**, 903–919.

New Economics Foundation (2006) *The Happy Planet Index*. London: New Economics Foundation.

Newitt, M. (1984) *The Comoro Islands. Struggle against Dependency in the Indian Ocean*. Boulder, CO: Westview.

Newitt, M. (1992) The perils of being a microstate: São Tomé and the Comoros Islands since independence. In H. Hintjens and M. Newitt, eds, *The Political Economy of Small Tropical Islands*. Exeter, UK: University of Exeter Press, pp. 76–92.

Nicholls, R. and Cazenave, A. (2010) Sea-level rise and its impact on coastal zones. *Science*, **328**(5985), 1517–1520.

Nietschmann, B. (1979) Ecological change, inflation and migration in the far western Caribbean. *Geographical Review*, **69**, 1–24.

Noble, C., Pereira, N. and Saune, N. (2011) *Urban Youth in the Pacific*. Suva, Fiji: UNDP.

Novaczek, J., Mitchell, J. and Veiteyaki, J., eds (2005) *Pacific Voices: Equity and Sustainability in Pacific Islands Fisheries*. Suva, Fiji: Institute of Pacific Studies.

Novelo-Casanova, D. and Suarez, G. (2010) Natural and man-made hazards in the Cayman Islands. *Natural Hazards*, **55**, 441–466.

Nowak, J.-J. and Sahli, M. (2007) Coastal tourism and 'Dutch disease' in a small island economy. *Tourism Economics*, **13**, 49–65.

Numbasa, G. and Koczberski, G. (2012) Migration, informal urban settlements and non-market land transactions on customary land: a case study of Wewak, East Sepik Province, Papua New Guinea. *Australian Geographer*, **43**(2), 143–162.

Nunn, P. (1990) *Oceanic Islands*. Oxford: Blackwell.

Nunn, P. (2007) *Climate, Environment and Society in the Pacific during the Last Millennium*. Amsterdam, Netherlands: Elsevier.

Nunn, P. (2009) *Vanished Islands and Hidden Continents of the Pacific*. Honolulu, Hawaii: University of Hawaii Press.

Nunn, P. (2010) Bridging the gulf between science and society: imperatives for minimizing societal disruption from climate change in the Pacific. In A. Sumi, K. Fukushi and A. Hiramatsu, eds, *Adaptation and Mitigation Strategies for Climate Change*. New York: Springer, pp. 233–248.

Nunn, P., Baniala, M., Harrison, M. and Geraghty, P. (2006) Vanished islands in Vanuatu: new research and a preliminary geohazard assessment. *Journal of the Royal Society of New Zealand*, **36**, 37–50.

Nunn, P. and Kumar, R. (2006) Coastal evolution in the Asia-Pacific region. In N. Harvey, ed., *Global Change and Integrated Coastal Management: The Asia Pacific Region*, Berlin: Springer, pp. 93–116.

Nurse, L. and Moore, R. (2005) Adaptation to global climate change: an urgent requirement for small island developing states. *Review of European Community and International Environmental Law*, **14**, 100–107.

O'Brien, K. and Leichenko, R. (2000) Double exposure: assessing the impacts of climate change within the context of economic globlization. *Global Environmental Change*, **10**, 221–232.

O'Dowd, D., Green, P. and Lake, P. (2003) Invasional 'meltdown' on an oceanic island. *Ecology Letters*, **6**, 812–817.

O'Garra, T. (2012) Economic valuation of a traditional fishing ground on the coral coast of Fiji. *Ocean and Coastal Management*, **56**, 44–55.

Ogden, M. (1994) MIRAB and the Republic of the Marshall Islands. *Isla*, **2**, 237–272.

Oles, B. (2007a) Access and alienation: the promise and threat of stewardship on Mokil Atoll. *Human Organization*, **66**, 78–89.

Oles, B. (2007b) Transformations in the sociocultural values and meanings of reefs and resources on Mwoakilloa. *Coral Reefs*, **26**, 971–981.

Olwig, K. (2003) *Global Culture, Island Identity. Continuity and Change in the Afro-Caribbean Community of Nevis*. Chur, Switzerland: Harwood.

Olwig, K. (2007) *Caribbean Journeys. An Ethnography of Migration and Home in Three Family Networks*. Durham, NC: Duke University Press.

O'Meara, T. (1990) *Samoan Planters: Tradition and Economic Development in Polynesia*. New York: Holt, Rinehart and Winston.

O'Meara, T. (1993) The cult of custom meets the search for money in Western Samoa. In V. Lockwood, T. Harding and B. Wallace, eds, *Contemporary Pacific Societies*. Englewood Cliffs, NJ: Prentice Hall, pp. 135–155.

Oostindie, G. (2005) *Paradise Overseas: The Dutch Caribbean*. Oxford: Macmillan.

Opeskin, B. (2009) Malaria in Pacific populations: seen but not heard? *Journal of Population Research*, **26**, 175–199.

Ormerod, B. (1981) Discourse and dispossession: Edouard Glissant's image of contemporary Martinique. *Caribbean Quarterly*, **27**, 1–12.

Osawa, Y., Fujita, K., Umezawa, Y., Kayanne, H. et al. (2010) Human impacts on large benthic foraminifers near a densely populated area of Majuro Atoll, Marshall Islands. *Marine Pollution Bulletin*, **60**, 1279–1287.

Osborne, D. (1966) *The Archaeology of the Palau Islands*. Bulletin No. 230. Honolulu, Hawaii: Bishop Museum.

Osti, R., Tanaka, S. and Tokioka, T. (2009) The importance of mangrove forest in tsunami disaster mitigation. *Disasters*, **33**, 203–213.

Owen, A. (2010) Guam culture, immigration and the US military build-up. *Asia Pacific Viewpoint*, **51**, 304–318.

Pabon, L., Umapathi, N. and Waqavonovono, E. (2012) How geographically concentrated is poverty in Fiji? *Asia Pacific Viewpoint*, **53**, 205–217.

Palan, R., Murphy, R. and Chavagneux, C. (2010) *Tax Havens, How Globalization Really Works*. Ithaca, NY: Cornell University Press.

Palmer, R. (2009) *The Caribbean Economy in the Age of Globalization*. New York: Palgrave Macmillan.

Pam, C. and Henry, R. (2012) Risky places. Climate change discourse and the transformation of place on Moch (Federated States of Micronesia), *Shima*, **6**, 30–46.

Parkes, J. (2011) Palau: managing macaque monkeys. *The Pacific Invasive Initiatives News*, December, 10–11.

Parris, H. (2010) Tuna dreams and tuna realities: defining the term 'maximising economic returns from the tuna fisheries' in six Pacific Island states. *Marine Policy*, **34**, 105–113.

Parry, M., Canziani, O., Palutikof, J., van der Linden, P. and Hanson, C., eds (2007) *Climate Change 2007: Impacts Adaptation and Vulnerability*. Cambridge: Cambridge University Press.

Parsons, E., Rose, N. and Telecky, T. (2010) The trade in live Indo-Pacific bottlenose dolphins from Solomon Islands: a CITES decision implementation case study. *Marine Policy*, **34**, 384–388.

Pascal, M., Richer de Forges, B., Le Guyader, H. and Simberloff, D. (2008) Mining and other threats to the New Caledonia biodiversity hotspot. *Conservation Biology*, **22**, 498–499.

Paton, K. and Fairbairn-Dunlop, P. (2010) Listening to local voices: Tuvaluans respond to climate change. *Local Environment*, **15**, 687–698.

Patterson, T., Gulden, T., Cousins, K. and Kraev, E. (2004) Integrating environmental, social and economic systems: a dynamic model of tourism in Dominica. *Ecological Modelling*, **175**, 121–136.

Paul, E. (1987) *Fisheries Development and the Food Needs of Mauritius*. Rotterdam, Netherlands: Balkema.

Paulay, G. and Benayahu, Y. (1999) Patterns and consequences of coral bleaching in Micronesia (Majuro and Guam) in 1992–94. *Micronesica*, **32**, 109–124.

Paulson, D. (1993) Hurricane hazard in Western Samoa. *Geographical Review*, **83**, 43–53.

Paulson, D. and Rogers, S. (1997) Maintaining subsistence security in Western Samoa. *Geoforum*, **28**, 173–187.

Payet, R. (2008) Climate change and tourism in the Seychelles. In N. Leary, J. Adewujon, V. Barros, I. Burton et al., eds, *Climate Change and Adaptation*. London: Earthscan, pp. 155–169.

Payne, A. (1987) Economic issues. In C. Clarke and T. Payne, eds, *Politics, Security and Development in Small States*. London: Allen and Unwin, pp. 50–61.

Payne, A. (2004) Small states in the global politics of development. *The Round Table*, **93**, 623–635.
Payne, A. (2006) The end of green gold? Comparative development options and strategies in the Eastern Caribbean banana-producing islands. *Studies in Comparative International Development*, **41**, 25–46.
Pelling, M. and Uitto, J. (2001) Small island developing states: natural disaster vulnerability and global change. *Environmental Hazards*, **3**, 49–62.
Penaia, A., Titimaea, A., Kohlhase, J., Nelson, F. et al. (2007) Reducing flood risk. From science to policy: the Samoa Process. Miscellaneous Report No. 693. Suva, Fiji: SOPAC.
Perch, L. and Roy, R. (2010) Social policy in the post-crisis context of small island developing states: a synthesis. Working Paper No. 67. Brazil: UNDP International Policy Centre for Inclusive Growth.
Perch-Nielsen, S. (2010) The vulnerability of beach tourism to climate change: an index approach. *Climatic Change*, **100**, 579–606.
Pereira, N. (2011) Return(ed) to paradise: the deportation experience in Samoa and Tonga. Most Policy Paper No. 21. Paris: UNESCO.
Perminow, A. (1993) Between the forest and the big lagoon: the microeconomy of Kotu Island in the Kingdom of Tonga. *Pacific Viewpoint*, **34**, 179–192.
Perry, C., Kench, P., Smithers, S., Riegl, B., Yamano, H. and O'Leary, M. (2011) Implications of reef ecosystem change for the stability and maintenance of coral reef islands. *Global Change Biology*, **17**, 3679–3696.
Petersen, E. (2003) The catch in trading fishing access for foreign aid. *Marine Policy*, **7**, 219–228.
Petersen, G. (1979) Breadfruit or rice? The political economics of a vote in Micronesia. *Science and Society*, **43**, 472–485.
Peterson, A. and Stead, S. (2011) Rule breaking and livelihood options in marine protected areas. *Environmental Conservation*, **38**, 342–352.
Peteru, S., Regan, S. and Klak, T. (2010) Local vibrancy in a globalizing world: evidence from Dominica. *Focus*, **53**(4), Winter, 125–133.
Philpott, S. (1968) Remittance obligations: social networks and choice among Montserratian migrants in Britain. *Man*, **3**, 465–476.
Philpott, S. (1999) The breath of the 'beast': migration, volcanic disaster, place and identity in Montserrat. In R. King and J. Connell, eds, *Small Worlds, Global Lives. Islands and Migration*. London: Pinter, pp. 137–160.
Pierce, P. (1999) Changing coastline at Mele Beach, Vanuatu. *Geography*, **84**, 149–163.
Pikacha, P. (2008) *Wild West. Rainforests of Western Solomon Islands*. Honiara, Solomon Islands: Melanesian Geo Publications.

Poirine, B. (1998) Should we hate or love MIRAB? *The Contemporary Pacific*, **10**, 65–105.

Pollard, S. (1987) The viability and vulnerability of a small island state. The case of Kiribati. Working Paper No. 87/14. Canberra, Australia: National Centre for Development Studies Islands/Australia.

Pollnac, R., Christie, P., Cinner, J., Dalton, T. et al. (2011) Marine reserves as linked social-ecological systems. *Proceedings of the National Academy of Sciences*, **107**, 18262–18265.

Pollock, N. (1978) Takapoto. La prosperité, retour aux îles. *Journal de le Société des Océanistes*, **34**, 133–135.

Pollock, N. (1996) Namu Atoll revisited: a follow-up study of 25 years of resource use. *Atoll Research Bulletin*, **441**, 1–11.

Pomponio, A. (1993). Education is development on a ten-acre island. In V.S. Lockwood, T.G. Harding and B.J. Wallace, eds, *Contemporary Pacific Societies.* Upper Saddle River, NJ: Prentice Hall, pp. 197–215.

Pop, G. (2011) Cape Verde. In S. Mohapatra and D. Ratha, eds, *Remittance Markets in Africa.* Washington: World Bank, pp. 91–111.

Potter, A. and Sluyter, A. (2010) Renegotiating Barbuda's commons: recent changes in Barbudan open-range cattle herding. *Journal of Cultural Geography*, **27**, 129–150.

Potter, R., Barker, D., Conway, D. and Klak, T. (2004) *The Contemporary Caribbean*. Harlow, UK: Pearson.

Powlesland, R., Butler, D. and Westbrooke, I. (2008) Was Tropical Cyclone Heta or hunting by people responsible for decline of the lupe (*Ducula pacifica*) (Aves: Columbidae) population on Niue during 1994–2004? *Pacific Science*, **62**, 461–471.

Pratchett, M., Trapon, M., Berumen, M. and Chong-Seng, K. (2011) Recent disturbances augment community shifts in coral assemblages in Moorea, French Polynesia. *Coral Reefs*, **30**, 183–193.

Pretes, M. and Petersen, E. (2004) Rethinking fisheries policy in the Pacific. *Marine Policy*, **28**, 297–309.

Price, R. (1985) The dark complete world of a Caribbean store: a note on the world-system. *Review*, **9**, 215–219.

Prideaux, B. and McNamara, K. (2013) Turning a global crisis into a tourism opportunity: the perspective from Tuvalu. *International Journal of Tourism Research.*

Pulsipher, L. (2010) Ash falls in Montserrat. *Focus*, **53**(1), Spring, 10–13.

Quammen, D. (1996) *The Song of the Dodo. Island Biogeography in an Age of Extinctions*. London: Hutchinson.

Rakoto, A. (1986) Can custom be custom built? Cultural obstacles to Fijian commercial enterprise. In C. Griffin and M. Monsell-Davis, eds, *Fijians in Town.* Suva, Fiji: Institute of Pacific Studies, pp. 4–17.

Ramsay, D. (1999) Economic aspects of sand-mining/Kosrae. http://www.csiwisepractices.org/?read=142.

Rankey, E. (2011) Nature and stability of atoll island shorelines: Gilbert Island chain, Kiribati, equatorial Pacific. *Sedimentology*, **58**, 1831–1859.

Rao, M. (2002) Challenges and issues for tourism in the South Pacific island states: the case of Fiji Islands. *Tourism Economics*, **8**, 401–429.

Rasanathan, J. and Bhushan, A. (2011) Gender-based violence in Solomon Islands. Suva, Fiji: WHO.

Rasmussen, K., May, W., Birk, T., Mataki, M. et al. (2009) Climate change on three Polynesian outliers in the Solomon Islands: impacts, vulnerability and adaptation. *Geografisk Tidsskrift*, **109**, 1–13.

Ratuva, S. (2011) The gap between global thinking and local living: dilemmas of constitutional reform in Nauru. *Journal of the Polynesian Society*, **120**, 241–268.

Rawlins, G. (2003) Barbados. In UNCTAD, *Turning Losses into Gains: SIDS and Multilateral Trade Liberalisation in Agriculture*. New York and Geneva: United Nations, pp. 35–96.

Read, R. (2004) The implications of increasing globalisation and regionalism for the economic growth of small island states. *World Development*, **32**, 365–378.

Read, R. (2010) Trade, economic vulnerability, resilience and the implications of climate change in small island and littoral developing economies. Issue Paper No. 12. Geneva: ICTSD.

Reaser, J., Meyerson, L., Cronk, Q., de Poorter, M. et al. (2007) Ecological and socioeconomic impacts of invasive alien species in island ecosystems. *Environmental Conservation*, **34**, 98–111.

Reenberg, A., Birch-Thomsen, T., Mertz, O., Fog, B. and Christiansen, S. (2008) Adaptation of human coping strategies in a small island society in the SW Pacific: 50 years of change in the coupled human-environment system on Bellona, Solomon Islands. *Human Ecology*, **36**, 807–819.

Regenvanu, R., Wyatt, S. and Tacconi, L. (1997) Changing forestry regimes in Vanuatu: is sustainable management possible? *The Contemporary Pacific*, **9**, 73–96.

Rego, M. (2008) Cape Verdean tongues: speaking of nation at home and abroad. In L. Batalha and J. Carling, eds, *Transnational Archipelago. Perspectives on Cape Verdean Migration and Diaspora*. Amsterdam: Amsterdam University Press, pp. 145–160.

Reyburn, R., Kim, D., Emch, M., Khatib, A., von Seidlin, L. and Ali, M. (2011) Climate variability and the outbreaks of cholera in Zanzibar, East Africa: a time series analysis. *American Journal of Tropical Medicine and Hygiene*, **84**, 862–869.

Richardson, B. (1975) The overdevelopment of Carriacou. *Geographical Review*, **65**, 390–399.

Richardson, B. (1986) *Panama Money in the Barbados, 1900–1920*. Knoxville, TN: University of Tennessee Press.

Richardson-Ngwenya, P. (2010) The EU sugar reform and the responses of Caribbean sugar producers. *Geography*, **95**, 70–79.

Richmond, O. (2011) De-romanticising the local, de-mystifying the international: hybridity in Timor Leste and the Solomon Islands. *The Pacific Review*, **24**, 115–136.

Rigamoto, R., Tyagi, A. and Thaman, R. (2004) Ethnobotanical importance of coastal plant species of Rotuma Island. *South Pacific Journal of Natural Science*, **22**, 23–29.

Rigg, J. (2006) Land, farming, livelihoods, and poverty: rethinking the links in the rural south. *World Development*, **34**, 180–202.

Rio, K. (2011) Policing the Holy Nation: the state and righteous violence in Vanuatu. *Oceania*, **81**, 51–71.

Roberts, C. and Hawkins, J. (1994) *Report on the Status of Bonaire's Coral Reefs*. St Thomas, Virgin Islands: University of the Virgin Islands Eastern Caribbean Center.

Roberts, S. (1995) Small place, big money: the Cayman Islands and the international financial system. *Economic Geography*, **71**, 237–256.

Roberts, S., Wright, S. and O'Neill, P. (2007) Good governance in the Pacific? Ambivalence and possibility. *Geoforum*, **38**, 967–984.

Robinson, J., Guillottreau, P., Jimenez-Toribio, R., Lantz, F. et al. (2010) Impacts of climate variability on the tuna economy of Seychelles. *Climate Research*, **43**, 149–162.

Rodda, G. and Savidge, J. (2007) Biology and impacts of Pacific Island invasive species. 2. *Boiga irregularis*, the brown tree snake (Reptilia: Colubridae). *Pacific Science*, **61**, 307–324.

Roderick, G. and Vernon, P. (2009) Invasion biology. In R. Gillespie and D. Clague, eds, *Encyclopedia of Islands*. Berkeley, CA: University of California Press, pp. 475–481.

Rodman, M. (1987a) *Masters of Tradition. Consequences of Customary Land Tenure in Longana, Vanuatu*. Vancouver, Canada: University of British Columbia Press.

Rodman, M. (1987b) Constraining capitalism? Contradictions of self-reliance in Vanuatu fisheries development. *American Ethnologist*, **14**, 712–726.

Rodman, M. (1993) Keeping options open: copra and fish in Rural Vanuatu. In T. Harding and B. Wallace, eds, *Contemporary Pacific Societies*. Englewood Cliffs, NJ: Prentice Hall, pp. 171–184.

Rogers, C. (2009) Coral bleaching and disease should not be underestimated as causes of Caribbean reef decline. *Proceedings of the Royal Society B*, **276**, 197–198.
Rogers, G. (1981) The evacuation of Niuafo'ou: an outlier in the Kingdom of Tonga. *Journal of Pacific History*, **16**, 149–163.
Rokoduru, A. (2008) Transient greener pastures in managed temporary labour migration in the Pacific: Fiji Nurses in the Marshall Islands. In J. Connell, ed., *The International Migration of Health Workers*. New York: Routledge, pp. 172–181.
Rolett, B. (2009) Deforestation. In R. Gillespie and D. Clague, eds, *Encyclopedia of Islands*. Berkeley, CA: University of California Press, pp. 221–224.
Rollason, W. (2011) We are playing football: seeing the game on Panapompom, PNG. *Journal of the Royal Anthropological Institute*, **17**, 481–503.
Rongo, T., Bush, M. and van Woesik, R. (2009) Did ciguatera prompt the late Holocene Polynesian voyages of discovery? *Journal of Biogeography*, **36**, 1423–1432.
Roopnarine, L. (2008) United States Virgin Islands migration. *Social and Economic Studies*, **57**(3 & 4), 131–156.
Roopnarine, L. (2011) St Croix's Secession movement in the United States Virgin Islands: sentimental or serious. *Journal of Eastern Caribbean Studies*, **36**, 43–66.
Roper, T. (2009) Implications for the Pacific. *Tiempo*, **70**, 3–7.
Rose, R. and Mathiason, N. (2011) Windwards' banana farmers face ruin. *Guardian Weekly*, 14 January, 44.
Rowe, M. (2008) *Bermuda's Water Supply*. Hamilton, Bermuda: Government of Bermuda.
Roy, P. and Connell, J. (1991) Climatic change and the future of atoll states. *Journal of Coastal Research*, **7**, 1057–1075.
Royle, S. (2001) *A Geography of Islands*. London: Routledge.
Royle, S. (2006) Extreme tourism: lessons from the world's cold water islands. In G. Baldacchino, ed., *Extreme Tourism in Cold Water Islands*. Oxford: Elsevier, pp. 181–191.
Rubenstein, H. (1983) Remittances and rural underdevelopment in the English-speaking Caribbean. *Human Organization*, **42**, 295–306.
Rubenstein, H. (1987) *Coping with Poverty, Adaptive Strategies in a Caribbean Village*. Boulder, CO: Westview.
Rubenstein, H. (1988) Ganja as a peasant resource in St Vincent: a preliminary analysis. In J. Brierley and H. Rubenstein, eds, *Small Farming and Peasant Resources in the Caribbean*, Geographical Studies No. 10. Winnipeg, Canada: University of Winnipeg Manitoba pp. 119–133.

Rubinstein, D. (2003) A tale of two islands. In T. Aoyama, ed., *Social Homeostasis of Small Islands in an Island Zone*. Occasional Paper No. 39, Kagoshima, Japan: Kagoshima Research University Center for the Pacific Islands.

Rubow, C. (2013) Enacting cyclones. The mixed response to climate change in the Cook Islands. In K. Hastrup and M. Skrydstrup, eds, *The Social Life of Climate Change Models: Anticipating Nature*. New York and Abingdon, UK: Routledge, pp. 57–76.

Ruddle, K. (1998) The context of policy design for existing community-based fisheries management systems in the Pacific islands. *Ocean and Coastal Management*, **40**, 105–126.

Rudiak-Gould, P. (2009) *Surviving Paradise. One Year on a Disappearing Island*. New York: Sterling.

Rudiak-Gould, P. (2010) *The Fallen Palm. Climate Change and Culture Change in the Marshall Islands*. Saarbrucken, Germany: VDM Verlag.

Rudiak-Gould, P. (2011) Climate change and anthropology. *Anthropology Today*, **27**(2), 9–12.

Rudiak-Gould, P. (2013) Promiscuous corroboration and climate change translation: a case study from the Marshall Islands. *Global Environmental Change*, **22**, 46–54.

Rull, V., Canellas-Bolta, N., Saez, A., Giralt, S., Pla, S. and Margalef, O. (2010) Paleoecology of Easter Island: evidence and uncertainties. *Earth-Science Reviews*, **99**, 50–60.

Russell, L. (2009) *Stayin' Alive. A Report on Urban Livelihoods in Honiara, Solomon Islands*. Sydney, Australia: APHEDA.

Rutz, H. (1982) Bureaucracy and brokerage: Fijian villages and public goods. In W. Rodman and D. Counts, eds, *Middlemen and Brokers in Oceania*. Ann Arbor, MI: University of Michigan Press, pp. 149–186.

Safina, C. and Klinger, D. (2008) Collapse of bluefin tuna in the Western Atlantic. *Conservation Biology*, **22**, 243–246.

Salesa, T. (2003) 'Travel-happy' Samoa: colonialism, Samoan migration and a 'brown Pacific'. *New Zealand Journal of History*, **37**, 172–187.

Sanchez, C., Lee, T., Young, S., Batts, D. et al. (2009) Risk factors for mortality during the 2002 landslides in Chuuk, Federated States of Micronesia. *Disasters*, **33**, 705–720.

Savidge, J. (1987) Extinction of an island forest avifauna by an introduced snake. *Ecology*, **68**, 660–668.

de Scally, F. (2008) Historical tropical cyclone activity and impacts in the Cook Islands. *Pacific Science*, **62**, 443–459.

Schaeffer, M. and Hare, B. (2010) *Persistence of Atoll Isslands Under Recent and Projected Sea-Level Rise*. Potsdam, Germany: Climate Analytics.

Schandl, H., Collins, K., Heyenga, S. and West, J. (2011) *Sustainability Assessment of Selected Countries in the Pacific Islands*. Canberra, Australia: CSIRO.

Scheffers, A. (2004) Tsunami imprints on the Leeward Netherlands Antilles (Aruba, Curaçao, Bonaire) and their relation to other coastal problems. *Quaternary International*, **120**, 163–172.

Scheffers, A. and Scheffers, S. (2006) Documentation of the impact of Hurricane Ivan on the coastline of Bonaire (Netherlands Antilles). *Journal of Coastal Research*, **22**, 1437–1450.

Scheyvens, R. (2007) Poor cousins no more: valuing the development potential of domestic and diaspora tourism. *Progress in Development Studies*, **7**, 307–325.

Scheyvens, R. (2008) On the beach: small scale tourism in Samoa. In J. Connell and B. Rugendyke, eds, *Tourism at the Grassroots*. London: Routledge, pp. 131–147.

Scheyvens, R. (2011) *Tourism and Poverty*. London: Routledge.

Scheyvens, R. and Russell, M. (2012) Tourism, land tenure and poverty alleviation in Fiji. *Tourism Geographies*, **14**, 1–25.

Schleupner, C. (2007) Spatial assessment of sea level rise on Martinique's coastal zone and analysis of planning frameworks for adaptation. *Journal of Coastal Conservation*, **11**, 91–103.

Schleupner, C. (2008) Evaluation of coastal squeeze and its consequences for the Caribbean island of Martinique. *Ocean and Coastal Management*, **51**, 383–390.

Schleyer, M. and Benayahu, Y. (2010) Pre- and post-1998 ENSO records of shallow-water octocorals (Alcyonacea) in the Chagos Archipelago. *Marine Pollution Bulletin*, **60**, 2197–2200.

Schumacher, E. (1973) *Small Is Beautiful*. London: Blond and Briggs.

Schutz, B. and Tenten, R. (1979) Adjustment. Problems of growth and change, 1892 to 1944. In L. Mason, ed., *Kiribati: Aspects of History*. Suva, Fiji: Institute of Pacific Studies, pp. 106–127.

Schwarz, A.-M., Béné, C., Bennett, G., Boso, D. et al. (2011) Vulnerability and resilience of remote rural communities to shocks and global changes: empirical analysis from Solomon Islands. *Global Environmental Change*, **21**, 1128–1140.

Scourse, A. and Wilkins, C. (2009) Impacts of modernisation on traditional food resource management and food security on Eauripik atoll, Federated States of Micronesia. *Food Security*, **1**, 169–176.

Sear, C., Hulme, M., Adger, N. and Brown, K. (2001) *The Impacts of Global Climate Change on the UK Overseas Territories*. Chatham and Norwich, UK: Natural Resources Institute, University of Greenwich and Tyndall Centre for Climate Change Research.

Seers, D. (1983) *The Political Economy of Nationalism*. Oxford: Oxford University Press.

Seiden, A., Hawley, N., Schulz, D., Raifman, S. and McGarvey, S. (2012) Long-term trends in food availability, food prices and obesity in Samoa. *American Journal of Human Biology*, **24**, 286–295.

Selig, E., Casey, K. and Bruno, J. (2012) Temperature-driven coral decline: the role of marine protected areas. *Global Change Biology*, **18**, 1561–1570.

Sem, G. (1986) Cooperation, management of fisheries and regionalism. *South Pacific Forum,* **3**, 113–135.

Serrano, E. (2007) Tourism in the South Shetland Islands: recent changes in local destinations and activities. *Tourism in Marine Environments*, **4**, 221–235.

Sevele, F. (1973) The importance of agriculture in Pacific Island Development. Unpublished paper to the Episcopal Conference of the Pacific, Wellington, New Zealand.

Sevele, F. (1983) Regional trade: limited potential. *Pacific Perspective*, **11**, 23–27.

Shakeela, A., Ruhanen, L. and Breakey, N. (2011) The role of employment in the sustainable development paradigm: the local tourism market in small island developing states. *Journal of Human Resources in Hospitality and Tourism*, **10**, 331–353.

Shakeela, A. and Weaver, D. (2012) Resident reactions to a tourism incident. *Annals of Tourism Research*, **39**, 1337–1358.

Shakow, D. and O'Keefe, P. (1981) Yes, we have no bananas: the economic effects of minor hazards in the Windward Islands. *Ambio*, **10**, 344.

Shankman, P. (1976) *Migration and Underdevelopment: the Case of Western Samoa*. Boulder, CO: Westview.

Shankman, P. (1993) The Samoan exodus. In V. Lockwood, T. Harding and B. Wallace, eds, *Contemporary Pacific Societies*. Englewood Cliffs, NJ: Prentice Hall, pp. 156–170.

Shankman, P. (1999) Development, sustainability and the deforestation of Samoa. *Pacific Studies*, **22**, 167–188.

Shanks, S. (2010) Introducing a transferable fishing day management regime for Pacific Island countries. *Marine Policy*, **34**, 988–994.

Sharman, J. (2005) South Pacific tax havens: from leaders in the race to the bottom to laggards in the race to the top? *Accounting Forum*, **29**, 311–323.

Sharman, J. (2011) *The Money Laundry. Regulating Criminal Finance in the Global Economy*. Ithaca, NY: Cornell University Press.

Shaxson, N. (2011) *Treasure Islands. Tax Havens and the Men Who Stole the World*. London: Bodley Head.

Shearman, P., Bryan, J. and Laurance, W. (2012) Are we approaching 'peak timber' in the tropics? *Biological Conservation*, **151**, 17–21.
Shen, S. and Binns, T. (2012) Pathways, motivations and challenges: contemporary Tuvaluan migration to New Zealand. *GeoJournal*, **77**, 63–82.
Shen, S. and Gemenne, F. (2011) Contrasted views on environmental change and migration: the case of Tuvaluan migration to New Zealand. *International Migration*, **49**, 224–242.
Sheppard, C. (1999) Coral decline and weather patterns over 20 years in the Chagos Archipelago, central Indian Ocean. *Ambio*, **28**, 472–478.
Sheppard, C. (2010) Marine protected areas and pelagic fishing: the case of the Chagos Archipelago. *Marine Pollution Bulletin*, **60**, 1899–1901.
Shore, B. (1978) Introduction. In C. Macpherson, B. Shore and B. Franco, eds, *New Neighbors: Islanders in Adaptation*. Santa Cruz, CA: Center for South Pacific Studies, pp. xiii–xiv.
Shorten, G. (2002) *Earthquake and Tsunami Damage Assessment in Port Vila*. Preliminary Report No. 135. Suva, Fiji: SOPAC.
Sidaway, J. (2010) 'One island, one team, one mission': geopolitics, sovereignty, 'race' and rendition. *Geopolitics*, **15**, 667–683.
Sieck, M., Ibisch, P., Moloney, K. and Jeltsch, F. (2011) Current models broadly neglect specific needs of biodiversity conservation in protected areas under climate change. *BMC Ecology*, **11**(12), 1–12 (online).
Sievanen, L., Gruby, R. and Campbell, L. (2013) Fixing marine governance in Fiji? The new scalar narrative of ecosystem-based management. *Global Environmental Change* (in press).
Simberloff, D. (2005) Why do introduced species appear to devastate islands more than mainland areas? *Pacific Science*, **49**, 87–97.
Simberloff, D. (2009) Introduced species. In R. Gillespie and D. Clague, eds, *Encyclopedia of Islands*. Berkeley, CA: University of California Press, pp. 469–475.
Siméoni, P. and Ballu, V. (2012) Le mythe des premiers réfugiés climatiques: mouvements de populations et changements environnementaux aux îles Torres. *Annales de Géographie*, **685**, 219–241.
Simpson, A., Johnson, R. and Cummins, P. (2011) Volcanic threat in developing countries of the Asia-Pacific region: probabilistic hazard assessment, population risks and information gaps. *Natural Hazards*, **57**, 151–165.
Simpson, G. (1990) Wallerstein's world systems theory and the Cook Islands: a critical examination. *Pacific Studies*, **4**, 73–94.
Sinan, H. and Whitmarsh, D. (2010) Wealth-based fisheries management and resource rent capture: an application to the Maldives marine fisheries. *Marine Policy*, **34**, 389–394.

Singh, R., Hales, S., de Wet, N., Raj, R., Hearnden, M. and Weinstein, P. (2001) The influence of climatic variation and change on diarrheal disease in the Pacific Islands. *Environmental Health Perspectives*, **109**, 155–159.

Skelton, T. (2000) Political uncertainties and natural disasters: Montserratian identity and colonial status. *Interventions: International Journal of Post-Colonial Studies*, **2**, 103–107.

Skelton, T. (2008) Globalization, cultures of land, home and natural hazard: the case of Montserrat in the Caribbean. In E. Kofman and G. Young, eds, *Globalization. Theory and Practice*. New York: Continuum, pp. 57–72.

Slatter, C. (2006) *The Con/Dominion of Vanuatu? Paying the Price of Investment and Land Liberalisation: A Case Study of Vanuatu's Tourism Industry*. Auckland, New Zealand: Oxfam New Zealand.

Slemrod, J. (2008) Why is Elvis on Burkina Faso postage stamps? Cross-country evidence on the commercialization of state sovereignty. *Journal of Empirical Legal Studies*, **5**, 683–712.

Slinger-Friedman, V. (2009) Ecotourism in Dominica: studying the potential for economic development, environmental protection and cultural conservation. *Island Studies Journal*, **4**, 3–24.

Small, C. (1997) *Voyages. From Tongan Villages to American Suburbs*. Ithaca, NY, and London: Cornell University Press.

Smith, M. and Shepherd, J. (1996) Tsunami waves generated by volcanic landslides: an assessment of the hazard associated with Kick 'em Jenny. In W. McGuire, A. Jones and J. Neuberg, eds, *Volcano Instability on the Earth and Other Planets*. Special Publication No. 110, London: Geological Society, pp. 115–123.

Smith, R. (2004) Access to healthcare via telehealth: experiences from the Pacific. *Perspectives on Global Development and Technology*, **3**, 197–211.

Smith, R. and Collen, J. (2004) Sand and gravel resources on Majuro Atoll, Marshall Islands. Technical Report No. 360. Suva, Fiji: SOPAC.

Snowdon, W., Moodie, M., Schultz, J. and Swinburn, B. (2011) Modelling of potential food policy interventions in Fiji and Tonga and their impacts on noncommunicable disease mortality. *Food Policy*, **36**, 597–605.

Sofer, M. (2007) Yaqona and the Fijian periphery revisited. *Asia Pacific Viewpoint*, **48**, 234–249.

Sofer, M. (2009) Twenty years of change in the Fijian periphery: the case of Kadavu Island. *Singapore Journal of Tropical Geography*, **30**, 343–357.

Solomon, S. (1994) A review of coastal processes and analysis of historical coastal change in the vicinity of Apia, Western Samoa. Technical Report No. 208. Suva, Fiji: SOPAC.

Sosis, R. (2002) Patch choice decisions among Ifaluk fishers. *American Anthropologist*, **104**, 583–598.

Sovacool, B. (2012a) Perceptions of climate change risks and resilient island planning in the Maldives. *Mitigation and Adaptation Strategies for Global Change*, **17**, 731–752.

Sovacool, B. (2012b) Expert views of climate change adaptation in the Maldives. *Climatic Change*, **114**, 295–300.

Spennemann, D. (1996) Nontraditional settlement patterns and typhoon hazard on contemporary Majuro Atoll, Republic of the Marshall Islands. *Environmental Management*, **20**, 337–348.

Spennemann, D. (2009) Water futures and their influence on sovereignty in the Marshall Islands. In D. Ghosh, H. Goodall and S. Donald, eds, *Water, Sovereignty and Borders in Asia and Oceania*. London: Routledge, pp. 50–71.

SPREP (1993) *South Pacific Biodiversity Conservation Programme*. Apia, Samoa: SPREP.

Spriggs, M. (1986) Landscape, land use and political transformation in southern Melanesia. In P. Kirch, ed., *Island Societies*. Cambridge: Cambridge University Press, pp. 6–19.

Stallings, C. (2009) Fishery-independent data reveal negative effect of human population density on Caribbean predatory fish communities. *PLoS One*, **4**(5) (online).

Steadman, D. (2006) *Extinction and Biogeography of Tropical Pacific Birds*. Chicago, IL, and London: University of Chicago Press.

Steeves, J. (2011) 'Unbounded politics' and the democratic model in Solomon Islands: the 2010 national elections. *Commonwealth and Comparative Politics*, **49**, 342–358.

Steinberg, P. and McDowell, S. (2003) Mutiny on the bandwidth: the semiotics of statehood in the internet domain name registries of Pitcairn Island and Niue. *New Media and Society*, **5**, 47–67.

Stenseth, N. and Voje, K. (2009) Easter Island: climate change might have contributed to past cultural and societal changes. *Climate Research*, **39**, 111–114.

Stevens, C. (1999) Defining and understanding sustainability in small island states. *Pacific Studies*, **22**, 1–12.

Stillman, S. (2011) The invisible army. *The New Yorker*, 6 June, 56–65.

Stoffle, B., Purcell, T., Stoffle, R., Arnett, K. and Minnis, J. (2009) Credit, identity and resilience in the Bahamas and Barbados. *Ethnology*, **48**, 71–84.

Stoffle, R. and Minnis, J. (2007) Marine protected areas and the coral reefs of traditional settlements in the Exumas, Bahamas. *Coral Reefs*, **26**, 1023–1032.

Stokes, M., Leichter, J. and Genovese, S. (2010) Long-term declines in coral cover at Bonaire, Netherlands Antilles. *Atoll Research Bulletin*, **582**, 1–19.

Storey, D. (2006) End of the line? Globalisation and Fiji's garment industry. In S. Firth, ed., *Globalisation and Governance in the Pacific Islands*. Canberra, Australia: ANU E Press, pp. 217–236.

Storey, D. and Hunter, S. (2010) Kiribati: an environmental 'perfect storm'. *Australian Geographer*, **41**, 167–181.

Strobl, E. (2012) The economic growth impact of natural disasters in developing countries: Evidence from hurricane strikes in the Central American and Caribbean regions. *Journal of Development Economics*, **97**, 130–141.

Sturton, M. (1992) *Tonga: Development Through Agricultural Exports*. Economic Report No. 4. Honolulu, Hawaii: Pacific Islands Development Program.

Sunshine, C. (1985) *The Caribbean. Survival, Struggle and Sovereignty*. Boston, MA: EPICA.

Sutherland, K., Smit, B., Wulf, K. and Nakalevu, T. (2005) Vulnerability in Samoa. *Tiempo*, **54**, 11–15.

Sutton, P. (2011) The concept of small states in the international political economy. *The Round Table*, **100**, 141–153.

Tabucanon, G. (2012) The Banaban resettlement: implications for Pacific environmental migration. *Pacific Studies*, **35**, 343–370.

Tabucanon, G. and Opeskin, B. (2011) The resettlement of Nauruans in Australia: an early case of failed environmental migration. *Journal of Pacific History*, **46**, 337–356.

Taleb, N. (2007) *The Black Swan*. Harmondsworth, UK: Penguin.

Tan, L. (2012) Flooded Fijians plead for residency. *New Zealand Herald*, 9 April, 9.

Taylor, R., Bampton, D. and Lopez, A. (2005) Contemporary patterns of Pacific island mortality. *International Journal of Epidemiology*, **18**, 634–646.

Taylor, M. (1986) Multinationals, business organisations and the development of the Fiji economy. In M. Taylor and N. Thrift, eds, *Multinationals and the Restructuring of the World Economy*. London: Croom Helm, pp. 49–85.

Taylor, M. (1987) Business organisations, the formal sector and development. In M. Taylor, ed., *Fiji: Future Imperfect?* Sydney, Australia: Allen and Unwin, pp. 1–13.

Teeuw, R., Rust, D., Solana, C. and Dewdney, C. (2009) Large coastal landslides and tsunami hazard in the Caribbean. *Eos*, **90**(10), 10 March, 81–82.

Teh, L., Teh, L., Starkhouse, B., Kuridrani, N. et al. (2008) Preliminary assessment of the socio-economic importance of export trade in coral reef resources on Fijian society. *Proceedings of the 11th International Coral Reef Symposium*, 1048–1052.

Teh, L., Teh, L., Starkhouse, B. and Sumaila, R. (2009) An overview of socio-economic and ecological perspectives of Fiji's inshore reef fisheries. *Marine Policy*, **3**, 807–817.

Teiwaki, R. and Associates (2005) Preliminary socio-economic baseline survey and waste stream analysis for Bikenibeu West, South Tarawa, Kiribati. Technical Report No. 14. Apia, Samoa: SPREP IWP-Pacific.

Terry, J. (2007) *Tropical Cyclones, Climatology and Impacts in the South Pacific*. New York: Springer.

Terry, J. and Chui, T. (2012) Evaluating the fate of freshwater lenses on atoll islands after eustatic sea-level rise and cyclone driven inundation: a modelling approach. *Global and Planetary Change*, **88–89**, 76–84.

Terry, J. and Etienne, S. (2010) Recent heightened tropical cyclone activity east of 180° in the South Pacific. *Weather*, **65**, 193–195.

Terry, J. and Falkland, A. (2010) Responses of atoll freshwater lenses to storm-surge overwash in the northern Cook Islands. *Hydrogeology Journal*, **18**, 749–759.

Terry, J. and Gienko, G. (2010) Climatological aspects of South Pacific tropical cyclones, based on analysis of the RSMC-Nadi (Fiji) regional archive. *Climate Research*, **42**, 223–233.

Terry, J. and Thaman, R. (2008) Physical geography of Majuro and the Marshall Islands. In J. Terry and F. Thomas, eds, *The Marshall Islands. Environment, History and Society in the Atolls*. Suva, Fiji: USP Faculty of Oceans and Islands, pp. 1–22.

Terry, J., Kisun, P., Qareqare, A. and Rajan, J. (2006) Lagoon degradation and management in Yanuca Channel on the Coral Coast of Fiji. *South Pacific Journal of Natural Science*, **24**, 1–10.

Thaman, R. (1995) Urban food gardening in the Pacific Islands: a basis for food security in rapidly urbanising small-island states. *Habitat International*, **19**, 209–224.

Thaman, R. (2008) The ecosystems and flora of Majuro. In J. Terry and F. Thomas, eds, *The Marshall Islands. Environment, History and Society in the Atolls*. Suva, Fiji: USP Faculty of Oceans and Islands, pp. 23–68.

Thaman, R. and Thomas, P. (1985) Cassava and change in Pacific Island food systems. In D. Cattle and K. Schwerin, eds, *Food Energy in*

Tropical Ecosystems. New York and London: Gordon and Breach, pp. 189–223.

Thaman, R., Puia, T., Wilson, T., Namona, A. and Fong, T. (2010) Marine biodiversity and ethnobiodiversity of Bellona (Mungiki) Island, Solomon Islands. *Singapore Journal of Tropical Geography*, **31**, 70–84

Thomas, C. and Hosein, R. (2006) Conceptualizing a strategic approach to mode 4 and the managed migration of nursing skills. *Social and Economic Studies*, **55**(4), 165–195.

Thomas, F. and Tonganibeia, K. (2007) Pacific island rural development. Challenges and prospects in Kiribati. In J. Connell and E. Waddell, eds, *Environment, Development and Change in Rural Asia-Pacific*. London: Routledge, pp. 38–55.

Thomassin, A., White, C., Stead, S. and David, G. (2010) Social acceptability of a marine protected area: the case of Réunion island. *Ocean and Coastal Management*, **53**, 169–179.

Thompson, R. (1985) Towards agricultural self-reliance in Grenada: an alternative model. In P. Gomes, ed., *Rural Development in the Caribbean*. London: Hurst, pp. 123–153,

Thorndike, T. (1985) *Grenada. Politics, Economy and Society*. London: Frances Pinter.

Thorndike, T. (1987) Antigua and Barbuda. In C. Clarke and T. Payne, eds, *Politics, Security and Development in Small States*. London: Allen and Unwin, pp. 96–112.

Thow, A., Swinburn, B., Colagiuri, S., Diligolevu, M., Quested, S., Vivili, P. and Leeder, S. (2010) Trade and food policy: case studies from three Pacific Island countries. *Food Policy*, **35**, 556–564.

Thur, S. (2010) User fees as sustainable financing mechanisms for marine protected areas: an application to the Bonaire National Marine Park. *Marine Policy*, **34**, 63–69.

von Tigerstrom, B. (2005) Small island developing states and international trade: special challenges in the global partnership for development. *Melbourne Journal of International Law*, **6**, 402–436.

Timms, B. and Conway, D. (2012) Slow tourism at the Caribbean's geographical margins. *Tourism Geographies*, **14**, 396–418.

Tomblin, J. and Aspinall, W. (1975) Reconnaissance report of the Antigua, West Indies, earthquake of 8th October 1974. *Bulletin of the Seismological Society of America*, **65**, 1553–1573.

Tompkins, E., Hurlston, L.-A. and Poortinga, W. (2009) Foreignness as a constraint on learning: the impact of migrants on disaster resilience in small islands. *Environmental Hazards*, **8**, 263–277.

Tonga Ministry of Finance (2011) Second National Millennium Development Goals report. Nuku'alofa, Kingdom of Tonga: Ministry of Finance.

Tonkinson, R. (1977) The exploitation of ambiguity: a New Hebrides case. In M. Lieber, ed., *Exiles and Migrants in Oceania*. Honolulu, Hawaii: University Press of Hawaii, pp. 269–295.

Tonkinson, R. (1979) The paradox of permanency in a resettled New Hebridean community. *Disasters*, **4**, 105–116.

Tonkinson, R. (2004) Spiritual prescription, social reality: reflections on religious dynamism. *Anthropological Forum*, **14**, 183–201.

Transparency International (2011) *Corruption Perceptions Index 2011*. Berlin, Germany: Transparency International.

Traufetter, G. (2012) Climate change or tectonic shifts? The mystery of the sinking South Pacific Islands. *Der Spiegel*, **24**, 11 June.

Trist, C. (1999) Recreating ocean space: recreational consumption and representation of the Caribbean marine environment. *Professional Geographer*, **51**, 376–387.

Tsing, S. (2005) *Friction. An Ethnography of Global Connection*. Princeton, NJ: Princeton University Press.

Tubuna, S., Haszler, H., Hone, P. and Gonemaituba, W. (2007) Policy responses to threats to rural household incomes, Seaqaqa, Northern Fiji. *Journal of Pacific Studies*, **30**, 87–117.

Tupper, M. and Rudd, M. (2002) Species-specific impacts of a small marine reserve on reef fish production and fishing productivity in the Turks and Caicos. *Environmental Conservation*, **29**, 484–492.

Turner, R., Cakacaka, A., Graham, N., Polunin, N., Pratchett, M., Stead, S. and Wilson, S. (2007) Declining reliance on marine resources in remote South Pacific societies: ecological versus socio-economic drivers. *Coral Reefs*, **26**, 997–1008.

Turvey, R. (2007) Vulnerability assessment of developing countries: the case of small-island developing states. *Development Policy Review*, **25**, 2423–264.

Underhill, Y. (1989) Population mobility as a household strategy: the case of Manihiki atoll, Cook Islands. Unpublished MA thesis, University of Hawaii, Honolulu.

Underwood, R. (1985) Excursions into inauthenticity: the Chamorros of Guam. In M. Chapman, ed., *Mobility and Identity in the Island Pacific*, *Pacific Viewpoint*, **26**, 160–184.

UNCTAD (1985) *The Least Developed Countries. 1984 Report*. New York: United Nations.

UNCTAD (2004) *Linking International Trade with Poverty Reduction. The Least Developed Countries Report 2004*. New York and Geneva: United Nations.

UNDP (2011) *Sustainability and Equity: a Better Future for All. Human Development Report 2011*. New York: UNDP.

UNDP (2012) *Human Development and the Shift to Better Citizen Security. Caribbean Human Development Report 2012*. New York: UNDP.
UNEP (1999) *Pacific Islands Environment Outlook*. Suva, Fiji: UNEP.
UNESCAP/POC (2002) *Squatter Settlement Assessment Nasinu*. Phase II Report. Suva, Fiji.
United Nations University (2011) *World Risk Report 2011*. Berlin, Germany: UNU.
Uyarra, M., Cote, I., Gill, J., Tinch, R. et al. (2005) Island-specific preferences of tourists for environmental features: implications of climate change for tourist- dependent states. *Environmental Conservation*, **32**, 11–19.
van der Grijp, P. (1993) *Islanders of the South*. Leiden: KITLV Press.
van der Grijp, P. (1999) Modernisation, mondialisation et identité Polynésienne. La modèle 'MIRAB' dans le Pacifique. *La Pensée*, **318**, 83–95.
van der Grijp, P. (2004) *Identity and Development. Tongan Culture, Agriculture and the Perenniality of the Gift*. Leiden, the Netherlands: KITLV Press.
van der Velde, M., Green, S., Vanclooster, M. and Clothier, B. (2007) Sustainable development in small island developing states: agricultural intensification, economic development, and freshwater resources management on the coral atoll of Tongatapu. *Ecological Economics*, **61**, 456–468.
Van Fossen, A. (2003) Money laundering, global financial instability and tax havens in the Pacific. *The Contemporary Pacific*, **15**, 237–275.
Van Fossen, A. (2007) Citizenship for sale: passports of convenience from Pacific Island tax havens. *Commonwealth and Comparative Politics*, **45**, 138–163.
Van Trease, H. (2012) Vanuatu. *The Contemporary Pacific*, **24**, 414–431.
Veitch, C., Clout, M. and Towns, D. (2011) *Island Invasives: Eradication and Management*. Gland, Switzerland: IUCN.
Veiteyaki, J. (1995) *Fisheries Development in Fiji. The Quest for Sustainability*. Suva, Fiji: USP.
Veiteyaki, J. (1997) Traditional marine resource management practices used in the Pacific Islands: an agenda for change. *Ocean and Coastal Management*, **30**, 123–136.
Veiteyaki, J. (2006) Caring for the environment and the mitigation of natural extreme events in Gau, Fiji Islands: a self-help community initiative. *Island Studies Journal*, **1**, 239–252.
Vergès, F. (2003) The island of wandering souls: processes of creolisation, politics of emancipation and the problematic of absence on Réunion Island. In R. Edmond and V. Smith, eds, *Islands in History and Representation*. London: Routledge, pp. 162–176.

Vianna, G., Meekan, M., Pannell, D., Marsh, S. and Meeuwig, J. (2012) Socio-economic value and community benefits from shark-diving tourism in Palau: a sustainable use of reef shark populations. *Biological Conservation*, **145**, 267–277.

Vicente, P. (2010) Does oil corrupt? Evidence from a natural experiment in West Africa, *Journal of Development Economics*, **92**, 28–38.

Vidal, J. (2011) Forecast: strange weather ahead. *Guardian Weekly*, 24 June, 28–29.

Vine, D. (2009) *Island of Shame. The Secret History of the US Military Base on Diego Garcia*. Princeton, NJ: Princeton University Press.

Vlcek, W. (2010) Money laundering prevention and small state development: insights from the case of the Bahamas. *Commonwealth and Comparative Politics*, **48**, 373–391.

Voight, B., Sparks, S., Miller, A., Stewart, R. et al. (1999) Magma flow instability and cyclic activity at Soufrière Hills volcano, Montserrat, British West Indies. *Science*, **283**, 1138–1142.

Voigt-Graf, C. (2003) Fijian teachers on the move: causes, implications and policies. *Asia Pacific Viewpoint,* **44**, 163–174.

Voigt-Graf, C. (2004) Towards a geography of transnational spaces: Indian transnational communities in Australia. *Global Networks*, **4**, 25–49.

Voigt-Graf, C. and Connell, J. (2006) Towards autonomy? Gendered migration in Pacific Island countries. In K. Ferro and M. Wallner, eds, *Migration Happens. Reasons, Effects and Opportunities of Migration in the South Pacific*. Vienna: Lit Verlag, pp. 43–62.

de Vries, B. (1985) Development aid to small countries. In P. Selwyn, ed., *Development Policy in Small Countries*. London: Croom Helm, pp. 164–184.

Wade, M. (2011) Now is not the time for donor fatigue. *Sydney Morning Herald*, 9 July, 17.

Walker, B. and Robinson, M. (2009) Economic development, marine protected areas and gendered access to fishing resources in a Polynesian lagoon. *Gender, Place and Culture*, **16**, 467–484.

Walker, L. and Bellingham, P. (2011) *Island Environments in a Changing World*. Cambridge: Cambridge University Press.

Wallis, J. (2010) 'Friendly islands' in an unfriendly system: examining the process of Tonga's WTO accession. *Asia Pacific Viewpoint*, **51**, 262–277.

Walmsley, S., Purvis, J. and Ninnes, C. (2006) The role of small-scale fisheries management in poverty reduction strategies in the Western Indian Ocean region. *Ocean and Coastal Management,* **49**, 812–833.

Walsh, A.C. (1982) *Migration Urbanisation and Development in South Pacific Countries*. Bangkok: ESCAP.

Walsh, K., McInnes, K. and McBride, J. (2012) Climate change impacts on tropical cyclones and extreme sea levels in the South Pacific: a regional assessment. *Global and Planetary Change*, **80**, 149–164.

Walters, H (1984) Fisheries development in St Lucia and the Lesser Antilles. *The Courier*, **95**, 94–95.

Ward, D. (2008) Ecological patterning and invasive ants (Hymenoptera: Formicidae) in a tropical rain forest ant community from Fiji. *Pacific Science*, **62**, 473–482.

Ward, R.G. (1959) The banana industry in Western Samoa. *Economic Geography*, **3**, 123–137.

Ward, R.G. (1961) Internal migration in Fiji. *Journal of the Polynesian Society*, **70**, 257–271.

Ward, R.G. (1982) The effects of scale on social and economic organisation. In R. May and H. Nelson, eds, *Melanesia; Beyond Diversity*. Canberra, Australia: ANU Research School of Pacific Studies, pp. 181–191.

Ward, R.G (1989) Earth's empty quarter? The Pacific Islands in a Pacific century. *Geographical Journal*, **155**, 235–246.

Ward, R.G. (1998) Remote runways: air transport and distance in Tonga. *Australian Geographical Studies*, **36**, 177–186.

Warren-Rhodes, K., Schwarz, A.-M., Boyle, L., Albert, J. et al. (2011) Mangrove ecosystem services and the potential for carbon revenue programmes in Solomon Islands. *Environmental Conservation*, **38**, 485–496.

Weaver, D. (1993) Model of urban tourism for small Caribbean islands. *Geographical Review*, **83**, 134–140.

Weaver, D. (2003) The political ecology of tourism in the Cayman Islands. In S. Gössling, ed., *Tourism and Development in Tropical Islands*. Cheltenham, UK and Northampton, MA: Edward Elgar, pp. 38–59.

Webb, A. (2005) An assessment of coastal processes, impacts, erosion mitigation options and beach mining: South Tarawa. Report No. 46. Suva, Fiji: SOPAC.

Webb, A. and Kench, P. (2010) The dynamic response of reef islands to sea level rise: evidence from multi-decadal analysis of island change in the Central Pacific. *Global and Planetary Change*, **72**, 234–246.

Webb, E., Seamon, J. and Fa'aumu, S. (2011) Frequent, low-amplitude disturbances drive high tree turnover rates on a remote, cyclone-prone Polynesian island. *Journal of Biogeography*, **38**, 1240–1252.

Webster, P., Holland, G., Curry, J. and Chang, H.-R. (2005) Changes in tropical cyclone number, duration and intensity in a warming environment. *Science*, **309**, 1844–1846.

Wesley-Smith, T. (2006) There goes the neighbourhood: the politics of failed states and regional intervention in the Pacific. In J. Bryant-Tokalau and I. Frazer, eds, *Redefining the Pacific? Regionalism Past, Present and Future*. Farnham, UK: Ashgate, pp. 121–126.

Wessen, A., Hooper, A., Huntsman, J., Prior, I. and Salmond, C. (1992) *Migration and Health in a Small Society*. Oxford: Clarendon Press.

White, I., Falkland, T., Metutera, T., Katatia, M. et al. (2008) Safe water for people in low, small island Pacific Nations: the rural–urban dilemma. *Development*, **51**, 282–287.

White, I., Falkland, T. and Scott, D. (1999) Droughts in small coral islands: case study, South Tarawa, Kiribati. IHP Technical Documents in Hydrology No. 26. Paris: UNESCO.

Wichman, V. (2012) *Fishing for Answers*. Rarotonga, Cook Islands: Ministry of Marine Resources.

Wilkinson, C. and Salvat, B. (2012) Coastal resource degradation in the tropics: does the tragedy of the commons apply for coral reefs, mangrove forests and seagrass beds? *Marine Pollution Bulletin*, **64**, 1096–1105.

Wilkinson, D. (2004) The parable of Green Mountain: Ascension Island, ecosystem construction and ecological fitting. *Journal of Bio geography*, **31**, 1–4.

Wilkinson, P. (1999) Caribbean cruise tourism: delusion? illusion? *Tourism Geographies*, **1**, 261–282.

Wilkinson, P. (2003) Tourism policy and planning in St Lucia. In S. Gössling, ed., *Tourism and Development in Tropical Islands*. Cheltenham, UK and Northampton, MA: Edward Elgar, pp. 88–120.

Williams, B. and Grottoli, A. (2010) Recent shoaling of the nutricline and thermocline in the western tropical Pacific. *Geophysical Research Letters*, **37**, L22601, doi:10.1029/2010GL044867.

Williams, G., Knapp, I., Maragos, J. and Davy, S. (2010) Modeling patterns of coral bleaching at a remote Central Pacific atoll. *Marine Pollution Bulletin*, **60**, 1467–1476.

Wilson, D. (1994) Unique by a thousand miles. Seychelles Tourism revisited. *Annals of Tourism Research*, **21**, 20–45.

Wilson, K. (2010) Wan Laki Aelan? Diverse development strategies on Aniwa, Vanuatu. Unpublished BA Honours thesis, University of Sydney, Australia.

Wiltshire, R. (2010) Caribbean migration: challenges to leadership. *Journal of Eastern Caribbean Studies*, **35**(2), 61–79.

Winters, A. (2005a) Small economies in a globalising world. *Pacific Economic Bulletin*, **20**(3), 94–102.

Winters, A. (2005b) Policy challenges for small economies in a globalising world. In S. Chand, ed., *Pacific Islands Regional Integration and Governance*. Canberra, Australia: Asia Pacific Press, pp. 7–25.

Winters, A. and Martins, P. (2005) *Beautiful but Costly: Business Costs in Small Remote Economies*. London: Commonwealth Secretariat.

Winthorpe, M. (2004) Words from Whitesands: reef management in Tanna, Vanuatu. Unpublished BSc Hons thesis, University of Sydney, Australia.

Wittersheim, E. (2011) Paradise for sale. The sweet illusions of economic growth in Vanuatu. *Journal de la Société des Océanistes*, **133**, 323–332.

Wong, P., Marone, E., Lana, P. and Fortes, M. (2005) Island systems. In R. Hassan, R. Scholes and N. Ash, eds, *Ecosystems and Human Wellbeing*. Washington, DC: Island Press, pp. 663–680.

Wood, C. and McCoy, T. (1985) Migration remittances and development: a study of Caribbean cane cutters in Florida. *International Migration Review*, **19**, 251–277.

Wood, D. (1967) The small territories: some political considerations. In B. Benedict, ed., *Problems of Smaller Territories*, London: Athlone Press, pp. 23–34.

Wood, R. (2004) Global currents: cruise ships in the Caribbean Sea. In D. Duval, ed., *Tourism in the Caribbean*. London: Routledge, pp. 152–171.

Woodroffe, C. (2008) Reef-island topography and the vulnerability of atolls to sea-level rise. *Global and Planetary Change*, **62**, 77–96.

Woodroffe, C. and Biribo, N. (2011) Atolls. In D. Hopley, ed., *Encyclopaedia of Modern Coral Reefs*. Dordrecht, Netherlands: Springer, pp. 51–71.

Wolff, H., Chong, H. and Auffhammer, M. (2011) Classification, detection and consequences of data error: evidence from the Human Development Index. *Economic Journal*, **121**, 843–870.

World Bank (1987) *Fiji. A Transition to Manufacturing*. Washington, DC: World Bank.

World Bank (2006) *At Home and Away. Expanding Job Opportunities for Pacific Islanders through Labour Mobility*. Washington, DC: World Bank.

World Bank (2007) *Crime, Violence and Development: Trends, Costs and Policy Options in the Caribbean*. Washington, DC: World Bank.

World Bank (2011) *Migration and Remittances Factbook 2011*. Washington, DC: World Bank.

World Bank (2012) *Statistics for Small States 2011*, Washington, DC: World Bank.

Xue, C. (2001) Coastal erosion and management of Majuro Atoll, Marshall Islands. *Journal of Coastal Research*, **17**, 909–918.

Yahya, F., Parameswaran, A. and Sebastian, R. (2005) Tourism and the South Asian Littoral: voices from the Maldives. *South Asia*, **28**, 457–480.

Yamano, H., Kayanne, H., Yamaguchi, T., Kuwahara, Y. et al. (2007) Atoll island vulnerability to flooding and inundation revealed by historical reconstruction: Fongafale Islet, Funafuti Atoll, Tuvalu. *Global and Planetary Change*, **57**, 407–416.

Yang, J. (2011) China in Fiji: displacing traditional players? *Australian Journal of International Affairs*, **65**, 305–321.

Yang, M. (2000) Putting global capitalism in its place. Economic hybridity, bataille and ritual expenditure. *Current Anthropology*, **41**, 477–510.

Yeo, S. and Blong, R. (2010) Fiji's worst natural disaster: the 1931 hurricane and flood. *Disasters*, **34**, 657–683.

Yoon, M. (2011) Factors hindering 'larger' representation of women in parliament: the case of Seychelles. *Commonwealth and Comparative Politics*, **49**, 98–114.

Zable, A. (1973) Neo-colonialism and race relations in New Guinea and the Pacific. *Race and Class*, **14**, 393–441.

Zann, S. and Zann, L. (2008) Poverty in paradise? Issues in poverty and development in Fijian fishing villages. *SPC Women in Fisheries Information Bulletin*, **18**, 36–41.

Zecca, A. and Chiari, L. (2012) Lower bounds to future sea-level rise. *Global and Planetary Change*, **98**, 1–5.

Index

Abaiang (Kiribati) 192
affluence 219, 238, 241, 252
Africa 20, 24, 25, 26, 28, 39, 171, 173
Agalega (Mauritius) 129, 134, 206
ageing 62, 125
agriculture 28, 55–69, 108, 128–9, 202, 260
  commercial 55, 58, 61–7, 91, 101, 102, 249
  subsistence 56–61, 133, 142, 250
Agrihan (Northern Marianas) 133
aid 19, 28, 36, 38, 39–43, 47, 54, 95, 116, 160, 175, 237, 239, 244, 265, 258, 259
air transport 37–8, 95, 98, 99, 113, 131, 152, 235
Aitutaki (Cook Islands) 15
Alamagan (Northern Marianas) 205
Alaska 152, 241
alcohol 119, 148, 253
Aldabra (Seychelles) 217
Alexander Selkirk Island (Chile) 216, 217
Ambae (Vanuatu) 222, 250, 251
Ambrym (Vanuatu) 68, 133–4, 249–50, 253
American Samoa 3, 8, 22, 76, 126, 131, 152, 153, 179, 204, 225, 235
Angaur (Palau) 94, 218
Anguilla 3, 7, 9, 17, 56, 98, 99, 100, 101, 108, 112, 118, 126, 130, 150, 153, 210–11, 235
Aniwa (Vanuatu) 9, 59, 69, 131, 162, 169, 208–9
Antigua 3, 5, 19, 22, 39, 44, 62, 134, 211, 222, 223, 239
  economy 29, 99–100, 104, 114
  environment 105, 223
  society 48, 120
ants 216, 218
Anuta (Solomon Islands) 17, 58, 130
AOSIS 2–3, 21, 26
Apia (Samoa) 72, 119, 140, 205, 206
aquaculture 81
Aranuka (Kiribati) 194
Aruba 3, 7, 38, 103, 113, 154
Ascension 3, 7, 15, 216, 217, 219
Asia 23, 48, 69, 74, 91, 92
Asian Development Bank (ADB) 20, 28, 36, 40, 258
asymmetry 31–2, 39, 116, 237, 257, 258–9
Atlantic Ocean 55, 73, 75, 77, 150, 199, 222
atolls 14–15, 61, 126, 127, 129, 131–2, 134–5, 188, 217, 227, 234, 257
  economies 57, 59, 63, 77, 228, 234, 243–4
  physical environment 135, 141–2, 144–6, 186–7, 189–95, 204, 207, 210, 211–13, 226, 227, 228–9, 231, 240, 242–3
AusAID 178
Australia 3, 34, 36, 93, 152, 154, 157, 163, 164, 170, 177, 178, 179, 180–81, 182

Bahamas 3, 4, 5, 14, 27, 36, 48, 129, 131, 151, 153, 170
  economy 83, 100, 103, 110, 111, 113, 118
  environment 204, 207, 220–21
  society 112, 147, 176, 245, 249
Bahrain 2, 21
Banaba (Ocean Island) 43, 94, 136, 151, 182, 235
bananas 19, 29, 32, 33, 34, 54, 61, 63, 125, 203, 208, 236, 238, 259

banks 19, 50, 80, 110–11, 113, 163, 182
Barbados 3, 5, 22, 24, 165, 170, 175–6, 177, 214
  economy 33, 34, 37, 57, 62, 63, 92, 93, 112, 113
  environment 59, 60, 71, 104–5, 106, 108, 242
  society 35, 47, 125, 147, 230, 249
Barbados Programme of Action 22
Barbuda 17, 56, 130, 134, 153, 161
Beaches 142, 143, 193–4, 196, 199, 227
bêche-de-mer 78, 234, 249
begging 119, 138
Belize 2
Bellona (Solomon Islands) 58, 60, 128, 130, 204, 251
Bequia (St Vincent) 209, 210
Bermuda 3, 7, 14, 19, 40, 98, 99, 103, 104, 110, 111, 112, 114, 122, 126, 154, 195, 196, 208, 210, 211, 236, 237, 241
Bhutan 25
Bikini (Marshall Islands) 132, 134, 135, 232, 235
biodiversity 14–15, 60, 66, 86, 107, 109, 215–21
biological invasions 216–19
Bird Island (Seychelles) 218
birds 69, 109, 214–21, 227
black swans 234–5, 258
Bonaire 3, 7, 18, 83, 108, 197, 219
Bora Bora (French Polynesia) 99, 109
boredom 129, 148, 232
Bougainville (PNG) 18, 95, 130, 136, 157, 190, 218, 235, 236
brain drain *see* migration, skilled
Brazil 90
British Indian Ocean Territory (BIOT) 4, 8, 83, 114, 132, 134, 136, 154, 157, 189, 191, 235
British Virgin Islands (BVI) 3, 7, 9, 56, 72, 99, 104, 110, 111, 112, 125, 131
brown tree snake 218
bureaucracy 31, 41, 46, 49–50, 53, 130, 139, 160, 171–2
business 27, 110, 167–8, 174, 184
  see also stores
cactus 61
cagou 217
Canada 151, 152, 154, 170, 177, 182, 257
Canary Islands 61
Cape Verde 3, 5, 15, 20, 75, 151, 152, 164, 165, 177, 182, 257
  economy 98, 117, 118, 159, 160, 166, 167–8, 234
  environment 157, 207
  society 44, 125, 155, 170–71, 174, 176, 249
capital 8, 47, 50, 81, 91, 93, 96, 110, 111, 133, 145, 167, 174–5, 228, 234, 237, 240, 254
capitalism 67, 68, 90, 130, 237, 245, 248, 250, 253
Caribbean 9, 17, 29, 33, 36, 40, 45, 46, 48–9, 52, 55–6, 58, 60, 61–2, 63–4, 65, 66, 69, 71, 72, 81, 94, 97–8, 102, 103–4, 106, 125, 133, 150, 154, 167, 170–73, 176, 177, 181, 186, 192, 197, 198, 199, 207–8, 214, 220, 222–3, 224, 226, 237, 238, 245, 249
Caribbean Basin Initiative 30
CARICOM 36, 37, 38, 181
Carriacou (Grenada) 101, 161, 186
Carteret Islands (Bougainville, PNG) 135, 190, 191, 193, 240
casino 47–8
cassava 58, 75, 118, 129, 202, 206
Castries (St Lucia) 140–41, 145
Cayman Islands 3, 7, 14, 83, 98, 100, 103, 104, 109, 110, 111–12, 113, 131, 154, 160, 170, 195, 202, 203, 228
Chagos Archipelago *see* BIOT
children 77, 82, 87, 127, 155, 169, 246
chillies 61
China 40–41, 90, 115
cholera 213–14
Christmas Island (Indian Ocean) 4, 8, 40, 94, 115, 153, 154, 218, 236
Chuuk (FSM) 82, 129, 224

ciguatera 215
Cikobia (Fiji) 204
civil society 46–8, 51–2, 111, 239
class 86, 140, 246, 252
climate 9, 183
climate change 9, 42, 74, 76, 85, 87, 108, 145, 186–201, 213–15, 216, 227, 239, 253
Clipperton (France) 94
coastal squeeze 16, 72, 86, 102, 106, 108, 196–7, 200, 221, 223, 228–9, 239–40, 243
coasts 23, 96, 106, 108, 140, 143–4, 190–93, 202, 205, 234, 242
cocaine 48, 65
cocoa 29, 61, 62, 65, 66
Cocos-Keeling Islands 4, 8, 14, 62, 99, 115
coffee 61, 65
coins 114
colonialism 21, 42, 129–30, 237, 245
commodities 22, 32–4, 87, 232
  niches 34, 35, 59, 63, 64–5,
Commonwealth Secretariat 2, 21, 22
Comoros Islands 4, 5, 6, 9, 15, 17, 18, 20, 23, 24, 25, 27, 40, 43, 130, 153, 154, 157, 164, 182, 258
  economy 28, 39, 59–60, 64, 100, 108, 159, 165
  environment 79, 218, 222–3
  society 4, 125, 126, 171, 173, 174, 238
conservation 73, 75–6, 82–6, 107, 196, 212, 221, 241
Cook Islands 3, 8, 57, 61, 77, 78, 87, 96, 109, 112, 114, 129, 130, 134–5, 151, 152, 153, 159, 160, 170, 172, 176, 202, 208, 233, 235, 236, 253
copra 61, 62, 63, 65, 131, 161, 162, 203, 220, 251, 258
coral atolls *see* atolls
coral reefs 70, 74, 77, 80, 82, 84, 105, 106, 142, 189–90, 194, 195–201, 204, 227, 240
  bleaching 189–90, 196, 197, 199–200, 228
  degradation 70, 82, 84, 186, 196–9, 204, 215, 219, 224
  sedimentation 80, 95, 107, 197, 198
corruption 24, 25, 28, 32, 44, 46, 47–9, 70, 111, 157, 244
cotton 166
coups 17, 32, 59, 157, 172, 207, 238
credit 33, 64, 78, 81, 249
crime 24, 47, 49, 63, 75, 86, 111, 119, 124, 138
cruise ships 102–5
Cuba 2, 3, 154, 172, 173
culture 17, 51, 100, 102, 115, 161, 165, 167, 168, 238, 245–54, 256–7
Curaçao 3, 7, 18, 153, 154
currency 19, 37
custom *see* tradition
cyclones 9, 58, 62, 71, 72, 113, 132, 134–5, 140, 143, 153, 162, 185–6, 189, 190, 198, 201–5, 208, 220, 228, 232, 240
Cyprus 2

data *see* statistics
debt 28–9
decentralization 132–3
decolonization 130, 236, 237
deforestation 60, 66, 67, 69–72, 86, 142, 187, 191, 206, 217, 220, 224, 243
democracy 44–6
demography *see* population
dengue 213, 214
Denmark 25
dependencies 52, 99, 110, 116, 125–6, 130–31, 153, 157, 160, 256
deregulation 28, 30
desalination 208–9, 211, 213, 236, 242
diabetes 95, 118, 147–8
diaspora 65, 159, 164, 165–6, 175–6, 184, 256, 260–61
Diego Garcia *see* BIOT
diet *see* nutrition
diplomacy 50
diseases 95, 108, 118, 140, 141–2, 147–8, 206, 211, 213–15, 217, 235
diversity 57, 58, 60, 232
  ecological 72, 73, 78, 240
  economic 29, 31, 32, 35, 67, 88, 128, 131, 134, 156, 158, 228, 232, 250

doctors, migration 154, 170–72
dodo 216
dolphins 49, 73, 220
Dominica 3, 5, 25, 29, 40, 44–5, 48, 83, 97, 233, 234, 235, 237
  economy 33, 47, 63, 79, 97, 113, 118, 159
  environment 71, 106, 210–11, 222, 224
  society 152, 170–71
Dominican Republic 2
drought 62, 93, 135, 157, 186, 188, 207–9, 212, 228, 250
drugs 34, 48–9, 65, 102, 119, 176, 235, 253
dugongs 73
Dutch disease 32, 47, 95, 102, 108, 160, 166–7, 169

earthquakes 23, 186, 192, 223–4, 240
Easter Island *see* Rapanui
East Timor (Timor Leste) 2, 172, 238
Eauripik (FSM) 79, 161
Ebeye (Marshall Islands) 140, 141
ecology 60, 133, 186, 215–21, 231, 240, 259
ecosystems 80, 196, 203–5, 242
ecotourism *see* tourism
education 24, 50–51, 55, 62, 79, 82, 119, 120, 126, 133, 156–7, 158, 166, 167, 169, 173–4, 180, 238, 246, 250, 253, 257, 259, 260
Egypt 29
elections 44, 45, 47, 87, 252
electronics 35, 93
elites 47, 83, 148, 246, 252
El Niño (ENSO) 9, 74, 187–90, 192, 194, 195, 199–200, 201, 208, 213, 240
emotion 183, 231, 247
employment 16, 92–3, 95, 99–100, 116–17, 128, 137–8, 170, 235, 236
  formal 66, 75, 76, 92, 110
  informal 119
  unemployment 16, 63, 76, 77, 116–17, 127, 138, 156, 243
endangered species 49, 73, 84, 216–21
energy 66, 70, 72, 76, 80, 96–7, 105, 109, 138, 166, 209, 238, 243
Enewetak (Marshall Islands) 135
Enron 111
environment viii, 14–15, 16, 22–3, 71–2
  degradation 16, 64, 66, 70, 73, 80, 85, 95–6, 98, 105–7, 136, 140–46, 157, 194–5, 196–206, 209–12, 220–21, 226–9, 235
  management 81–5, 107, 109, 204, 211–12, 221, 222–3, 229, 233–4, 241–2
equity 29, 43, 59, 68, 180, 248, 252
  see also uneven development
erosion 59, 66, 70, 72, 95, 116, 142, 143–4, 187, 191, 193, 197, 202, 211
Etal (FSM) 79, 82
ethnicity 4, 234, 238
European Union (EU) 28, 29, 36, 63, 74, 76, 78, 111, 181, 259
  Lomé Convention 29, 34, 93
Exclusive Economic Zones (EEZs) 72, 74–5
expatriates 31, 111, 113
export processing zones 92–3
exports 19, 27, 29, 30–31, 32, 33, 35, 48–9, 62, 63–5, 75, 77, 92–4, 114, 131, 235, 236, 253, 259

failed states 17, 24, 43, 236
Fais (FSM) 94
Fakaofo (Tokelau) 130, 161
Falahola(Tonga) 161, 167
Falkland Islands 3, 7, 40, 76, 96, 99, 103, 125, 153, 217, 236, 238
family 46, 126, 128, 129, 136, 137, 149, 155, 156, 158, 182, 234
Faroe Islands 3, 73, 99, 241
Farquhar (Seychelles) 94
fauna 15, 73, 83, 140, 188, 213, 215–21
Federated States of Micronesia (FSM) 3, 4, 5, 15, 40, 74–5, 79, 118, 129, 131, 152–3, 155, 170, 258–9
  economy 39, 130
  environment 15, 73, 191, 193–4, 224
  society 79, 126, 232
fertility 124–6

Fiji 3, 5, 17, 27, 32, 33, 49, 59, 77, 84, 85, 92, 131, 134, 137, 142, 155, 157, 163, 170, 177–9, 182, 214, 232, 259
economy 19–20, 31, 33, 34, 36, 39, 59, 63, 64, 65, 72, 76, 80, 88, 90, 91, 93, 94, 96–7, 99, 101, 109, 113–14, 118, 121, 124, 155, 169
environment 59, 70, 72, 78, 87, 106–7, 108, 109, 141, 204, 205–7, 208, 209, 217, 218
society 51, 138–9, 147, 154, 170–74, 238, 245, 248
finance centres *see* tax havens
fishing 28, 48, 65, 70, 72–81, 128, 129, 196, 198, 200–201, 237, 243, 258, 259
agreements 47, 72, 74–5
commercial 72–81, 259
flags of convenience 113
floods 59, 62, 72, 95, 140, 141, 142, 187, 188, 190–91, 193, 202, 205–7, 224, 240
flora 15, 188, 203, 215–21
food 56, 58, 59, 62, 64, 85, 86–7, 92, 138, 253
food security 80, 85, 86–7, 119, 135, 157, 190, 228, 230, 232, 257
see also nutrition
France 3, 18, 40, 154, 177, 182, 237
French Polynesia 3, 8, 45, 77, 81, 131, 132, 151, 160, 202, 235
fuel (*see* energy)
Funafuti (Tuvalu) 131, 134, 140, 142, 144, 190–91, 192, 210, 211
Futuna (Wallis and Futuna) 204–5, 225

Galapagos Islands (Ecuador) 3, 18, 77, 78, 79, 80, 83, 103, 108, 128, 217, 219, 237, 239, 240
gambling 39, 113, 259
gender 24, 43, 155, 252
General Agreement on Trade in Services (GATS) 181
Georgia 41
Gibraltar 3
ginger 34, 59, 64
Gizo (Solomon Islands) 80, 198, 225
global financial crisis (GFC) vii, 98, 108, 111–12, 120, 163, 183, 235
global warming 16, 186–95
globalization 19–20, 22, 30, 87, 155, 181–2, 240, 245, 247–8, 251, 253, 256–7, 258
see also transnationalism
Gold Ridge (Solomon Islands) 96, 136
golf 105
governance 36, 41–2, 43–52, 70, 82, 94, 252
government islands 57, 99, 110, 153, 241
Grand Comore 79, 162
Greece vii, 235
Greenland 3
Grenada 3, 4, 5, 15, 37, 61, 68, 99, 141, 170, 203, 236
economy 29, 110, 113, 116–17, 118, 159
environment 141, 203, 210–11, 222
society 120, 170–71
Grenadines (St Vincent) 99, 209
Guadalcanal (Solomon Islands) 68, 136, 236
Guadeloupe 3, 7, 18, 48, 57, 97, 103, 118, 226
Guam 3, 8, 99, 115, 153, 154, 155, 204, 210, 218, 235, 236, 238
Guinea-Bissau 2
Gulf 170
Guyana 2, 154

Ha'ano (Tonga) 132
Ha'apai (Tonga) 128–9
Haiti 2, 81, 154, 203, 223
handicrafts 65, 92, 104, 258
happiness 24–5
Hawai'i 217
hazards, natural 9, 14, 18, 22–3, 56, 58, 59, 87, 106, 133–4, 137, 141, 185–6, 208, 213, 223, 226, 227–9, 232, 234, 239, 240, 253
health 126, 147–8, 169, 172–3, 206, 208, 211, 213–5
health care provision, national 50–51, 61, 214–15, 248–9
Henderson Island (Pitcairn) 217, 221

HIV/AIDS 102, 147
Honiara (Solomon Islands) 50, 68, 117, 119, 128, 130, 134, 137, 138, 139, 142
housing 102, 111, 136, 137–9, 143, 165, 166, 167, 180, 191, 203, 223, 224, 225
see also urban settlements
human capital 24, 50, 155, 169, 170, 174, 175–6, 261
human development index (HDI) 24
human resources 31–2, 50–51, 52–3, 93, 99, 168–77, 234
Hunter Island (Vanuatu) 122
hunting 60, 73, 204, 217, 220
hurricanes *see* cyclones
hybridity 245–54, 256
hydropower 97

Iceland 21
identity 136, 175, 176–7, 182–3, 231–2, 238, 241, 246–54, 256–7
national 17, 38, 40, 130, 248, 252, 257
Ifaluk (FSM) 59
Iguanas 217, 220
imports 18, 31, 35, 39, 56, 57, 64, 72, 86–7, 97, 101, 143
incomes 39, 52, 59, 65, 67, 72, 78, 79, 85, 86, 88, 92, 100, 101, 170
India 90, 172
Indian Ocean 55–6, 61, 75, 115, 150, 167, 186, 191, 199, 200, 222–3, 245
indigenous technical knowledge 55, 58, 60, 79, 82, 84, 85, 121, 221, 225, 228, 248, 253
individualism 64, 66, 82, 86, 102, 159, 164, 245, 247, 253
Indonesia 64
industry 18, 29, 32, 35, 76, 82, 90–94
inequality 9, 47, 68, 87–8, 90, 118–19, 139–40, 169
informal sector 124, 138–9
information technology (IT) 155, 172, 230, 257
infrastructure 50, 137, 139, 146
insularity 3, 22, 23
insurance 113, 237
International Monetary Fund (IMF) 20, 28, 34–5
Internet 39, 113, 114, 259, 261
invasive alien species 216–20
IPCC 191
Iran 113
islands, outer 79, 81, 97, 110, 121, 128, 129–32, 144, 152, 160, 161, 167, 209, 224, 231, 243
Isle of Pines (New Caledonia) 99
isolation 21, 59, 76, 87, 98, 103, 130–31, 183, 262
Israel 41

Jackson Report 244, 260
Jamaica 2, 48, 65, 176, 177, 179, 203, 214, 233
Japan 40, 41, 64, 154, 178, 210, 227, 235

Kabara (Fiji) 161
Kadavu (Fiji) 64, 162, 232, 237
Kapingamarangi (FSM) 79, 82, 258–9
Karthala (Comoros) 222–3
kava 30, 34, 64, 66, 206
Kick 'em Jenny (Grenada) 222, 226
kinship *see* family
Kiribati 3, 4, 5, 14, 19, 20, 22, 24, 38, 40, 41, 43, 46, 61, 131, 132, 133, 137, 148, 151, 174, 177–8, 180, 182, 225, 229, 231, 233, 243–4, 258, 260
economy 39, 77–8, 100, 155, 159
environment 134, 141–2, 145–6, 188, 191, 193–4, 199, 207
society 87, 117, 119, 125, 126, 128–9, 248
Kissinger, Henry 116
knowledge 53, 55, 58, 145, 175–6, 234, 252, 259
Kolombangara (Solomon Islands) 60, 70
Korea, North 41
Korea, South 178
Kosrae (FSM) 143, 166, 195
Kotu (Tonga) 64, 88, 231
Kwajalein (Marshall Islands) 114, 235

Labasa (Fiji) 141
labour 92, 94
lagoons 95, 105–6, 107, 142, 143, 144, 145, 146, 194, 198, 204, 205, 210, 243
Lamotrek (FSM) 246
land 56, 70, 101, 106, 113, 118, 167, 183, 190, 224, 255–6
disputes 67–8, 102, 127, 140, 144, 212
reclamation 19, 140, 191
tenure 56, 61, 67–9, 105, 136–7, 138, 207, 212, 246, 247, 252
use 66, 70, 115, 139, 142–3, 197, 229
landslides 141, 193, 206, 223–5
language 4, 46, 51, 53, 87–8, 155, 164, 176, 249, 252
Latin America 23, 63, 150
Lau islands (Fiji) 61, 65, 85, 88, 169, 209, 252, 254
Lau lagoon (Solomon Islands) 77, 79
leadership 44–6, 67, 82, 171, 252–3
legal systems 49, 50, 51, 92, 249
Lesotho 2
less developed countries (LDCs) 17, 20, 23
Lewis, Sir Arthur 91
Libya 41
Lihir (PNG) 96, 136
Line Islands (Kiribati) 62, 207
literacy 24, 51, 122
livelihoods 65–6, 67, 77, 83–5, 88, 118, 156, 182, 250–51, 254, 260
livelihood portfolio 88, 128, 131, 134, 156, 250–51, 255
Lofanga (Tonga) 82, 131, 161
logging *see* timber industry
Lopevi (Vanuatu) 134
Lord Howe Island (Australia) 217, 219

Macquarie Island 217
Madagascar 64, 80, 198
Maiana (Kiribati) 194
Majuro (Marshall Islands) 76, 80, 140, 144, 145, 191, 193, 194, 198, 208, 210, 211, 212, 220, 242
Makatea (French Polynesia) 94
Makira (Solomon Islands) 66, 20, 241–2
Malaita (Solomon Islands) 66, 68, 169, 250
malaria 106, 148, 213, 214, 217
Malden (Kiribati) 230
Maldives 4, 6, 14, 17, 20, 44, 100, 115, 129, 132–3, 143–4, 148, 243–4
economy 31, 38, 74, 75, 98, 99, 108
environment 73, 105, 106, 109, 136–7, 140, 143–4, 187, 192, 194, 225, 226, 240
society 47, 99, 102, 119, 146
Male (Maldives) 130, 140, 144, 146
Malta 2
management 77, 78, 79, 81–5, 86, 130, 136, 137, 139, 145, 198, 201, 211–12, 221, 224, 228, 233–4, 241–2, 252–3
Manam (PNG) 132, 133–4, 222
Mandok (PNG) 169, 249
Mangaia (Cook Islands) 233
mangroves 70, 71–2, 80, 95, 105, 106, 143, 195, 224, 225, 242
Manihiki (Cook Islands) 77, 161, 163
manufacturing *see* industry
Manus (PNG) 115
Marianas, Northern (*see* Northern Marianas)
marijuana 34, 48–9, 65
marine protected areas (MPAs) 81–5, 86, 198, 233, 252, 255
markets 62, 64, 65–6, 81, 84, 88, 92, 94, 118, 138–9
Marquesa Islands (French Polynesia) 15, 217
Marshall Islands 3, 5, 14, 40, 44, 45, 131, 152–3, 154, 155, 170, 178, 208, 233, 235, 243–5
economy 39, 47, 111, 113, 114–6
environment 144, 188, 191, 193, 194, 198, 208, 227, 242
society 125, 127, 135, 147, 176
Martinique 3, 7, 48, 57, 87, 97, 106, 138, 153, 222, 249, 257
materialism 42, 102, 245–6
Matthew Island (Vanuatu) 122
Mauke (Cook Islands) 161

Mauritius 2, 3, 9, 73, 87, 94, 129, 131, 134, 136, 235
Mayotte 4, 8, 18, 40, 126, 130, 153, 234, 236
McDonald's 245
mechanisation 66–7
media 51–2
Melanesia 17, 44, 58, 60, 182
Melanesian Spearhead Group 36
Merir (Palau) 132
Mexico 34, 179
Micronesia 17, 79, 147, 150, 166, 214, 236, 257
Middle East 155, 178–9
migration 19, 40, 41, 42, 56, 65, 78, 84, 93, 94, 148–84, 207, 232, 236–7, 239, 240, 244, 245–6, 255–6, 259, 261
  culture of 149, 151, 152, 232, 256, 261–2
  impact 148–84
  policy 128, 174, 177–81, 183, 244, 260
  return 128, 173–7, 183
  rural–urban 108, 117, 121, 126, 127–33, 206, 207, 243, 250
  seasonal 168, 177, 179–81
  skilled 46–7, 120, 154–5, 157, 158, 169–74, 183
military 42, 48, 114–15, 154, 155, 158, 178–9, 236
millennium development goals 24, 25, 243
Minerva Reef 122
mining 43, 94–7, 115, 136, 151, 210
MIRAB 41, 160–63, 166, 256
mobility 185, 217, 231–2, 254
Moch (FSM) 131, 191, 228
modernity 33, 58, 62, 67–9, 101, 122–3, 129, 245, 247, 251, 253, 256, 257
Mokil (FSM) 57
monetary policy 19
money 45, 156, 163–4, 246, 247
monkeys 218
Montserrat 3, 7, 19, 133–4, 136, 143, 151, 152, 153, 166, 203, 222–3, 235, 236
Moorea (French Polynesia) 199, 200, 219
moral economy 229, 244, 253, 254
mortality 124–6
multinational corporations 31, 74, 92, 100, 101, 104, 113, 149, 245
music 102, 113, 248, 253
Mustique (St Vincent) 101

Nadi (Fiji) 140, 206
Naikeleyaga (Fiji) 255
Nanumea (Tuvalu) 132, 152, 161
Naqalevu (Fiji) 132
Nassau (Bahamas) 104
natural disasters 9, 22–3, 27, 50, 58, 60, 87, 140–41, 185–213
Nauru 3, 5, 15, 17, 21, 45, 111–12, 235, 236
  economy 48, 59, 94–5, 115, 151
  environment 75, 191, 209
  society 126
Navua (Fiji) 141, 206
neo-colonialism 38
neo-liberalism 35, 258
nepotism 46, 49, 53, 157, 252
Netherlands 154, 237
Netherlands Antilles 18, 40, 235
Nevis 17, 88, 120, 130, 150, 152, 155, 158, 164, 166, 177, 225
New Caledonia 3, 8, 18, 57, 58, 81, 103, 122, 177, 204, 210, 217, 218, 235
Newfoundland 73
New Georgia (Solomon Islands) 58, 70, 78
New Guinea 14
  see also Papua New Guinea
New Zealand 3, 36, 93, 131, 135, 152, 153, 154, 159, 163, 164, 170, 179–81, 182, 207, 222, 244, 257
Ngulu (FSM) 132
Niuas (Tonga) 134, 160, 225
Niue 3, 8, 15, 26, 40, 44, 45–6, 65, 98, 110, 111, 114, 125, 127, 134, 152, 153, 160, 172, 203, 220
non-communicable diseases (NCDs) 95, 147–8
non-governmental organizations (NGOs) 51, 121

noni 34, 64
Norfolk Island 3, 8, 40, 50, 99, 153, 219
Noro (Solomon Islands) 76
Northern Marianas 3, 8, 32, 47, 48, 94, 115, 131, 132, 153, 155, 222, 235
Noumea (New Caledonia) 80, 106
nuclear power vii, 114, 133–4, 230
Nuku'alofa (Tonga) 72, 117, 140
Nukumanu (Bougainville, PNG) 79
nurses 154, 158, 164, 170–72, 181
nutrition 27, 35, 57, 79, 86–7, 102, 118, 129, 140, 142, 147, 166, 207

obesity 95, 147–8
oceans 188, 189–91
  acidification 73, 196, 197, 228
occupational multiplicity 88, 138, 158, 250, 254, 255, 260
offshore finance 47–8, 110–13, 154
oil 31–2, 37, 96, 109, 111, 144, 236, 238, 258
oil palm 62
Ontong Java (Solomon Islands) 59, 78, 220, 260
Ovalau (Fiji) 76

Paama (Vanuatu) 162, 205, 231
Pacific 50, 55, 59, 68, 111, 117, 129, 139, 191, 197, 200, 201, 214, 220, 222, 243
Pacific Agreement on Closer Economic Relations (PACER) 36
Pacific Forum 36, 38
  Pacific Forum Fisheries Agency 37
Pacific Islands Countries Trade Agreement (PICTA) 36
'Pacific Solution' 115, 235
Palau 3, 5, 40, 46, 131, 132, 153, 154, 155, 179, 218
  economy 39, 65, 107
  environment 70, 189, 218
  society 127
Palmyra (US) 189, 227, 230
Panapompom (PNG) 253
papayas 64
Papua New Guinea (PNG) 2, 3, 36, 75, 76, 82, 83, 96
passports 47, 113
pearls 77, 163
Pentecost (Vanuatu) 64
Philippines 64, 172, 177, 178
Phoenix Islands (Kiribati) 132, 135, 189, 207
pigs 60, 190, 210, 226
piracy 48, 98
Pitcairn Islands 4, 8, 28, 47, 98, 99, 103, 114, 124, 125, 132, 153, 160, 211, 216, 217, 230, 238, 254, 255
plantations 29, 55, 58, 61–2
plate tectonics 186, 187, 190, 191–2
Pohnpei (FSM) 66, 70, 224, 245
Poland 76
police 42, 48, 49, 50
politics 17, 18, 24, 28, 43–9, 124, 157, 252
pollution *see* environmental degradation
Polynesia 17, 38, 129, 147, 150, 182, 233
Ponam (PNG) 161, 162, 168, 255
population 3, 9, 16, 24, 58, 67, 73, 78, 82, 83, 87, 124–7, 131–2, 191, 211, 224, 239–40, 244, 260
pornography 114, 116
poverty 9, 24, 27, 28, 63, 85, 86, 87, 97, 102, 117–21, 124, 135, 137, 141, 143, 147, 165, 169, 173, 179, 206, 209, 223, 229, 234, 239, 240–41, 243, 252, 254, 255, 257
prices 66, 79, 86, 240
private sector 46, 116, 130–31, 239, 242
privatization 28, 111
prostitution 102, 119, 138
protectionism 29, 30, 32–3, 35, 78, 81, 92–4, 122, 259
public sector 49, 53, 99, 116, 130, 183, 232–3 *see also* privatization
Puerto Rico 3, 48, 206, 235
Pukapuka (Cook Islands) 212

Rabi (Fiji) 94, 136, 182
Raine Island (Australia) 194
rainfall 56, 59, 60, 97, 146, 187, 189, 197, 205–6, 207–9, 211, 213, 224–5
Rakahanga (Cook Islands) 134–5

RAMSI 28, 37, 236
Ranongga (Solomon Islands) 68, 134–5
Rapanui (Easter Island) 3, 18, 69, 99, 102, 103, 122, 216, 219, 230, 234
rats 217, 218, 219
reciprocity 68, 127
recycling 145
Redonda (Antigua) 114, 151
reefs *see* coral reefs
Reef Islands (Solomon Islands) 130–31, 162, 193, 194
refugees 115, 154
  environmental 178, 190, 195, 227, 239, 244, 261
religion 51, 155, 183, 234, 238, 247, 251
remittances 16, 19, 41, 66, 121, 129, 153, 167, 168, 208, 249–50, 258
remoteness 19, 22, 23, 107, 129, 150, 167, 168, 208, 249–50, 258
Rendova (Solomon Islands) 56–7, 129, 225
repo taro 57, 87, 252
resettlement 133–7, 190, 250
resilience 1, 27, 71, 85, 97–8, 132, 144, 186, 193–4, 198–200, 204–5, 247, 248, 257
resorts 100, 101
resource curse *see* Dutch disease
return migration *see* migration, return
Réunion 3, 55, 83, 237, 257
rice 56, 62, 86, 118, 251
rivers 141, 145, 209, 212
  sedimentation 67, 70, 95
Rodrigues (Mauritius) 78, 80, 83–4, 128, 131, 175, 186, 206, 233, 234
Rongelap (Marshall Islands) 153, 191
Rotuma (Fiji) 17, 130, 161, 169
Roviana (Solomon Islands) 118
rugby 171, 178
Russia 41, 90, 111, 115

Saba 3, 7, 18, 71, 99, 107–8, 151
safety nets 35, 60, 88, 117, 249, 254
St Barts (St Barthélemy) 18, 99
St Croix (USVI) 17, 109
St Helena 3, 7, 15, 99, 103, 125, 153, 160, 217, 236, 238
St Kitts and Nevis 3, 5, 22, 130, 214
  economy 29, 32, 33, 62, 63, 93, 97, 113, 118, 159
  environment 150, 225
  society 120, 125, 127, 150, 152, 170–71
St Lucia 3, 5, 22, 25, 44, 45, 48, 154, 155, 223, 224
  economy 32, 33, 63, 65, 92–3, 97, 99, 100, 104, 117, 118, 159–60
  environment 107, 224–5
  society 87–8, 120, 170–71
Saint Martin 18, 233
Saint Pierre et Miquelon 3, 73, 153, 241
St Thomas (USVI) 104, 109
St Vincent 3, 5, 25, 37, 48–9, 71, 170, 178, 220
  economy 33, 63, 65, 99, 114, 118, 159, 236
  environment 222
  society 125, 170–71
Saipan (Northern Marianas) 94
Salinity 142, 187, 206, 207, 208, 210, 211
Samoa 4, 6, 20, 22, 38, 152, 165, 158, 164, 180, 258
  economy 35, 42, 57–8, 63, 65, 66, 75, 76, 79, 80, 92, 93, 96, 97, 107, 114, 121, 159, 160, 163
  environment 70, 72, 107, 134, 205, 225, 226, 235
  society 37, 44, 46, 51, 117–18, 125, 126, 147, 155–6, 157, 170–71, 173–6, 245–6
sand 200, 223
  mining 106, 142, 143–4, 187, 191, 194
sandalwood 69
sanitation 119, 124, 141–2, 146
Santo (Vanuatu) 15
São Tomé et Principe 3, 5, 15, 17, 20, 22, 23, 24, 43, 44, 47, 226, 238, 258
  economy 28, 29, 31–2, 39, 58, 59–60, 61–2, 96, 98, 100, 113, 114, 117, 118
  society 44, 47, 125, 170
Savai'i (Samoa) 160

science 83, 187, 228, 253, 259
sea-level rise 4, 14, 16, 146, 148, 187–95, 227–8, 242–3
seaweed 77–8, 198, 243
secession 17–18, 130
second homes 106, 113
Secretariat of the Pacific Regional Environmental Program (SPREP) 37
security 17, 19, 28, 40, 42, 114–16, 155, 236, 259
service sector 29, 110–16
sewage 105, 140, 141, 142, 146, 197, 198, 210, 211, 212
Seychelles 4, 6, 9, 19, 22, 28, 45, 48, 115, 129
  economy 28, 34, 74, 75, 76, 77, 98, 100, 102, 108, 111, 113
  environment 105, 109, 225, 241
  society 37, 44, 51, 101, 125
shipping 37, 131, 151, 155
Sigatoka (Fiji) 65
Sikaiana (Solomon Islands) 121, 232
Simberi (PNG) 96
Simbo (Solomon Islands) 225
Singapore 2, 21
Sint Eustatius 3, 7, 18, 153
Sint Maarten 3, 7, 18, 28, 48, 56, 98, 99, 103, 104, 108, 154, 233, 262
Siwai (Bougainville, PNG) 128–9
skills 93, 99, 110, 111, 113, 117, 151, 154, 169–74, 175–6, 229, 244, 254
slavery 55, 61, 150
Small Island Developing States (SIDS) 2–3, 21–3, 26, 29, 32, 39
smallness 2, 4, 17, 18, 21–3, 26, 46, 52, 130, 132, 208, 230, 262
smoking 148
snails 216
social capital 167–8, 240, 246–8
social obligations 66, 68, 127, 138, 156, 161, 164, 165, 172, 175, 229, 231, 240, 246–51
socialism 44
soils 9, 14, 56, 60, 62, 64, 95, 106, 141, 185
Solomon Islands 4, 6, 14, 17, 20, 24, 28, 41, 43, 45, 48, 130, 132, 135, 148, 157, 180, 181, 220, 224, 228, 233, 236, 240, 252, 258
  economy 31, 36, 47, 49, 62, 69–70, 75, 78, 95–6, 100, 128
  environment 15, 23, 29, 69, 134, 188, 193, 198, 215, 224, 225, 241–2
  society 49, 83, 125, 126, 127, 129, 138–9, 238, 247, 250
sorcery 119
Sorol (FSM) 132
South Africa 90
South Georgia 3, 7, 103, 217
South Tarawa *see* Tarawa
SPARTECA 37, 93
sport 37, 105, 154, 155, 171, 173, 178, 238, 253
squash (pumpkin) 64, 163, 168
squatter settlements *see* urbanisation, settlements
stamps 114, 116, 256
statistics 1, 9–14, 25, 26, 53, 84, 229, 233–4
status 57, 86, 156, 157, 165, 183, 245–6, 252, 254, 256
stores 86–7, 129, 138, 165, 167, 249, 251
storms 190, 191, 193, 195, 197, 205–6, 219, 226
strategic services 114–16, 235, 236
structural adjustment programmes 28, 239
subsidy 59, 78, 80, 81, 91, 92, 99, 102, 122, 125, 130, 132, 147, 153, 241, 256, 258–9
sugar 19, 29, 32, 33, 34, 54, 57, 59, 61, 62, 63, 72, 87, 121, 141, 186, 206–7, 209, 238
suicide 147, 155–6
Surinam 2
sustainability 60, 75–6, 82–5, 107, 127, 182, 255
Suva (Fiji) 65, 137, 138, 139, 140, 141, 145
sweet potato 58, 129, 206

Tahiti (French Polynesia) 49, 88, 98, 122, 219
Taiwan 40–41, 76, 178

Takapoto (French Polynesia) 132
Takuu (Mortlock Islands) (Bougainville, PNG) 135, 190, 193
Tamana (Kiribati) 207, 250
Tanga (PNG) 234–5
Tanna (Vanuatu) 82, 125, 162, 206, 222, 223, 231, 238
Tarawa (Kiribati) 140, 141–2, 143, 144, 150, 166, 191, 192, 194, 210, 211–12
taro 34, 58, 65, 87, 132, 166, 206, 211, 212, 235, 250, 251
Taveuni (Fiji) 68
tax havens 110–13
teachers 154–5, 170–71
technocracy 41, 53, 252
technology 74, 78, 82, 91, 175, 246
telephones 37, 114, 131, 163, 182, 261
television 37, 113–14, 238
Tetepare (Solomon Islands) 84
textiles 32, 35, 93–4
Thailand 76
Tikopia (Solomon Islands) 58, 127, 143, 204, 231–2, 233
timber industry 31, 47, 69–72, 92, 224, 242, 255, 258
Tinian (Northern Marianas) 115
Tokelau 4, 8, 14, 40, 125, 129, 130, 133, 135, 152, 153, 160, 208, 209, 232, 244
Tonga 4, 6, 19, 22, 38, 49, 129, 130, 152, 156, 159, 160, 164, 177–8, 180, 214
  economy 31, 35, 57, 63, 64, 66–7, 75, 78, 82, 88, 92, 93, 97, 107, 113, 114, 117, 118, 120–21, 138, 162, 166–8, 169
  environment 23, 72, 134, 191, 195, 203, 209–10, 215, 222, 225
  society 101, 121, 125, 127, 128–9, 157, 158, 170–71, 173, 174, 176
Torres Islands (Vanuatu) 190, 192, 225
tourism 19, 28, 29, 37, 57, 62, 77, 90, 97–109, 117, 121, 125, 130, 160, 170, 218, 219, 223, 235, 237, 238, 242, 258, 262
  economic impact 56, 81, 83, 99–101, 103–4
  ecotourism 101, 107, 255
  environmental impact 68, 72, 83, 104–8 , 143, 190, 197, 210
  medical 113
  social impact 102, 147, 154
trade 18, 19, 27–39, 53, 54, 64–5, 181, 238
  fair trade 33, 65
trade unions 45, 51, 92
tradition 51, 66, 67–9, 82, 167, 204, 231, 232, 246–54, 256
'tragedy of the commons' 79
transnational corporation of kin 156, 158, 159, 195, 256
transnational corporations *see* multinational corporations
transnationalism 28, 41, 149, 150, 164, 261, 262
transport 18, 19, 31, 37–8, 64, 81, 87, 92, 100, 109, 131
  see also air transport
Trinidad 2, 48, 94
Tristan da Cunha 3, 4, 7, 55, 59, 76, 125, 133–4, 153, 160, 222, 235
Trobriand Islands (PNG) 231
trochus 77
tsunami 107, 134, 143–4, 163, 225–6, 227, 235, 240
Tuamotus (French Polynesia) 57
Tubuai (French Polynesia) 204–5
tuna 73–6, 259
Turks and Caicos Islands 3, 8, 14, 28, 47, 56, 101, 103, 111–12, 118, 125, 131, 153–4, 197, 211, 220
turtles 73, 83, 188, 215, 217
Tuvalu 4, 6, 14, 18, 20, 21, 22, 27, 40, 41, 43, 61, 108, 129, 130, 131, 134, 148, 151, 152, 157, 158, 174, 177–8, 180, 182, 187, 195, 236, 239, 243–4, 258
  economy 39, 75, 92, 100, 113, 114, 116, 159, 160, 161
  environment 142, 144, 188, 190–91, 192, 193–4, 202, 208, 209, 240, 243
  society 125, 128–9, 130, 157, 161

Ujae (Marshall Islands) 130
UKOTS viii, 40, 47, 110, 111–12, 125, 153, 237, 239
Ulithi (FSM) 107
UNCLOS 14
unemployment *see* employment
uneven development 39, 102, 183–4, 239, 240, 252
Union Island 130
United Arab Emirates (UAE) 154
United Kingdom 3, 47
United Nations 2, 19, 23, 24, 40, 50, 235, 237
  UNCTAD 20, 23
  UNDP 24
United States of America 3, 34, 39, 40, 48, 63, 74, 75, 94, 113, 114–16, 151, 152–3, 154, 163, 176, 182, 236, 258–9
United States Virgin Islands (USVI) 3, 8, 9, 56, 99, 109, 151, 153, 154, 179
universities 19, 37, 51
urban bias 49, 101, 120, 130, 161
urbanisation 72, 105, 119, 123, 124–49, 205, 210, 213–14, 223, 228, 239, 243
  economy 117, 138, 175
  planning 139, 205, 223–4, 229
  settlements 50, 119, 121, 124, 137–40, 141, 146, 212, 223, 224
  (*see also* housing)
Urelapa (Vanuatu) 4

Vanatinai (PNG) 249
Vanikoro (Solomon Islands) 193
vanilla 64
Vanua Lava (Vanuatu) 131, 251
Vanua Levu (Fiji) 212
Vanuatu 4, 5, 17, 20, 22, 23, 25, 27, 30, 43, 52, 122, 129, 178–80, 181, 214, 219, 238, 258
  economy 35, 36, 47, 57, 65–6, 96, 97, 98, 99, 103, 111–12, 113, 114
  environment 23, 69, 82, 106, 190, 192, 198, 203, 206, 222–3, 225, 241
  society 25, 51, 117, 119, 120, 125, 126, 128, 213, 238, 249–50
Vatukoula (Fiji) 96
Vatulele (Fiji) 87
vegetables 34
Victoria (Seychelles) 140
Vila (Vanuatu) 117, 119, 128, 231, 236, 252
violence 51, 95, 119, 128, 231, 236, 252
  domestic 50, 119
volcanism 4, 14–15, 132, 133–4, 162, 186, 222–3
vulnerability 1, 17, 21, 22–3, 63, 196, 203–5, 223, 232

Wagina (Solomon Islands) 78
Wakaya (Fiji) 101
Wallis and Futuna 4, 8, 9, 114, 125, 127, 130, 164, 174
waste disposal 72, 141, 142, 144–6, 210, 211
  see also sewage
water 66, 75, 81, 105–6, 119, 132, 137, 141–3, 180, 193, 207, 209–13, 224, 238, 243
  lenses 144, 191, 207–8, 209, 210, 211–12, 258
Western Union 163
whales 73
Woleai (Yap, FSM) 57, 77, 82, 86, 130, 191, 204, 208, 215, 228, 232, 262
women 46, 50, 79, 92–3, 100, 126, 151, 155, 162, 163, 170, 172, 218
World Bank 2, 20, 21, 23, 25, 28, 40, 42, 91, 94, 163
World Health Organization (WHO) 154
World Trade Organization (WTO) 21, 30, 31–6, 44, 181

yams 58
Yap (FSM) 57, 78, 129, 130, 231, 232
ylang-ylang 64
youth 16, 117, 119, 126–7, 148, 155–6, 172, 253

Zanzibar 214